Power Generation, Operation, and Control

Power Generation, Operation, and Control

Allen J. Wood
Bruce F. Wollenberg

Power Technologies, Inc.
Schenectady, New York
and
Rensselaer Polytechnic Institute
Troy, New York

John Wiley & Sons

New York Chichester Brisbane Toronto Singapore

Library of Congress Cataloging in Publication Data

Wood, Allen J.
  Power generation, operation, and control.
  Includes index.
  1. Electric power systems.   I. Wollenberg, Bruce F.
II. Title.
TK1001.W64   1983   621.31   83-1172
ISBN 0-471-09182-0

Printed in the United States of America

10 9 8 7 6 5 4 3 2 1

Preface

The fundamental purpose of this text is to introduce and explore a number of engineering and economic matters involved in planning, operating, and controlling power generation and transmission systems in electric utilities. It is intended for first-year graduate students in electric power engineering. We believe that it will also serve as a suitable self-study text for anyone with an undergraduate electrical engineering education and an understanding of steady-state power circuit analysis.

This text brings together material that has evolved since 1966 in teaching a graduate-level course in the electric power engineering department at Rensselaer Polytechnic Institute (RPI). The topics included serve as an effective means to introduce graduate students to advanced mathematical and operations research methods applied to practical electric power engineering problems. Some areas of the text cover methods that are currently being applied in the control and operation of electric power generation systems. The overall selection of topics, undoubtedly, reflects the interests of the authors.

In a one-semester course it is, of course, impossible to consider all the problems and "current practices" in this field. We can only introduce the types of problems that arise, illustrate theoretical and practical computational approaches, and point the student in the direction of seeking more information and developing advanced skills as they are required.

The material has regularly been taught in the second semester of a first-year graduate course. Some acquaintance with both advanced calculus methods (e.g., LaGrange multipliers) and basic undergraduate control theory is needed. Optimization methods are introduced as they are needed to solve practical problems and used without recourse to extensive mathematical proofs. This material is intended for an engineering course: mathematical rigor is important but is more properly the province of an applied or theoretical mathematics course. With the exception of Chapter 12, the text is self-contained in the sense that the various applied mathematical techniques are presented and developed as they are utilized.

Chapter 12, dealing with state estimation, may require more understanding of statistical and probabilistic methods than is provided in the text.

The first seven chapters of the text follow a natural sequence, with each succeeding chapter introducing further complications to the generation scheduling problem and new solution techniques. Chapter 8 treats methods used in generation system planning and introduces probabilistic techniques in the computation of fuel consumption and energy production costs. Chapter 8 stands alone and might be used in any position after the first seven chapters. Chapter 9 introduces generation control and discusses practices in modern U.S. utilities and pools. We have attempted to provide the "big picture" in this chapter to illustrate how the various pieces fit together in an electric power control system.

The topics of energy and power interchange between utilities and the economic and scheduling problems that may arise in coordinating the economic operation of interconnected utilities are discussed in Chapter 10. Chapters 11 and 12 are a unit. Chapter 11 is concerned with power system security and develops the analytical framework used to control bulk power systems in such a fashion that security is enhanced. Everything, including power systems, seems to have a propensity to fail. Power system security practices try to control and operate power systems in a defensive posture so that the effects of these inevitable failures are minimized. Finally, Chapter 12 is an introduction to the use of state estimation in electric power systems. We have chosen to use a maximum likelihood formulation since the quantitative measurement-weighting functions arise in a natural sense in the course of the development.

Each chapter is provided with a set of problems and an annotated reference list for further reading. Many (if not most) of these problems should be solved using a digital computer. At RPI we are able to provide the students with some fundamental programs (e.g., a load flow, a routine for scheduling of thermal units). The engineering students of today are well prepared to utilize the computer effectively when access to one is provided. Real bulk power systems have problems that usually call forth Dr. Bellman's curse of dimensionality—computers help and are essential to solve practical-sized problems.

The authors wish to express their appreciation to K. A. Clements, H. H. Happ, H. M. Merrill, C. K. Pang, M. A. Sager, and J. C. Westcott, who each reviewed portions of this text in draft form and offered suggestions. In addition, Dr. Clements used earlier versions of this text in graduate courses taught at Worcester Polytechnic Institute and in a course for utility engineers taught in Boston, Massachusetts.

Much of the material in this text originated from work done by our past and current associates at Power Technologies, Inc., the General Electric Company, and Leeds and Northrup Company. A number of IEEE papers have been used as primary sources and are cited where appropriate. It is not possible to avoid omitting references and sources that are considered to be significant by one group or another. We make no apology for omissions and only ask for indulgence from those readers whose favorites have been left out. Those interested may easily trace the references back to original sources.

We would like to express our appreciation for the fine typing job done on the original manuscript by Liane Brown and Bonnalyne MacLean.

This book is dedicated in general to all of our teachers, both professors and associates, and in particular to Dr. E. T. B. Gross.

Allen J. Wood
Bruce F. Wollenberg

Contents

Chapter 6

Generation with Limited Energy Supply   155

Chapter 7

Hydrothermal Coordination   189

Chapter 8

Energy Production Cost Models for Fuel Budgeting and Planning 239

Chapter 9

Control of Generation 291

Chapter 12

An Introduction to State Estimation in Power Systems

Power Generation, Operation, and Control

chapter 1

Introduction

1.1 PURPOSE OF COURSE

The objectives of a first-year, one-semester graduate course in electric power generation, operation, and control include the desire to

1. Acquaint electric power engineering students with power generation systems, their operation in an economic mode, and their control.
2. Introduce students to the important "terminal" characteristics for thermal and hydroelectric power generation systems.
3. Introduce mathematical optimization methods and apply them to practical operating problems.
4. Introduce methods for solving complicated problems involving both economic analysis and network analysis and illustrate these techniques with relatively simple problems.
5. Introduce methods that are used in modern control systems for power generation systems.

1.2 COURSE SCOPE

Topics to be addressed include

1. Power generation characteristics.
2. Economic dispatch and the general economic dispatch problem.
3. The economic dispatch problem for thermal units.

4. Methods of solution of the thermal dispatch problem.
5. Transmission losses.
6. Introduction to network transformation methods.
7. Loss formula in the economic dispatch, coordination equations.
8. Methods for solution of the coordination equations.
9. Unit commitment problem and an introduction to dynamic programming.
10. Economic dispatching in systems with limited energy availability.
11. The hydrothermal coordination problem.
12. Examples of solution methods for hydrothermal systems.
13. Probabilistic production cost calculations.
14. Power system control.
15. Capacity and energy interchange pricing analysis.
16. Techniques of analyzing power system security.
17. An introduction to least-squares techniques for power system state estimation.

In many cases this will be only an introduction to the topic area. Many additional problems and topics that represent important, practical problems would require more time and space than is available. Still other problems, such as those involving light-water nuclear reactors, would require several chapters to lay a firm foundation. These topics can only be the subject of a brief overview.

1.3 ECONOMIC IMPORTANCE

The efficient and optimum economic operation and planning of electric power generation systems have always occupied an important position in the electric power industry. Prior to 1973 and the oil embargo that signaled the rapid escalation in fuel prices, electric utilities in the United States spent about 20% of their total revenues on fuel for the production of electrical energy. By 1980, that figure had risen to more than 40% of total revenues. In the 5 years after 1973, U.S. electric utility fuel costs escalated at a rate that averaged 25% compounded on an annual basis. The efficient use of the available fuel is growing in importance both monetarily and because most of the fuel used represents irreplaceable natural resources.

An idea of the magnitude of the amounts of money under consideration can be obtained by considering the annual operating expenses of a large utility for purchasing fuel. Assume the following parameters for a moderately large system.

Annual peak load: 10,000 MW

Annual load factor: 60%

Average annual heat rate for converting fuel to electric energy: 10,500 Btu/kWh

Average fuel cost: $2.00 per million Btu (MBtu)

With these assumptions the total annual fuel cost for this system is as follows.

Annual energy produced: 10^7 kW $\times$ 8760 h/yr $\times$ 0.60 = 5.256 $\times$ 10^{10} kWh

Annual fuel consumption: 10,500 Btu/kWh $\times$ 5.256 $\times$ 10^{10} kWh

$$= 55.188 \times 10^{13} \text{ Btu}$$

Annual fuel cost: 55.188 $\times$ 10^{13} Btu $\times$ 2 $\times$ 10^{-6} \$/Btu = \$1.104 billion

To put this cost in perspective, it represents a direct requirement for revenues from the average customer of this system of 2.1 cents per kWh just to recover the expense for fuel.

A savings in the operation of this system of a small percent represents a significant reduction in operating cost as well as in the quantities of fuel consumed. It is no wonder that this area has warranted a great deal of attention from engineers through the years.

Periodic increases in basic fuel price levels serve to accentuate the problem and increase its economic significance. Inflation also causes problems in developing and presenting methods, techniques, and examples of the economic operation of electric power generating systems. Actually, recent fuel costs always seem to be ancient history and entirely inappropriate to current conditions. To avoid leaving false impressions about the actual value of the methods to be discussed, all the examples and problems that are in the text are expressed in a nameless, fictional monetary unit to be designated as an "R".

1.4 PROBLEMS SOLVED AND UNSOLVED

This text represents a progress report in an engineering area that has been and is still undergoing rapid change. It concerns established engineering problem areas (i.e., economic dispatch and control of interconnected systems) that have taken on new importance in recent years. The original problem of economic dispatch for thermal systems was solved by numerous methods years ago. Recently there has been a rapid growth in applied mathematical methods and the availability of computational capability for solving problems of this nature so that more involved problems have been successfully solved.

The classic problem is the economic dispatch of fossil-fired generation systems to achieve minimum operating cost. This problem area has taken on a subtle twist as the public has become increasingly concerned with environmental matters, so that "economic dispatch" now includes the dispatch of systems to minimize pollutants and conserve various forms of fuel as well as to achieve minimum costs. In addition, there is a need to expand the limited economic optimization problem to incorporate constraints on system operation to ensure the "security" of the system, thereby preventing the collapse of the system due to unforeseen conditions. The hydrothermal coordination problem is another optimum operating problem area that has received a great deal of attention. Even so, there are difficult problems involving hydrothermal coordination that cannot be solved in a theoretically satisfying fashion in a rapid and efficient computational manner.

The post World War II period saw the increasing installation of pumped-storage hydroelectric plants in the U.S. and a great deal of interest in energy storage systems. These storage systems involve another difficult aspect of the optimum economic operating problem. Methods are available for solving coordination of hydroelectric, thermal, and pumped-storage electric systems. However, closely associated with this economic dispatch problem is the problem of the proper commitment of an array of units out of a total array of units to serve the expected load demands in an "optimal" manner. This unit commitment problem can only be solved approximately for practical-sized systems at the present time.

There are other unsolved problems in the area of power generation, operation, and control. For example, we would like to be able to schedule maintenance outages of units on an "optimum" basis. However, at this time, we are not even sure what we mean by the term "optimum." In interconnected system operations there are control, economic dispatch, and accounting problems that need further work and resolution. Many researchers are still working on solutions to the "optimal load flow" problem that promise to extend economic dispatch to include the optimal setting of generator voltages, transformer taps, and reactive power sources. The list is long and undoubtedly will expand as new problems arise and will contract as new methodologies are developed, tested, and utilized on a routine basis.

FURTHER READING

The following books are suggested as sources of information for this general area. The first four are the "classics"; the remainder are either specialized to a limited number of topics or else represent collections of articles or chapters on varied topics related to this broad subject.

1. Steinberg, M. J., Smith, T. H., *Economy Loading of Power Plants and Electric Systems*, Wiley, New York, 1943.

2. Kirchmayer, L. K., *Economic Operation of Power Systems*, Wiley, New York, 1958.

3. Kirchmayer, L. K., *Economic Control of Interconnected Systems*, Wiley, New York, 1959.

4. Cohn, N., *Control of Generation and Power Flow On Interconnected Systems*, Wiley, New York, 1961.

5. Hano, I., *Operating Characteristics of Electric Power Systems*, Denki Shoin, Tokyo, 1967.

6. Handschin, E. (ed.), *Real-Time Control of Electric Power Systems*, Elsevier, Amsterdam, 1972

7. Savulescu, S. C. (ed.), *Computerized Operation of Power Systems*, Elsevier, Amsterdam, 1976.

8. Sterling, M. J. H., *Power System Control*, Peregrinus, London, 1978.

9. El-Hawary, M. E., Christensen, G. S., *Optimal Economic Operation of Electric Power Systems*, Academic, New York, 1979.

chapter 2

Characteristics of Power Generation Units

2.1 CHARACTERISTICS OF STEAM UNITS

In analyzing the problems associated with the controlled operation of power systems, there are many possible parameters of interest. Fundamental to the economic operating problem is the set of input-output characteristics of a thermal power generation unit. A typical boiler-turbine-generator unit is sketched in Figure 2.1. This unit consists of a single boiler that generates steam to drive a single turbine generator set. The electrical output of this set is connected not only to the electric power system but also to the auxiliary power system in the power plant. A typical steam turbine unit may require 2 to 6% of the gross output of the unit for the auxiliary power requirements necessary to drive boiler feed pumps, fans, condenser circulating water pumps, and so on. In defining the unit characteristics, we will talk about *gross* input versus *net* output. That is, gross input to the plant represents the total input whether measured in terms of dollars per hour or tons of coal per hour or millions of cubic feet of gas per hour, or any other units. The net output of the plant is the electrical power output available to the electric utility system. Occasionally engineers

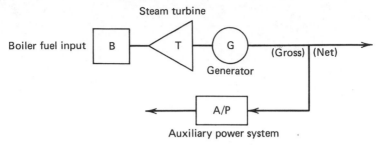

FIG. 2.1 Boiler-turbine-generator unit.

will develop gross input-gross output characteristics. In such situations, the data should be converted to net output to be more useful in scheduling the generation.

In defining the characteristics of steam turbine units, the following terms will be used.

H: Btu per hour heat input to the unit (or MBtu/h)

F: Fuel cost times H is the ℝ per hour (ℝ/h) input to the unit for fuel.

Occasionally the dollar per hour operating cost of a unit will include prorated operation and maintenance costs. That is, the labor cost for the operating crew will be included as part of the operating cost if this cost can be expressed directly as a function of the power output of the unit. The output of the generation unit will be designated by P, the megawatt net output of the unit.

Figure 2.2 shows the input-output characteristic of a steam unit in idealized form. The input to the unit shown on the ordinate may be either in terms of heat energy requirements [millions of Btu per hour (MBtu/h)] or in terms of total

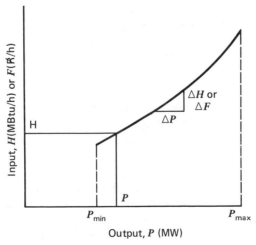

FIG. 2.2 Input-output curve of a steam turbine generator.

cost per hour (dollars per hour). The output is normally the net electrical output of the unit. The characteristic shown is idealized in that it is presented as a smooth, convex curve.

These data may be obtained from design calculations or from heat rate tests. When heat rate test data are used, it will usually be found that the data points do not fall on a smooth curve. Steam turbine generating units have several critical operating constraints. Generally, the minimum load at which a unit can operate is influenced more by the steam generator and the regenerative cycle than by the turbine. The only critical parameters for the turbine are shell and rotor metal differential temperatures, exhaust hood temperature, and rotor and shell expansion. Minimum load limitations are generally caused by fuel combustion stability and inherent steam generator design constraints. For example, most supercritical units cannot operate below 30% of design capability. A minimum flow of 30% is required to cool the tubes in the furnace of the steam generator adequately. Turbines do not have any inherent overload capability, so that the data shown on these curves normally do not extend much beyond 5% of the manufacturer's stated valve-wide-open capability.

The incremental heat rate characteristic for a unit of this type is shown in Figure 2.3. This incremental heat rate characteristic is the slope (the derivative) of the input-output characteristic, ($\Delta H/\Delta P$ or $\Delta F/\Delta P$). The data shown on this curve are in terms of Btu per kilowatt hour (or dollars per kilowatt hour) versus the net power output of the unit in megawatts. This characteristic is widely used in economic dispatching of the unit. It is converted to an incremental fuel cost characteristic by multiplying the incremental heat rate in Btu per kilowatt hour by the equivalent fuel cost in terms of dollars per Btu. Frequently this characteristic is approximated by a sequence of straight-line segments.

The last important characteristic of a steam unit is the unit (net) heat rate characteristic shown in Figure 2.4. This characteristic is H/P versus P. It is the reciprocal of the usual efficiency characteristic developed for machinery. The unit

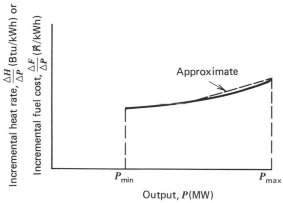

FIG. 2.3   Incremental heat (cost) rate characteristic.

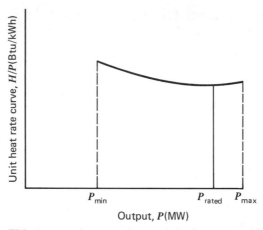

FIG. 2.4 Net heat rate characteristic of a steam turbine generator unit.

heat rate characteristic shows the heat input per kilowatt hour of output versus the megawatt output of the unit. Conventional steam turbine units are between 30 and 35% efficient, so that their unit heat rates range between approximately 11,400 Btu/kWh and 9800 Btu/kWh. (A kilowatt hour has a thermal equivalent of approximately 3412 Btu.) Unit heat rate characteristics are a function of unit design parameters such as initial steam conditions, stages of reheat and the reheat temperatures, condenser pressure, and the complexity of the regenerative feed-water cycle. These are important considerations in the establishment of the unit's efficiency. For purposes of estimation, a typical heat rate of 10,500 Btu/kWh may be used occasionally to approximate actual unit heat rate characteristics.

Many different formats are used to represent the input-output characteristic shown in Figure 2.2. The data obtained from heat rate tests or from the plant design engineers may be fitted by a polynomial curve. In many cases quadratic characteristics have been fit to these data. A series of straight-line segments may also be used to represent the input-output characteristics. The different representations will, of course, result in different incremental heat rate characteristics. Figure 2.5 shows two such variations. The solid line shows the incremental heat rate characteristic that results when the input versus output characteristic is a quadratic curve or some other continuous, smooth, convex function. This incremental heat rate characteristic is monotonically increasing as a function of the power output of the unit. The dashed lines in Figure 2.5 show a stepped incremental characteristic that results when a series of straight-line segments are used to represent the input-output characteristics of the unit. The use of these different representations may require that different scheduling methods be used for establishing the optimum economic operation of a power system. Both formats are useful, and both may be represented by tables of data. Only the first, the solid line, may be represented by a continuous analytic function, and only the first

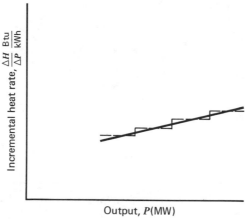

FIG. 2.5 Approximate representations of the incremental heat rate curve.

has a derivative that is nonzero. (That is, d^2F/dP^2 equals zero if dF/dP is constant.)

2.2 VARIATIONS IN STEAM UNIT CHARACTERISTICS

A number of different steam unit characteristics exist. For large steam turbine generators the input-output characteristics shown on Figure 2.2 are not always as smooth as indicated there. Large steam turbine generators will have a number of steam admission valves that are opened in sequence to obtain ever-increasing output of the unit. Figure 2.6 shows both an input-output and an incremental heat rate characteristic for a unit with four valves. As the unit loading increases, the input to the unit increases and the incremental heat rate decreases between opening points for any two valves. However, when a valve is first opened, the throttling losses increase rapidly and the incremental heat rate rises suddenly. This gives rise to the discontinuous type of incremental heat rate characteristic shown in Figure 2.6. It is possible to use this type of characteristic in order to schedule steam units, although it is usually not done. This type of input-output characteristic is nonconvex; hence optimization techniques that require convex characteristics may not be used with impunity.

Another type of steam unit that may be encountered is the *common-header plant*, which contains a number of different boilers connected to a common steam line (called a common header). Figure 2.7 is a sketch of a rather complex common-header plant. In this plant there are not only a number of boilers and turbines each connected to the common header but also a "topping turbine" connected to the common header. A *topping turbine* is one in which steam is exhausted from the turbine and fed not to a condenser but to the common steam header.

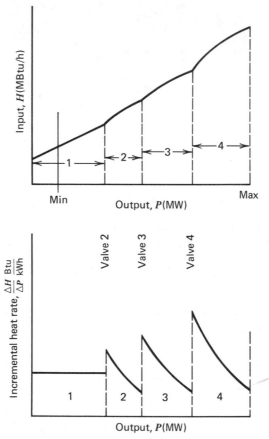

FIG. 2.6 Characteristics of a steam turbine generator with four steam admission valves.

A common-header plant will have a number of different input-output characteristics that result from different combinations of boilers and turbines connected to the header. Steinberg and Smith (*Economy Loading of Power Plants and Electric Systems*, Wiley, 1943) treats this type of plant quite extensively. Common-header plants were constructed originally not only to provide a large electrical output from a single plant but also to provide steam sendout for the heating and cooling of buildings in dense urban areas. After World War II, a number of these plants were modernized by the installation of the type of topping turbine shown in Figure 2.7. For a period of time during the 1960s, these common-header plants were being dismantled and replaced by modern, efficient plants. However, as urban areas began to reconstruct, a number of metropolitan utilities found that their steam loads were growing and that the common-header plants could not be dismantled but had to be expanded to provide steam supplies to new buildings.

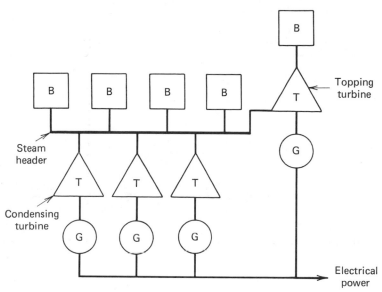

FIG. 2.7 A common-header steam plant.

In the late 1960s, a new type of steam plant configuration began to find favor, the *combined cycle plant*. A simple cycle gas turbine unit consists of a gas turbine and compressor connected on a single shaft to a generating unit. Simple cycle gas turbines have efficiencies in the range of 25 to 30% (i.e., unit heat rate of 13,600 to 11,400 MBtu/kWh based on the higher heating value of the fuel), require oil or gas for fuel, and are used primarily for peaking duty in electric power systems. The high temperature exhaust from a simple cycle unit is discharged to the atmosphere. The combined cycle plant utilizes the high temperature exhaust of gas turbines in heat recovery steam generators (HRSG) to generate steam to drive a separate steam turbine generating unit. The advantage of this type of plant is that it has a high thermal cycle efficiency. Figure 2.8 is a schematic diagram of this type of plant with four gas turbines and a single steam turbine. Figure 2.9 shows the resulting unit heat rate characteristic for the combined cycle plant.

It appears from data available from various manufacturers of combined cycle units that unit heat rates of 8500 Btu/kWh may be achievable with this type of plant. The incremental heat rate characteristics are very difficult to ascertain from the available data. It appears that the incremental heat rate characteristics are monotonically decreasing (or at least nonincreasing) as a function of the power output of the plant. The existence of this type of discontinuous characteristic as well as the negative-sloping incremental characteristic make the optimum economic scheduling of this type of unit somewhat difficult. When conventional economic dispatching methods are utilized to schedule this type of plant, it may be found that these plants are operating at full load all the time that they are available. It is an open question as to whether or not this type of plant is

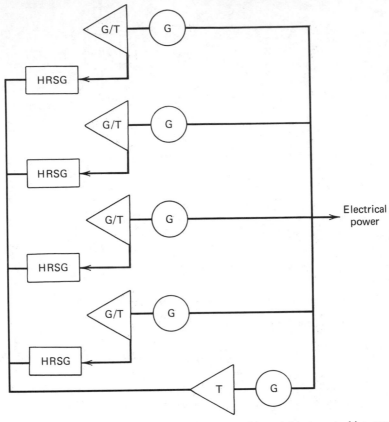

FIG. 2.8 A combined cycle plant with four gas turbines and a steam turbine generator.

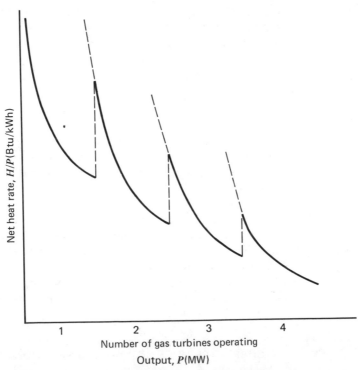

FIG. 2.9 Combined cycle plant heat rate characteristic.

designed to operate at full load for 6000 to 7000 h/yr. Of course, the actual operation of this type of plant will depend on relative fuel costs as well as on the unit generating mix available in the system.

Variations in the design of combined cycle plants have been utilized in recent years. There is, for example, a design configuration using a single shaft that contains two gas turbines driving a single electric generator. The exhaust gases from these two jet engines are then utilized in a heat recovery steam generator. Designs of heat recovery steam generators have been proposed that make use of additional fuel supplies to augment the steam-generating capability of the HRSG. Most combined cycle plants can utilize the gas turbines as simple cycle units when the HRSG and/or steam turbine generating units are out of service. These plants require a bypass around the HRGS for the exhaust gases.

2.3 LIGHT-WATER MODERATED NUCLEAR REACTOR UNITS

U.S. utilities have adopted the light-water moderated reactor as the "standard" type of nuclear steam supply system. These reactors are either pressurized water reactors (PWRs) or boiling water reactors (BWRs) and use slightly enriched uranium as the basic energy supply source. The uranium that occurs in nature contains approximately seven-tenths of 1% by weight of U_{235}. This natural uranium must be enriched so that the content of U_{235} is in the range of 2 to 4% for use in either a PWR or a BWR.

The enriched uranium must be fabricated into fuel assemblies by various manufacturing processes. At the time the fuel assemblies are loaded into the nuclear reactor core there has been a considerable investment made in this fuel. During the period of time in which fuel is in the reactor and is generating heat and steam, and electrical power is being obtained from the generator, the amount of usable fissionable material in the core is decreasing. At some point the reactor core is no longer able to maintain a critical state at a proper power level, so the core must be removed and new fuel reloaded into the reactor. Commercial power reactors are normally designed to replace one third to one fifth of the fuel in the core during reloading.

At this point the nuclear fuel assemblies that have been removed are highly radioactive and must be treated in some fahion. Originally it was intended that these assemblies would be reprocessed in commercial plants and that valuable materials would be obtained from the reprocessed core assemblies. At this time (1983) it is questionable if the U.S. reactor industry will develop an economically viable reprocessing system that is acceptable to the public in general. If this is not done, either these radioactive cores will need to be stored for some indeterminate period of time or the U.S. government will have to take over these fuel assemblies for storage and eventual reprocessing. In any case, an additional amount of money will need to be invested either in reprocessing the fuel or in storing it for some period of time.

The calculation of "fuel cost" in a situation such as this involves economic and accounting considerations and is really an investment analysis. Simply speaking, there will be a total dollar investment in a given core assembly. This dollar investment includes the cost of mining the uranium, milling the uranium core, converting it into a gaseous product that may be enriched, fabricating fuel assemblies, and delivering them to the reactor plus the cost of removing the fuel assemblies after they have been irradiated and either reprocessing them or storing them. Each of these fuel assemblies will have generated a given amount of electrical energy. A pseudo fuel cost may be obtained by dividing the total investment in dollars by the total amount of electrical energy generated by the assembly. Of course there are refinements that may be made in this simple computation. For example, it is possible using nuclear physics calculations to compute more precisely the amount of energy generated by a specific fuel assembly in the core in a given stage of operation of a reactor.

In the remainder of this text nuclear units will be treated as if they are ordinary thermal-generating units fueled by a fossil fuel. The considerations and computations of exact fuel reloading schedules and enrichment levels in the various fuel assemblies are beyond the scope of a one-semester graduate course because they require a background in nuclear engineering as well as detailed understanding of the fuel cycle and its economic aspects.

2.4 HYDROELECTRIC UNITS

Hydroelectric units have input-output characteristics similar to steam turbine units. The input is in terms of volume of water per unit time; the output is in terms of electrical power. Figure 2.10 shows a typical input-output curve for a hydroelectric plant where the net hydraulic head is constant. This characteristics

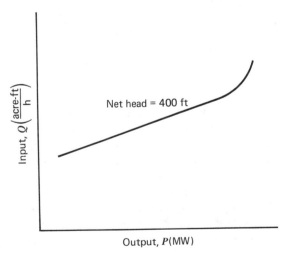

FIG. 2.10 Hydroelectric unit input-output curve.

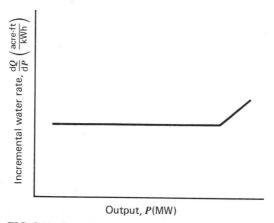

FIG. 2.11 Incremental water rate curve for hydroelectric plant.

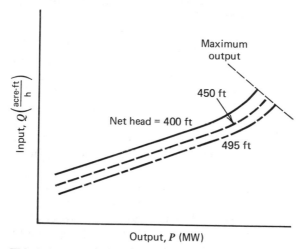

FIG. 2.12 Input-output curves for hydroelectric plant with a variable head.

shows an almost linear curve of input water volume requirements per unit time as a function of power output as the power output increases from minimum to rated load. Above this point the volume requirements increase as the efficiency of the unit falls off. The incremental water rate characteristics are shown in Figure 2.11. The units shown on both these curves are English units. That is, volume is shown as acre-feet (an acre of water a foot deep). If necessary, net hydraulic heads are shown in feet. Metric units are also used as are thousands of cubic feet per second (kft^3/sec) for the water rate.

Figure 2.12 shows the input-output characteristics of a hydroelectric plant with variable head. This type of characteristic occurs whenever the variation in the storage pond (i.e., forebay) and/or afterbay elevations is a fairly large percentage of the overall net hydraulic head. Scheduling hydroelectric plants with

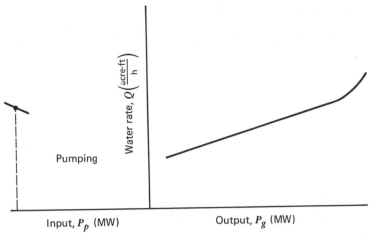

FIG. 2.13 Input-output characteristics for a pumped storage hydro plant with a fixed, net hydraulic head.

variable head characteristics is more difficult than scheduling hydroelectric plants with fixed heads. This is true not only because of the multiplicity of input-output curves that must be considered, but also because the maximum capability of the plant will also tend to vary with the hydraulic head. In Figure 2.12 the volume of water required for a given power output decreases as the head increases. (That is, $\partial Q/\partial$head or $\partial Q/\partial$volume are negative for a fixed power.) In a later section methods are discussed that have been proposed for the optimum scheduling of hydrothermal power systems where the hydroelectric systems exhibit variable head characteristics.

Figure 2.13 shows the type of characteristics exhibited by pumped-storage hydroelectric plants. These plants are designed so that water may be stored by pumping it against a net hydraulic head for discharge at a more propitious time. This type of plant was originally installed with separate hydraulic turbines and electric motor driven pumps. In recent years reversible, hydraulic pump turbines have been utilized. These reversible pump turbines exhibit normal input-output characteristics when utilized as turbines. In the pumping mode, however, the efficiency of operation tends to fall off when the pump is operated away from the rating of the unit. For this reason most plant operators will only operate these units in the pumping mode at a fixed pumping load. The incremental water characteristics when operating as a turbine are, of course, similar to the conventional units illustrated previously.

The scheduling of pumped-storage hydroelectric plants may also be complicated by the necessity of recognizing the variable-head effects. These effects may be most pronounced in the variation of the maximum capability of the plant rather than in the presence of multiple input-output curves. This variable maximum capability may have a significant effect on the requirements for selecting capacity to run on the system, since these pumped-storage hydro plants may usually be considered as spinning-reserve capability. That is, they will be used only during periods of highest cost generation on the thermal units; at other times they may be considered as readily available ("spinning reserve"). That is, during periods when they would normally be pumping, they may be shut off to reduce the demand. When idle, they may be started rapidly. In this case the maximum capacity available will have a significant impact on the requirements for having other units available to meet the systems total spinning-reserve requirements.

These hydroelectric plants and their characteristics (both the characteristics for the pumped-storage and the conventional storage hydroelectric plants) are affected greatly by the hydraulic configuration that exists where the plant is installed and by the requirements for water flows that may have nothing to do with power production. The characteristics just illustrated are for single, isolated plants. In many river systems plants are connected in both series and in parallel (hydraulically speaking). In this case the release of an upstream plant contributes to the inflow of downstream plants. There may be tributaries between plants that contribute to the water stored behind a downstream dam. The situation becomes even more complex when pumped-storage plants are constructed in conjunction with conventional hydroelectric plants. The problem of the optimum utilization of these resources involves the complicated problems associated with the scheduling of water as well as the optimum operation of the electric power system to minimize production cost. We can only touch on these matters in this text and introduce the subject. Because of the importance of the hydraulic coupling between plants, it is safe to assert that no two hydroelectric systems are exactly the same.

APPENDIX
Typical Generation Data

Up until the early 1950s, most U.S. utilities installed units of less than 100 MW. These units were relatively inefficient (about 950 psi steam and no reheat cycles). During the early 1950s, the economics of reheat cycles and advances in materials technology encouraged the installation of reheat units having steam temperatures of 1000°F and pressures in the range of 1450 to 2150 psi. Unit sizes for the new design reheat units ranged to 225 MW. In the late 1950s and early 1960s, U.S. utilities began installing larger units ranging up to 300 MW in size. In the late 1950s, U.S. utilities began installing larger, more efficient units (about 2400 psi with single reheat) ranging in size to 700 MW. In addition, in the late 1960s, some U.S. utilities began installing more efficient supercritical units (about 3500 psi, some with double reheat) ranging in size to 1300 MW. The bulk of these supercritical units range in size from 500 to 900 MW. However, many of the newest supercritical units range in size from 1150 to 1300 MW.

Typical heat rate data for these classes of fossil generation are shown in Table 2.1. These data are based on Federal Power Commission reports and other design data for U.S. utilities (see *Heat Rates for General Electric Steam*

TABLE 2.1 Typical Fossil Generation Unit Heat Rates

Fossil unit-description	Unit rating (MW)	100% output (Btu/kWh)	80% output (Btu/kWh)	60% output (Btu/kWh)	40% output (Btu/kWh)	25% output (Btu/kWh)
Steam—coal	50	11000	11088	11429	12166	13409[a]
Steam—oil	50	11500	11592	11949	12719	14019[a]
Steam—gas	50	11700	11794	12156	12940	14262[a]
Steam—coal	200	9500	9576	9871	10507	11581[a]
Steam—oil	200	9900	9979	10286	10949	12068[a]
Steam—gas	200	10050	10130	10442	11115	12251[a]
Steam—coal	400	9000	9045	9252	9783	10674[a]
Steam—oil	400	9400	9447	9663	10218	11148[a]
Steam—gas	400	9500	9548	9766	10327	11267[a]
Steam—coal	600	8900	8989	9265	9843	10814[a]
Steam—oil	600	9300	9393	9681	10286	11300[a]
Steam—gas	600	9400	9494	9785	10396	11421[a]
Steam—coal	800–1200	8750	8803	9048	9625[a]	
Steam—oil	800–1200	9100	9155	9409	10010[a]	
Steam—gas	800–1200	9200	9255	9513	10120[a]	

[a] For study purposes, units should not be loaded below the points shown.

TABLE 2.2 Approximate Unit Heat Rate Increase Over Valve-Best-Point Turbine Heat Rate In %

Unit size (MW)	Coal	Oil	Gas
50	22	28	30
200	20	25	27
400	16	21	22
600	16	21	22
800–1200	16	21	22

Turbine-Generators 100,000 kW and Larger, Large Steam Turbine Generator Department, G.E.).

The shape of the heat rate curves is based on the locus of design, "valve-best-points" for the various sizes of turbines. The magnitude of the turbine heat rate curve has been increased to obtain the unit heat rate, adjusting for the mean of the valve loops, boiler efficiency, and auxiliary power requirements. The resulting approximate increase from design turbine heat rate to obtain the generation unit heat rate in Table 2.1 is summarized in Table 2.2 for the various types and sizes of fossil units. Typical fossil unit maintenance requirements and forced outage rate (F.O.R.) data are shown in Table 2.3.

The annual maintenance requirements, based on a fairly recent report (1975) of the Edison Electric Institute on equipment availability, includes both scheduled and unscheduled maintenance (or maintenance outages). A maintenance outage occurs when a unit is taken out of service to repair a problem. Supercritical units have more complicated equipment than subcritical units; therefore

TABLE 2.3 Typical Fossil Unit Maintenance Requirements and Forced Outage Rates

Unit size (MW)	Maintenance requirements (Day/Yr)			Full outage		Partial outage data			Effective outage	
	Scheduled	Unscheduled	Total	F.O.R. (%)	Avg. repair time (Day)	Avg. derating (% Unit)	Avg. partial F.O.R. (%)	Avg. repair time (Day)	Effective F.O.R. (%)	Avg. repair time (Day)
50	17	7	24	2.3	2.2	15	2.4	0.9	2.7	2.4
200	24	12	36	5.3	2.2	24	8.9	3.2	7.4	3.0
400	31	14	45	9.5	2.5	22	14.0	4.9	13.0	3.6
600	31	14	45	16.0	2.7	21	25.0	8.8	21.0	4.5
800–1200	34	15	49	18.0	3.0	21	29.0	9.5	24.0	5.0

TABLE 2.4 Typical Nuclear Generation Unit Heat Rates

	100% Output (Btu/kWh)	75% Output (Btu/kWh)	50% Output (Btu/kWh)
Light-water reactor	10400	10442[a]	10951[a]

[a] Nuclear units are usually dispatched at fixed output.

the maintenance requirements have been increased 19% over the EEI data for units larger than 600 MW, which is based mainly on subcritical unit maintenance requirements.

The forced outage rate and outage duration data for the fossil units are also based on the cited EEI report. Full and partial outage data are presented as well as the resulting effective outage rate data. The forced outage rates increase significantly as the unit size and initial steam conditions increase. The EEI report does not separate the forced outage rates of supercritical units from the subcritical units greater than 600 MW. Since the steam pressure is much higher and the unit size will be larger for supercritical units, it is assumed that the forced outage data would be 10% higher than the 600 MW subcritical unit data, which represent larger subcritical unit forced outage rates.

Typical nuclear generation heat rate data are shown in Table 2.4. One set of typical characteristics can represent both types of light-water reactors for generation studies. The typical shape of the heat rate curve for light-water reactors is based on the locus of design valve-best-points, adjusted upward 8% to obtain the unit net heat rates. The additional 8% accounts for auxiliary power requirements and heat losses in the reactor auxiliary equipment. Maintenance and forced outage rate data shown in Table 2.5 are based on EEI data for the existing nuclear units.

Gas turbines range in size from a few megawatts to 100 MW for some of the newer units. Gas turbines are designed to burn liquid or gaseous fuels, mostly light distillate oil or natural gas. Typical gas turbine heat rates are shown in Table 2.6. Heat rate data based on the same sources as the fossil and nuclear units are presented only for full-load operation because these units are not normally operated at partial loads.

Typical gas turbine maintenance and forced outage data are presented in Table 2.7. The maintenance requirements are approximately the same for both industrial and jet engine plants, based on the EEI report. The effective forced outage rates shown in Table 2.7 are also about the same for both types of gas turbines if the EEI definition of forced outage rate is modified as follows.

$$\text{EEI F.O.R.} = \frac{\text{Time on forced outage}}{\text{Time in service } + \text{ Time on forced outage}}$$

$$\text{Modified F.O.R} = \frac{\text{Time on forced outage}}{\text{Time available for service} + \text{Time on forced outage}}$$

TABLE 2.5 Typical Nuclear Unit Maintenance Requirements and Forced Outage Rates

Nuclear unit size	Maintenance requirements (Day/Yr)			Full outage		Partial outage data			Effective outage	
	Scheduled	Unscheduled	Total	F.O.R. (%)	Avg. repair time (Day)	Avg. derating (% Unit)	Avg. partial F.O.R. (%)	Avg. repair time (Day)	Effective F.O.R. (%)	Avg. repair time (Day)
All	35	14	49	11	3.7	23	16	5	15	5

TABLE 2.6 Typical Gas Turbine Generation Unit Heat Rates

Unit type	Heat rate (Btu/kWh)
Industrial	13,600
Jet	16,000

TABLE 2.7 Typical Gas Turbine Maintenance Requirements and Forced Outage Rates

Unit type	Maintenance requirements (Day/Yr)			Full outage		Partial outage data			Effective outage	
	Scheduled	Unscheduled	Total	F.O.R. (%)	Avg. repair time (Day)	Avg. derating (% Unit)	Avg. partial F.O.R. (%)	Avg. repair time (Day)	Effective F.O.R. (%)	Avg. repair time (Day)
All	8	10	18	—	—	—	—	—	24	2

Since gas turbines are generally operated as peaking units due to high fuel cost and high unit heat rates, their actual running time is very small. The conventionally calculated forced outage rates for gas turbines are approximately 33% for industrial units and 48% for jet engines. However, if the service time is replaced by the time the peaking units are available, the forced outage rates reduce to about 24% for both types of gas turbines. It is recomended that the modified forced outage rate data be used for these peaking units. It is almost impossible to specify "typical" data for a hydroelectric plant since the essential characteristics depend on the hydraulic parameters associated with each plant site.

chapter 3

Economic Dispatch of Thermal Units and Methods of Solution

This chapter introduces techniques of power system optimization. For a complete understanding of how optimization problems are carried out, first read the appendix to this chapter where the concepts of the LaGrange multiplier and the Kuhn-Tucker conditions are introduced.

3.1 THE ECONOMIC DISPATCH PROBLEM

Figure 3.1 shows the configuration that will be studied in this section. This system consists of N thermal-generating units connected to a single bus-bar serving a received electrical load P_R. The input to each unit, shown as F_i, represents the cost rate* of the unit. The output of each unit, P_i, is the electrical power generated by that particular unit. The total cost rate of this system is, of course, the sum of the costs of each of the individual units. The

---

* Generating units consume fuel at a specific rate (e.g., MBtu/h), which as noted in Chapter 2 can be converted to ℞/h, which represents a cost rate. Starting in this chapter and throughout the remainder of the text, we will simply use the term generating unit "cost" to refer to ℞/h

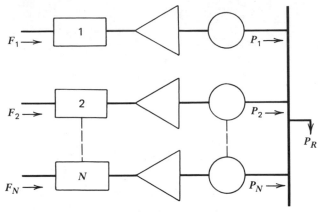

FIG. 3.1 N thermal units committed to serve a load of P_R.

essential constraint on the operation of this system is that the sum of the output powers must equal the load demand.

Mathematically speaking, the problem may be stated very concisely. That is, an objective function, F_T, is equal to the total cost for supplying the indicated load. The problem is to minimize F_T subject to the constraint that the sum of the powers generated must equal the received load. Note that any transmission losses are neglected and any operating limits are not explicitly stated when formulating this problem. That is,

$$F_T = F_1 + F_2 + F_3 + \cdots + F_N$$

$$= \sum_{i=1}^{N} F_i(P_i) \tag{3.1}$$

$$\phi = 0 = P_R - \sum_{i=1}^{N} P_i \tag{3.2}$$

This is a constrained optimization problem that may be attacked formally using advanced calculus methods that involve the LaGrange function.

In order to establish the necessary conditions for an extreme value of the objective function, add the constraint function to the objective function after the constraint function has been multiplied by an undetermined multiplier. This is known as the *LaGrange function* and is shown in Eq. 3.3.

$$\mathscr{L} = F_T + \lambda\phi \tag{3.3}$$

The necessary conditions for an extreme value of the objective function result when we take the first derivative of the LaGrange function with respect to each of the independent variables and set the derivatives equal to zero. In this case there are $N + 1$ variables, the N values of power output, P_i, plus the undetermined LaGrange multiplier, λ. The derivative of the LaGrange function with respect to the undetermined multiplier merely gives back the constraint equation. On the other hand, the N equations that result when we take the partial derivative of the LaGrange function with respect to the power output values one at a time give the set of equations shown as Eq. 3.4.

$$\frac{\partial \mathscr{L}}{\partial P_i} = \frac{\mathrm{d}F_i(P_i)}{\mathrm{d}P_i} - \lambda = 0$$

or
(3.4)

$$0 = \frac{\mathrm{d}F_i}{\mathrm{d}P_i} - \lambda \quad ; \;\Rightarrow\; \frac{\mathrm{d}F_i}{\mathrm{d}P_i} = \lambda$$

That is, the necessary condition for the existence of a minimum cost-operating condition for the thermal power system is that the incremental cost rates of all the units be equal to some undetermined value, λ. Of course, to this necessary condition we must add the constraint equation that the sum of the power outputs must be equal to the power demanded by the load. In addition, there are two inequalities that must be satisfied for each of the units. That is, the power output of each unit must be greater than or equal to the minimum power permitted and must also be less than or equal to the maximum power permitted on that particular unit.

These conditions and inequalities may be summarized as shown in the set of equations making up Eq. 3.5.

$$\frac{\mathrm{d}F_i}{\mathrm{d}P_i} = \lambda \qquad\qquad N \text{ equations}$$

$$P_{i,\,\min} \leq P_i \leq P_{i,\,\max} \qquad 2N \text{ inequalities}$$
(3.5)

$$\sum_{i=1}^{N} P_i = P_R \qquad\qquad 1 \text{ constraint}$$

When we recognize the inequality constraints, then the necessary conditions may be expanded slightly as shown in the set of equations making up Eq. 3.6.

$$\frac{\mathrm{d}F_i}{\mathrm{d}P_i} = \lambda \qquad \text{for } P_{i,\,\min} < P_i < P_{i,\max}$$

$$\frac{\mathrm{d}F_i}{\mathrm{d}P_i} \leq \lambda \qquad \text{for } P_i = P_{i,\,\max}$$
(3.6)

$$\frac{\mathrm{d}F_i}{\mathrm{d}P_i} \geq \lambda \qquad \text{for } P_i = P_{i,\,\min}$$

Several of the examples in this chapter use the following three generator units.

Unit 1: Coal-fired Steam Unit:
Max output = 600 MW
Min output = 150 MW
Input-output curve:

$$H_1 \left(\frac{\mathrm{MBtu}}{\mathrm{h}}\right) = 510.0 + 7.2\,P_1 + 0.00142\,P_1^2$$

Unit 2: Oil-fired Steam Unit:
Max output = 400 MW
Min output = 100 MW
Input-output curve:

$$H_2 \left(\frac{\text{MBtu}}{\text{h}} \right) = 310.0 + 7.85\, P_2 + 0.00194\, P_2^2$$

Unit 3: Oil-fired Steam Unit:
Max output = 200 MW
Min output = 50 MW
Input-output curve:

$$H_3 \left(\frac{\text{MBtu}}{\text{h}} \right) = 78.0 + 7.97\, P_3 + 0.00482\, P_3^2$$

EXAMPLE 3A

Suppose that we wish to determine the economic operating point for these three units when delivering a total of 850 MW. Before this problem can be solved, the fuel cost of each unit must be specified. Let the following fuel costs be in effect.

Unit 1: Fuel cost = 1.1 ₹/MBtu
Unit 2: Fuel cost = 1.0 ₹/MBtu
Unit 3: Fuel cost = 1.0 ₹/MBtu

Then

$$F_1(P_1) = H_1(P_1) \times 1.1 = 561 + 7.92\, P_1 + 0.001562\, P_1^2 \;\text{₹/h}$$
$$F_2(P_2) = H_2(P_2) \times 1.0 = 310 + 7.85\, P_2 + 0.00194\, P_2^2 \;\text{₹/h}$$
$$F_3(P_3) = H_3(P_3) \times 1.0 = \;\;78 + 7.97\, P_3 + 0.00482\, P_3^2 \;\text{₹/h}$$

Using Eq. 3.5, the conditions for an optimum dispatch are

$$\frac{dF_1}{dP_1} = 7.92 + 0.003124\, P_1 = \lambda$$

$$\frac{dF_2}{dP_2} = 7.85 + 0.00388\, P_2 = \lambda$$

$$\frac{dF_3}{dP_3} = 7.97 + 0.00964\, P_3 = \lambda$$

and

$$P_1 + P_2 + P_3 = 850 \text{ MW}$$

Solving for λ, one obtains

$$\lambda = 9.148 \ \text{R/MWh}$$

and then solving for P_1, P_2, and P_3,

$$P_1 = 393.2 \ \text{MW}$$

$$P_2 = 334.6 \ \text{MW}$$

$$P_3 = 122.2 \ \text{MW}$$

Note that all constraints are met, that is, each unit is within its high and low limit and the total output when summed over all three units meets the desired 850 MW total.

EXAMPLE 3B

Suppose the price of coal decreased to 0.9 R/MBtu. The fuel cost function for unit 1 becomes

$$F_1(P_1) = 459 + 6.48 \ P_1 + 0.00128 \ P_1^2$$

If one goes about the solution exactly as done here, the results are

$$\lambda = 8.284 \ \text{R/MWh}$$

and

$$P_1 = 704.6 \ \text{MW}$$

$$P_2 = 111.8 \ \text{MW}$$

$$P_3 = 32.6 \ \text{MW}$$

This solution meets the constraint requiring total generation to equal 850 MW, but units 1 and 3 are not within limit. To solve for the most economic dispatch while meeting unit limits, use Eq. 3.6.

Suppose unit 1 is set to its maximum output and unit 3 to its minimum output. The dispatch becomes

$$P_1 = 600 \ \text{MW}$$

$$P_2 = 200 \ \text{MW}$$

$$P_3 = 50 \ \text{MW}$$

From Eq. 3.6, we see that λ must equal the incremental cost of unit 2 since it is not at either limit. Then

$$\lambda = \left. \frac{dF_2}{dP_2} \right|_{P_2 = 200} = 8.626 \ \text{R/MWh}$$

Next, calculate the incremental cost for units 1 and 3 to see if they meet the conditions of Eq. 3.6.

$$\left.\frac{dF_1}{dP_1}\right|_{P_1=600} = 8.016 \text{ R/MWh}$$

$$\left.\frac{dF_3}{dP_3}\right|_{P_3=50} = 8.452 \text{ R/MWh}$$

Note that the incremental cost for unit 1 is less than λ, so unit 1 should be at its maximum. However, the incremental cost for unit 3 is not greater than λ, so unit 3 should not be forced to its minimum. Thus, to find the optimal dispatch, allow the incremental cost at units 2 and 3 to equal λ as follows.

$$P_1 = 600 \text{ MW}$$

$$\frac{dF_2}{dP_2} = 7.85 + 0.00388 \, P_2 = \lambda$$

$$\frac{dF_3}{dP_3} = 7.97 + 0.00964 \, P_3 = \lambda$$

$$P_2 + P_3 = 850 - P_1 = 250 \text{ MW}$$

which results in

$$\lambda = 8.576 \text{ R/MWh}$$

and

$$P_2 = 187.1 \text{ MW}$$

$$P_3 = 62.9 \text{ MW}$$

Note that this dispatch meets the conditions of Eq. 3.6 since

$$\left.\frac{dF_1}{dP_1}\right|_{P_1=600 \text{ MW}} = 8.016 \text{ R/MWh}$$

which is less than λ, while

$$\frac{dF_2}{dP_2} \quad \text{and} \quad \frac{dF_3}{dP_3}$$

both equal λ.

3.2 THERMAL SYSTEM DISPATCHING WITH NETWORK LOSSES CONSIDERED

Figure 3.2 shows symbolically an all thermal power generation system connected to an equivalent load bus through a transmission network. The economic-dispatching problem associated with this particular configuration is slightly more complicated to set up than the previous case. This is because the constraint

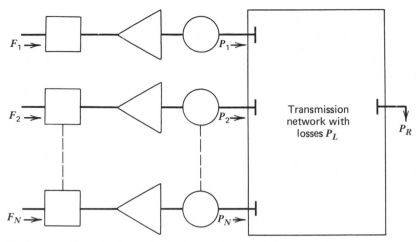

FIG. 3.2 *N* thermal units serving load through transmission network.

equation is now one that must include the network losses. The objective function, F_T, is the same as that defined for Eq. 3.1. However, the constraint equation previously shown in Eq. 3.2 must now be expanded to the one shown in Eq. 3.7.

$$P_R + P_L - \sum_{i=1}^{N} P_i = \phi = 0 \tag{3.7}$$

The same procedure is followed in the formal sense to establish the necessary conditions for a minimum cost operating solution, The LaGrange function is shown in Eq. 3.8. In taking the derivative of the LaGrange function with respect to each of the individual power outputs, P_i, it must be recognized that the loss in the transmission network, P_L, is a function of the network impedances and the currents flowing in the network. For our purposes, the currents will be considered only as a function of the independent variables P_i and the load P_R. Taking the derivative of the LaGrange function with respect to any one of the N values of P_i results in Eq. 3.9. There are N equations of this type to be satisfied along with the constraint equation shown in Eq. 3.7. This collection, Eq. 3.9 plus Eq. 3.7, is known collectively as the *coordination equations*.

$$\mathcal{L} = F_T + \lambda\phi \tag{3.8}$$

$$\frac{\partial\mathcal{L}}{\partial P_i} = \frac{dF_i}{dP_i} - \lambda\left(1 - \frac{\partial P_L}{\partial P_i}\right) = 0 \tag{3.9}$$

or

$$\frac{dF_i}{dP_i} + \lambda\frac{\partial P_L}{\partial P_i} = \lambda \qquad ; \text{ or } \qquad \frac{dF_i}{dP_i} = \lambda\left(1 - \frac{\partial P_L}{\partial P_i}\right)$$

$$P_R + P_L - \sum_{i=1}^{N} P_i = 0$$

It is much more difficult to solve this set of equations than the previous set with no losses since this second set involves the computation of the network loss in order to establish the validity of the solution in satisfying the constraint equation. There have been two general approaches to the solution of this problem. The first is the development of a mathematical expression for the losses in the network solely as a function of the power output of each of the units. This is the loss-formula method discussed at some length in Kirchmayer (*Economic Operation of Power Systems*, Wiley, 1958). The other basic approach to the solution of this problem is to incorporate the load-flow equations as essential constraints in the formal establishment of the optimization problem. This general approach is known as the *optimal load flow*.

EXAMPLE 3C

Starting with the same units and fuel costs as in Example 3A, we will include a simplified loss expression.

$$P_L = 0.00003\, P_1^2 + 0.00009\, P_2^2 + 0.00012\, P_3^2$$

This simplified loss formula will suffice to show the difficulties in calculating a dispatch for which losses are accounted. Note that real-world loss formulas are more complicated than the one used in this example (See Chapter 4).

Applying Eqs. 3.8 and 3.9,

$$\frac{dF_1}{dP_1} = \lambda\left(1 - \frac{\partial P_L}{\partial P_1}\right)$$

becomes

$$7.92 + 0.003124\, P_1 = \lambda[1 - 2(0.00003)P_1]$$

Similarly for P_2 and P_3,

$$7.85 + 0.00388\, P_2 = \lambda[1 - 2(0.00009)P_2]$$

$$7.97 + 0.00964\, P_3 = \lambda[1 - 2(0.00012)P_3]$$

and

$$P_1 + P_2 + P_3 - 850 - P_L = 0$$

We no longer have a set of linear equations as in Example 3A. This necessitates a more complex solution procedure as follows.

Step 1 Pick a set of starting values for P_1, P_2, and P_3 that sum to the load.

Step 2 Calculate the incremental losses $\partial P_L/\partial P_i$ as well as the total losses P_L. The incremental losses and total losses will be considered constant until we return to step 2.

Step 3 Calculate the value of λ that causes P_1, P_2, and P_3 to sum to the total load plus losses. This is now as simple as the calculations in Example 3A since the equations are again linear.

Step 4 Compare the P_1, P_2, and P_3 from step 3 to the values used at the start of step 2. If there is no significant change in any one of the values, go to step 5, otherwise go back to step 2.

Step 5 Done.

Using this procedure, we obtain

Step 1 Pick the P_1, P_2, and P_3 starting values as

$$P_1 = 400.0 \text{ MW}$$

$$P_2 = 300.0 \text{ MW}$$

$$P_3 = 150.0 \text{ MW}$$

Step 2 Incremental losses are

$$\frac{\partial P_L}{\partial P_1} = 2(0.00003)400 = 0.0240$$

$$\frac{\partial P_L}{\partial P_2} = 2(0.00009)300 = 0.0540$$

$$\frac{\partial P_L}{\partial P_3} = 2(0.00012)150 = 0.0360$$

Total losses are 15.6 MW

Step 3 We can now solve for λ using the following.

$$7.92 + 0.003124 \, P_1 = \lambda(1 - 0.0240) = \lambda(0.9760)$$

$$7.85 + 0.00388 \, P_2 = \lambda(1 - 0.0540) = \lambda(0.9460)$$

$$7.97 + 0.00964 \, P_3 = \lambda(1 - 0.0360) = \lambda(0.9640)$$

and

$$P_1 + P_2 + P_3 - 850 - 15.6 = P_1 + P_2 + P_3 - 865.6 = 0$$

These equations are now linear, so we can solve for λ directly. The results are

$$\lambda = 9.5252 \text{ R/MWh}$$

and the resulting generator outputs are

$$P_1 = 440.68$$

$$P_2 = 299.12$$

$$P_3 = 125.77$$

Step 4 Since these values for P_1, P_2, and P_3 are quite different from the starting values, we will return to step 2.

Step 2 The incremental losses are recalculated with the new generation values.

$$\frac{\partial P_L}{\partial P_1} = 2(0.00003)440.68 = 0.0264$$

$$\frac{\partial P_L}{\partial P_2} = 2(0.00009)299.12 = 0.0538$$

$$\frac{\partial P_L}{\partial P_3} = 2(0.00012)125.77 = 0.0301$$

Total losses are 15.78 MW.

Step 3 The new incremental losses and total losses are incorporated into the equations, and a new value of λ and P_1, P_2, and P_3 are solved for

$$7.92 + 0.003124\, P_1 = \lambda(1 - 0.0264) = \lambda(0.9736)$$

$$7.85 + 0.00388\, P_2 = \lambda(1 - 0.0538) = \lambda(0.9462)$$

$$7.97 + 0.00964\, P_3 = \lambda(1 - 0.0301) = \lambda(0.9699)$$

$$P_1 + P_2 + P_3 - 850 - 15.78 = P_1 + P_2 + P_3 - 865.78 = 0$$

resulting in $\lambda = 9.5275$ R/MWh and

$$P_1 = 433.94 \text{ MW}$$

$$P_2 = 300.11 \text{ MW}$$

$$P_3 = 131.74 \text{ MW}$$

$$\vdots$$

Table 3.1 summarizes the iterative process used to solve this problem.

TABLE 3.1 Iterative Process Used to Solve Example 3.

Iteration	P_1(MW)	P_2(MW)	P_3(MW)	Losses (MW)	Lambda (R/MWh)
Start	400.00	300.00	150.00	15.60	9.5252
1	440.68	299.12	125.77	15.78	9.5275
2	433.94	300.11	131.74	15.84	9.5285
3	435.87	299.94	130.42	15.83	9.5283
4	435.13	299.99	130.71	15.83	9.5284

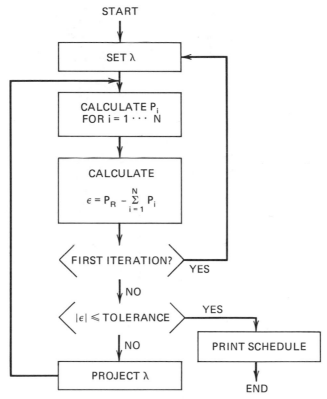

FIG. 3.3 Economic dispatch by the Lambda-iteration method.

3.3 THE LAMBDA-ITERATION METHOD

Figure 3.3 is a block diagram of the lambda-iteration method of solution for the all thermal, dispatching problem-neglecting losses. We can approach the solution to this problem by considering a graphical technique for solving the problem and then extending this into the area of computer algorithms.

Suppose we had a three-machine system and wish to find the optimum economic operating point. One approach would be to plot the incremental cost characteristics for each of these three units on the same graph, such as sketched in Figure 3.4. In order to establish the operating points of each of these three units such that we have minimum cost and at the same time satisfy the specified demand, we could use this sketch and a ruler to find the solution. That is, we could assume an incremental cost rate (λ) and find the power outputs of each of the three units for this value of incremental cost.

Of course, our first estimate will be incorrect. If we have assumed the value of incremental cost such that the total power output is too low, we must increase the λ value and try another solution. With two solutions, we can extrapolate (or

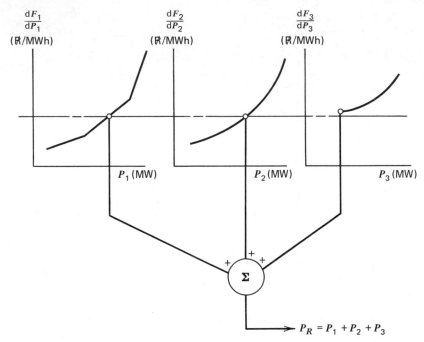

FIG. 3.4 Graphical solution to economic dispatch.

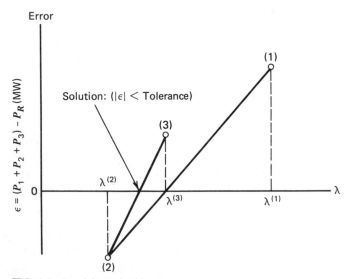

FIG. 3.5 Lambda projections.

interpolate) the two solutions to get closer to the desired value of total received power (see Figure 3.5).

By keeping track of the total demand versus the incremental cost, we can rapidly find the desired operating point. If we wished, we could manufacture a whole series of tables that would show the total power supplied for different incremental cost levels and combinations of units.

This same procedure can be adopted for a computer implementation as shown in Figure 3.3. That is, we will now establish a set of logical rules that would enable us to accomplish the same objective as we have just done with ruler and graph paper. The actual details of how the power output is established as a function of the incremental cost rate are of very little importance. We could, for example, store tables of data within the computer and interpolate between the stored power points to find exact power output for a specified value of incremental cost rate. Another approach would be to develop an analytical function for the power output as a function of the incremental cost rate, store this function (or its coefficients) in the computer, and use this to establish the output of each of the individual units.

This procedure is an iterative type of computation, and we must establish stopping rules. Two general forms of stopping rules seem appropriate for this application. The first is shown in Figure 3.3 and is essentially a rule based on finding the proper operating point within a specified tolerance. The other, not shown in Figure 3.3, involves counting the number of times through the iterative loop and stopping when a maximum number is exceeded.

The lambda-iteration procedure converges very rapidly for this particular type of optimization problem. The actual computational procedure is slightly more complex than that indicated in Figure 3.3 since it is necessary to observe the operating limits on each of the units during the course of the computation. The well-known Newton-Raphson method may be used to project the incremental cost value to drive the error between the computed and desired generation to zero.

EXAMPLE 3D

Assume that one wishes to use cubic functions to represent the input/output charactcristics of generating plants as follows.

$$H \text{ (MBtu/h)} = A + BP + CP^2 + DP^3 \qquad (P \text{ in MW})$$

For the three units, find the optimum schedule using the lambda-iteration method.

	A	B	C	D
Unit 1	749.55	6.95	9.68×10^{-4}	1.27×10^{-7}
Unit 2	1285.0	7.051	7.375×10^{-4}	6.453×10^{-8}
Unit 3	1531.0	6.531	1.04×10^{-3}	9.98×10^{-8}

Assume the fuel cost to be 1.0 ₽/MBtu for each unit and unit limits as follows.

$$320 \text{ MW} \le P_1 \le 800 \text{ MW}$$

$$300 \text{ MW} \le P_2 \le 1200 \text{ MW}$$

$$275 \text{ MW} \le P_3 \le 1100 \text{ MW}$$

Two sample calculations are shown, both using the flowchart in Figure 3.3. In this calculation, the value for λ on the second iteration is always set at 10% above or below the starting value depending on the sign of the error; for the remaining iterations, lambda is projected as in Figure 3.5.

The first example shows the advantage of starting λ near the optimum value.

$$P_R = 2500 \text{ MW}$$

$$\lambda_{start} = 8.0 \text{ ₽/MWh}$$

The second example shows the oscillatory problems that can be encountered with a lambda-iteration approach.

$$P_R = 2500 \text{ MW}$$

$$\lambda_{start} = 10.0 \text{ ₽/MWh}$$

Iteration	Lambda	Total generation (MW)	P_1	P_2	P_3
1	8.0000	1731.5	494.3	596.7	640.6
2	8.8000	2795.0	800.0	1043.0	952.0
3	8.5781	2526.0	734.7	923.4	867.9
4	8.5566	2497.5	726.1	911.7	859.7
5	8.5586	2500.0	726.9	912.7	860.4

Iteration	Lambda	Total generation (MW)	P_1	P_2	P_3
1	10.0000	3100.0	800.0	1200.0	1100.0
2	9.0000	2974.8	800.0	1148.3	1026.5
3	5.2068	895.0	320.0	300.0	275.0
4	8.1340	1920.6	551.7	674.5	694.4
5	9.7878	3100.0	800.0	1200.0	1100.0
6	8.9465	2927.0	800.0	1120.3	1006.7
7	6.8692	895.0	320.0	300.0	275.0
8	8.5099	2435.0	707.3	886.1	841.7
9	8.5791	2527.4	735.1	924.0	868.3
10	8.5586	2500.1	726.9	912.8	860.4

3.4 FIRST-ORDER GRADIENT METHOD

Many optimization procedures are nothing more than directed enumeration, or search, procedures. Such is the case with the methods that are known as *gradient methods.* These methods are well suited for the type of problem where there is a great deal of physical understanding of the nature of the problem and the behavior of the objective function as the independent variables are adjusted. Gradient methods have one very strong desirable characteristic. That is, gradient search techniques always start off with a feasible solution and search for the optimum solution along a trajectory that maintains a feasible solution at all times. By "feasible solution" is meant a solution in which all the constraint conditions are met.

There may be other methods of gradient search that start from nonfeasible solutions and search for solutions that are both optimal and feasible. However, the most commonly used gradient search techniques always provide feasible answers. This is a very desirable characteristic since the computational procedure may be interrupted at any time and the most recent solution point will still be a reasonable operating point of which to make use. If the method is such that the objective function is monotonically changing as each new operating point is established, then the most recent solution is the best one found to that particular stage of the computation. The disadvantage of this type of optimization procedure is that there is no clear "stopping rule." That is, the establishment of "the" optimum point is difficult. Usually, the gradient search method is allowed to continue until the search procedure can find no additional "significant" gain in the objective function or until the number of iterations has exceeded some reasonably high value.

To start the gradient technique, begin with the objective function of Eq. 3.1 and/or the constraint equation (Eq. 3.2). Next, assume that the power system is operating at a feasible operating point. That is, the sum of the power outputs of the units is equal to the load demand. Next, assume that the power outputs of each of these units is changed by some small amount and that a new feasible operating point is found.

Equation 3.10 is the Taylor-series expansion of the objective function of Eq. 3.1, expanded about the initial feasible operating point. Terms are shown in this Taylor-series expansion up to and including the second order. To a first order the change in the objective function is shown in Eq. 3.11. We have obtained this relationship by assuming that all second-order and higher terms may be neglected relative to the first-order terms and by subtracting the initial operating cost from the perturbed equation.

$$F_T + \Delta F_T = F_1(P_1) + F_2(P_2) + \cdots + F_N(P_N)$$

$$+ \frac{dF_1}{dP_1} \Delta P_1 + \frac{dF_2}{dP_2} \Delta P_2 + \cdots + \frac{dF_N}{dP_N} \Delta P_N$$

$$+ \frac{1}{2} \left(\frac{d^2F_1}{dP_1^2} [\Delta P_1]^2 + \frac{d^2F_2}{dP_2^2} [\Delta P_2]^2 + \cdots \right) + \cdots \qquad (3.10)$$

$$\Delta F_T = \frac{dF_1}{dP_1} \Delta P_1 + \frac{dF_2}{dP_2} \Delta P_2 + \cdots + \frac{dF_N}{dP_N} \Delta P_N \qquad (3.11)$$

Next, consider the constraint equation (Eq. 3.2) and allow each of the powers to be perturbed some small amount. The constraint equation is as follows.

$$\sum_{i=1}^{N} \Delta P_i = 0 \tag{3.12}$$

In this relationship again the initial values have been subtracted so that the constraint equation now merely states that the sum of the changes in all the power outputs must be zero. This constraint relationship removes one degree of freedom from the problem so that at least one unit must be selected as a dependent variable. Let us call this dependent unit the xth unit. In that case, the change in power output for this dependent unit is a negative sum of the change in power outputs of the remaining $N - 1$ units as shown in Eq. 3.13.

$$\Delta P_x = -\sum_{i \neq x} \Delta P_i \tag{3.13}$$

These two equations may be combined to establish an equation that gives the change in the objective function, F_T, as a function of the change in the power output of each of the $N - 1$ independent machines. This is shown in Eq. 3.14. The procedure for using this equation is indicated in Figure 3.6, which is a functional block diagram of a first-order gradient search method.

$$\Delta F_T = \sum_{i \neq x} \left[\frac{dF_i}{dP_i} - \frac{dF_x}{dP_x} \right] \Delta P_i = \sum_{i \neq x} \frac{\partial F_T}{\partial P_i} \Delta P_i \tag{3.14}$$

The technique starts with any feasible solution, that is, a solution that satisfies the constraint equation (Eq. 3.2). Next, the coefficients in Eq. 3.14 are evaluated for this operating point after the dependent machine, x, has been selected. This results in the numerical expression of Eq. 3.14, and the coefficients of each of the $N - 1$ power variables indicate the gain in objective function that may be achieved for a unit move in the power of that particular machine. Obviously, the machine that gains the most reduction in the operating cost function is the one to move first. This unit can be obtained by finding the maximum of the absolute values of the various coefficients of the changes in the power output.

When a move is selected, the amount to move the power must be checked carefully so that neither the independent unit (the ith unit) nor the dependent unit (the xth unit) is moved beyond its operating constraints. Both units must be moved in an equal and opposite manner so that the new operating point is still feasible.

As Figure 3.6 shows, the operating points should be checked after each step, or preferably before each step is made permanently. If the dependent unit comes up against an operating limit or an operating limit for the dependent unit is violated, then it is time to return to that stage of the computation where the dependent unit is selected. This procedure is quite simple and straightforward. It does require, however, a large number of iterations to come to a satisfactory convergence.

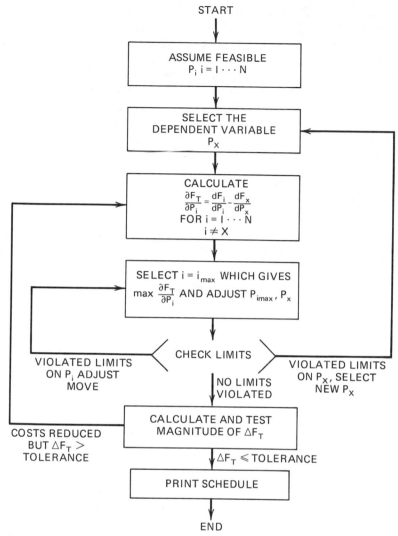

START

ASSUME FEASIBLE
$P_i \ i = 1 \cdots N$

SELECT THE
DEPENDENT VARIABLE
P_X

CALCULATE
$$\frac{\partial F_T}{\partial P_i} = \frac{dF_i}{dP_i} - \frac{dF_x}{dP_x}$$
FOR $i = 1 \cdots N$
$i \neq X$

SELECT $i = i_{max}$ WHICH GIVES
max $\frac{\partial F_T}{\partial P_i}$ AND ADJUST $P_{i_{max}}$, P_x

CHECK LIMITS

VIOLATED LIMITS
ON P_i ADJUST
MOVE

NO LIMITS
VIOLATED

VIOLATED LIMITS
ON P_X, SELECT
NEW P_X

CALCULATE AND TEST
MAGNITUDE OF ΔF_T

COSTS REDUCED
BUT $\Delta F_T >$
TOLERANCE

$\Delta F_T \leqslant$ TOLERANCE

PRINT SCHEDULE

END

FIG. 3.6   Economic dispatch by first-order gradient search.

Other procedures may be used to augment this first-order gradient search method. For example, in the coefficients shown in Eq. 3.14 the difference in incremental costs is a function of the operating point of the units taken two at a time. That is, given an initial set of operating points, these coefficients may be established. They will change, however, as the operating points of the unit are varied so that they are in fact functions of the changes in the operating power output, the ΔP_i values. Regarded in this way, you may look at the changes in the operating power point of each of the units.

If we now select a particular pair of units to move (that is, both i and x), then we can drive the coefficient as close to zero as is possible. Neglecting any operating limits of the units, it would be possible to move the power of the ith unit and x^{th} unit such that the two incremental costs will be equal and that the coefficient of the change in the output would vanish. This may easily be done for practical scheduling algorithms, particularly if the incremental cost characteristics are simple analytic functions or simple tabulated values. In that case, we can easily determine the "optimum step size" to take in each step of the gradient method. We would, however, restrict our movements so that we only moved in one direction at a time.

EXAMPLE 3E

Again, taking the data of Example 3A, solve for an economic dispatch of generation by the first-order gradient technique shown in Figure 3.6. The search will be started as follows.

$$\left. \begin{array}{l} P_1 = 400 \text{ MW} \\ P_2 = 300 \text{ MW} \\ P_3 = 150 \text{ MW} \end{array} \right\} \text{ Iteration 1}$$

We can use this starting point since it meets the condition that $P_1 + P_2 + P_3 = 850$ MW. The dependent variable, x, will be unit 3 output P_3. Then

$$\Delta F = \left(\frac{dF_1}{dP_1} - \frac{dF_3}{dP_3} \right) \Delta P_1 + \left(\frac{dF_2}{dP_2} - \frac{dF_3}{dP_3} \right) \Delta P_2$$

$$= (-0.2464) \Delta P_1 + (-0.4020) \Delta P_2$$

and

$$F_T = 8200.47 \text{ R/h.}$$

Since the larger coefficient appears with ΔP_2, we will move P_2 up. That is, we wish to decrease F (ΔF negative), so increasing P_2 (ΔP_2 positive) will achieve this since its coefficient is negative. The next iteration's conditions are (after increasing P_2 by 50 MW and decreasing P_3 by 50 MW)

$$\left. \begin{array}{l} P_1 = 400 \text{ MW} \\ P_2 = 350 \text{ MW} \\ P_3 = 100 \text{ MW} \end{array} \right\} \text{ Iteration 2}$$

which results in

$$\Delta F = (0.2356) \Delta P_1 + (0.2740) \Delta P_2$$

$$F_T = 8197.27 \text{ R/h}$$

Succeeding iterations are shown in the following table.

Iteration	P_1(MW)	P_2(MW)	P_3(MW)	F(R/h)	ΔF	ΔP_1(MW)	ΔP_2(MW)	ΔP_3(MW)
1	400	300	150	8200.47	$(-0.2464)\Delta P_1 + (-0.4020)\Delta P_2$	0	+50	−50
2	400	350	100	8197.27	$(0.2356)\Delta P_1 + (0.2740)\Delta P_2$	0	−25	+25
3	400	325	125	8194.65	$(-0.0054)\Delta P_1 + (-0.064)\Delta P_2$	0	+12.5	−12.5
4	400	337.5	112.5	8194.90	$(0.1151)\Delta P_1 + (0.105)\Delta P_2$	−10	0	+10
5	390	337.5	122.5	8194.38	$(-0.01254)\Delta P_1 + (-0.00861)\Delta P_2$	+5	0	−5
6	395	337.5	117.5	8194.48	$(0.05128)\Delta P_1 + (0.0568)\Delta P_2$	0	−2.5	+2.5
7	395	335	120	8194.38	$(0.0272)\Delta P_1 + (0.023)\Delta P_2$			

3.5 SECOND-ORDER GRADIENT METHOD

Let us return to the problem of finding the economic schedule for a given single level of demand. The gradient method shown in Section 3.4 may be refined and made more efficient by using the second-order terms of the Taylor series expansion of the total generation cost given in Eq. 3.10.

$$F_T + \Delta F_T = F_1(P_1) + F_2(P_2) + \cdots + F_N(P_N)$$

$$+ \frac{dF_1}{dP_1} \Delta P_1 + \frac{dF_2}{dP_2} \Delta P_2 + \cdots + \frac{dF_N}{dP_N} \Delta P_N$$

$$+ \frac{1}{2}\left(\frac{d^2F_1}{dP_1^2} (\Delta P_1)^2 + \frac{d^2F_2}{dP_2^2} (\Delta P_2)^2 + \cdots + \frac{d^2F_N}{dP_N^2} (\Delta P_N)^2 \right)$$

$$+ \cdots \tag{3.15}$$

Recall again that there are no mixed, second partial derivatives in the usual case. That is, the second derivative of the fuel cost rate of a given unit is normally only explicitly dependent on the power output of that unit, itself. That is,

$$\frac{\partial^2 F_i}{\partial P_i \partial P_j} = 0$$

for $i \neq j$. Also, the constraint requiring the sum of the power outputs of the individual units to equal the total demand must be treated as in Eqs. 3.12 and 3.13 in order that the schedule developed will not alter the frequency of the system. These are

$$\sum_{i=1}^{N} \Delta P_i = 0 \tag{3.16}$$

$$\Delta P_x = -\sum_{i \neq x} \Delta P_i \tag{3.17}$$

If we substitute the relationship of Eq. 3.17 into Eq. 3.15 keeping *both* the first- and second-order terms involving the perturbed powers, then we obtain

$$\Delta F_T = \sum_{i \neq x} \left(\frac{dF_i}{dP_i} - \frac{dF_x}{dP_x} \right) \Delta P_i$$

$$+ \frac{1}{2} \left\{ \sum_{i \neq x} \left[\frac{d^2F_1}{dP_1^2} (\Delta P_1)^2 + \frac{d^2F_2}{dP_2^2} (\Delta P_2)^2 + \cdots \right] \right.$$

$$\left. + \frac{d^2F_x}{dP_x^2} (\Delta P_1^2 + \Delta P_2^2 + \cdots + 2\Delta P_1 \Delta P_2 + 2\Delta P_1 \Delta P_3 + \cdots) \right\} \tag{3.18}$$

The change in the total operating cost, ΔF_T, can be treated using ordinary calculus methods since it is a function of the $N - 1$ independent changes in the power output levels, ΔP_i. There are no constraint conditions other than the limits on the plant outputs, which we will ignore at present. The best operating point will be achieved when the partial derivative of ΔF_T with respect to each independent variable, ΔP_i, is zero. That is, the partial derivatives $\partial \Delta F_T / \partial \Delta P_i$ must be zero for all i, $i \neq x$. These derivatives result in a set of simultaneous equations.

$$\frac{\partial \Delta F_T}{\partial \Delta P_1} = 0 = \left(\frac{dF_1}{dP_1} - \frac{dF_x}{dP_x}\right) + \frac{d^2 F_1}{dP_1^2} \Delta P_1 + \frac{d^2 F_x}{dP_x^2} \sum_{i \neq x} \Delta P_i$$

$$\frac{\partial \Delta F_T}{\partial \Delta P_2} = 0 = \left(\frac{dF_2}{dP_2} - \frac{dF_x}{dP_x}\right) + \frac{d^2 F_2}{dP_2^2} \Delta P_2 + \frac{d^2 F_x}{dP_x^2} \sum_{i \neq x} \Delta P_i \qquad (3.19)$$

$$\vdots$$

The presentation may be made somewhat clearer by letting

$$F_i' = \frac{dF_i}{dP_i}$$

and $\qquad (3.20)$

$$F_i'' = \frac{d^2 F_i}{dP_i^2}$$

both be evaluated for the initial operating point. Then the $N - 1$ simultaneous equations may be written in matrix form as

$$\begin{bmatrix} F_1'' + F_x'' & F_x'' & F_x'' & \cdots \\ F_x'' & F_2'' + F_x'' & F_x'' & \cdots \\ F_x'' & F_x'' & F_3'' + F_x'' & \cdots \\ \vdots & \vdots & \vdots & \end{bmatrix} \begin{bmatrix} \Delta P_1 \\ \Delta P_2 \\ \Delta P_3 \\ \vdots \end{bmatrix} = - \begin{bmatrix} F_1' - F_x' \\ F_2' - F_x' \\ F_3' - F_x' \\ \vdots \end{bmatrix} \qquad (3.21)$$

The second-order gradient search method yields the multidimensional components of the gradient of the cost function by the following steps.

1. Start with a feasible solution and compute the elements of the matrix equation.
2. Invert the matrix to solve for the amounts to shift each generator.
3. Verify that the solution does not violate any constraint.
4. Check the values of the vector $(F_i' - F_x')$ at the new operating point to see that all the incremental costs are equal given the operating limitations of each unit.

If the incremental costs are not equal, the procedure is repeated.

With a quadratic objective function, the second-order method will converge in one step assuming an unconstrained situation. The examples used in this text may lead you to the impression that the second-order method is one that always converges in a single step and does not seem to have any of the computational problems associated with the first-order methods. Don't you believe it! These methods all have their difficulties. Generally, in using gradient techniques it is the constraints that give rise to computational problems. (In some other optimization techniques these very same constraints are a blessing and serve to reduce the dimensionality of the problem.)

EXAMPLE 3F

We will solve the same dispatch problem (Example 3A) from the same starting conditions as in Example 3E using the second-order gradient method. The starting conditions are

$$P_1 = 400 \text{ MW}$$

$$P_2 = 300 \text{ MW}$$

$$P_3 = 150 \text{ MW}$$

$$F'_1 = 9.1696 \text{ R/MWh} \qquad F''_1 = 0.003124 \text{ (R/MWh)/MW}$$

$$F'_2 = 9.0140 \text{ R/MWh} \qquad F''_2 = 0.00388 \text{ (R/MWh)/MW}$$

$$F'_3 = 9.4160 \text{ R/MWh} \qquad F''_3 = 0.00964 \text{ (R/MWh)/MW}$$

Then using Eq. 3.21,

$$
\begin{bmatrix} F''_1 + F''_3 & F''_3 \\ F''_3 & F''_2 + F''_3 \end{bmatrix} \begin{bmatrix} \Delta P_1 \\ \Delta P_2 \end{bmatrix} = - \begin{bmatrix} F'_1 - F'_3 \\ F'_2 - F'_3 \end{bmatrix}
$$

or

$$
\begin{bmatrix} 0.012764 & 0.00964 \\ 0.00964 & 0.01352 \end{bmatrix} \begin{bmatrix} \Delta P_1 \\ \Delta P_2 \end{bmatrix} = - \begin{bmatrix} -0.2464 \\ -0.4020 \end{bmatrix}
$$

The solution of which is

$$
\begin{bmatrix} \Delta P_1 \\ \Delta P_2 \end{bmatrix} = \begin{bmatrix} -6.8301 \\ 34.603 \end{bmatrix}
$$

Then

$$P_1 = 400 - 6.83 = 393.17 \text{ MW}$$

$$P_2 = 300 + 34.6 = 334.6 \text{ MW}$$

and by definition $P_3 = 850 - P_1 - P_2 = 122.23$ MW.

3.6 BASE POINT AND PARTICIPATION FACTORS

This method assumes that the economic dispatch problem has to be solved repeatedly by moving the generators from one economically optimum schedule to another as the load changes by a reasonably small amount. We start from a given schedule—the *base point*. Next, the scheduler assumes a load change and investigates how much each generating unit needs to be moved (i.e., "participate" in the load change) in order that the new load be served at the most economic operating point.

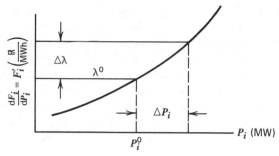

FIG. 3.7 Relationship of $\Delta\lambda$ and ΔP_i.

Assume that both the first and second derivatives in the cost versus power output function are available (i.e., both F_i' and F_i'' exist). The incremental cost curve of the ith unit is given in Figure 3.7. As the unit load is changed by an amount ΔP_i, the system incremental cost moves from λ^0 to $\lambda^0 + \Delta\lambda$. For a small change in power output on this single unit,

$$\Delta\lambda_i = \Delta\lambda \cong F_i''(P_i^0)\Delta P_i \tag{3.22}$$

This is true for each of the N units on the system so that,

$$\Delta P_1 = \frac{\Delta\lambda}{F_1''}$$

$$\Delta P_2 = \frac{\Delta\lambda}{F_2''}$$

$$\vdots$$

$$\Delta P_N = \frac{\Delta\lambda}{F_N''}$$

The total change in generation (= change in total system demand) is, of course, the sum of the individual unit changes. Let P_D be the total demand on the generators (where $P_D = P_R + \text{losses}$), then

$$\Delta P_D = \Delta P_1 + \Delta P_2 + \cdots + \Delta P_N$$

$$= \Delta\lambda \sum_i \left(\frac{1}{F_i''}\right) \tag{3.23}$$

The earlier equation, 3.22, can be used to find the *participation factor* for each unit as follows.

$$\left(\frac{\Delta P_i}{\Delta P_D}\right) = \frac{(1/F_i'')}{\sum_i \left(\frac{1}{F_i''}\right)} \tag{3.24}$$

The computer implementation of such a scheme of economic dispatch is straightforward. It might be done by provision of tables of the values of F_i'' as a

---

function of the load levels and devising a simple scheme to take the existing load plus the projected increase to look up these data and compute the factors.

A somewhat less elegant scheme to provide participation factors would involve a repeat economic dispatch calculation at $P_D^0 + \Delta P_D$. The base-point economic generation values are then subtracted from the new economic generation values and the difference divided by ΔP_D to provide the participation factors. This scheme works well in computer implementations where the execution time for the economic dispatch is short and will always give consistent answers when units reach limits or pass through break points on piecewise linear incremental cost functions.

EXAMPLE 3G

Starting from the optimal economic solution found in Example 3A, use the participation factor method to calculate the dispatch for a total load of 900 MW.
Using Eq. 3.24,

$$\frac{\Delta P_1}{\Delta P_D} = \frac{(0.003124)^{-1}}{(0.003124)^{-1} + (0.00388)^{-1} + (0.00964)^{-1}} = \frac{320.10}{681.57} = 0.47$$

Similarly,

$$\frac{\Delta P_2}{\Delta P_D} = \frac{(0.00388)^{-1}}{681.57} = 0.38$$

$$\frac{\Delta P_3}{\Delta P_D} = \frac{103.73}{681.57} = 0.15$$

$$\Delta P_D = 900 - 850 = 50$$

The new value of generation is calculated using

$$P_{new_i} = P_{base_i} + \left(\frac{\Delta P_i}{\Delta P_D}\right)\Delta P_D \qquad \text{for } i = 1, 2, 3$$

Then for each unit

$$P_{new_1} = 393.2 + (0.47)(50) = 416.7$$

$$P_{new_2} = 334.6 + (0.38)(50) = 353.6$$

$$P_{new_3} = 122.2 + (0.15)(50) = 129.7$$

3.7 ECONOMIC DISPATCH VERSUS UNIT COMMITMENT

At this point it may be well to emphasize the essential difference between the unit commitment and economic dispatch problem. The economic dispatch problem *assumes* that there are N units already connected to the system. The purpose of the economic dispatch problem is to find the optimum operating policy for these N units. This is the problem that we have been investigating so far in this text.

On the other hand, the unit commitment problem is more complex. We may assume that we have N units available to us and that we have a forecast of the demand to be served. The question that is asked in the unit commitment problem area is approximately as follows.

> Given that there are a number of subsets of the complete set of N generating units that would satisfy the expected demand, which of these subsets should be used in order to provide the minimum operating cost?

This unit commitment problem may be extended over some period of time, such as the 24 h of a day or the 168 h of a week. The unit commitment problem is a much more difficult problem to solve. The solution procedures involve the economic dispatch problem as a subproblem. That is, for each of the subsets of the total number of units that are to be tested, for any given set of them connected to the load the particular subset should be operated in optimum economic fashion. This will permit finding the minimum operating cost for that subset, but it does not establish which of the subsets is in fact the one that will give minimum cost over a period of time.

A later chapter will consider the unit commitment problem in some detail. The problem is more difficult to solve mathematically since it involves integer variables. That is, generating units must be either all on or all off. (How can you turn a switch half on?) Methods of solving optimization problems with integer variables have developed in recent years. The two most widely used techniques are *integer programming* and *dynamic programming*. A later section will explore the use of dynamic programming for the solution of a restricted range of unit commitment problems.

At this point (1983), there is no satisfactory, theoretically correct solution procedure for solving the unit commitment problem for a large system.

APPENDIX
Optimization Within Constraints

Suppose you are trying to maximize or minimize a function of several variables. It is relatively straightforward to find the maximum or minimum using rules of calculus. First, of course, you must find a set of values for the variables where the first derivative of the function with respect to each variable is zero. In addition, the second derivatives should be used to determine whether the solution found is a maximum, minimum, or a saddle point.

In optimizing a real-life problem one is usually confronted with a function to be maximized or minimized as well as numerous constraints that must be met. The constraints, sometimes called *side conditions*, can be other functions with conditions that must be met or they can be simple conditions such as limits on the variables themselves.

Before we begin this discussion on constrained optimization, we will put down some definitions. Since the objective is to maximize or minimize a mathematical function, we will call this function the *objective function*. The constraint functions and simple variable limits will be lumped under the term *constraints*. The region defined by the constraints is said to be the *feasible region* for the independent variables. If the constraints are such that no such region exists, that is, there are no values for the independent variables that satisfy all the constraints, then the problem is said to have an *infeasible* solution. When an optimum solution to a constrained optimization problem occurs at the boundary of the feasible region defined by a constraint, we say the constraint is *binding*. If the optimum solution lies away from the boundary, the constraint is *nonbinding*.

To begin, let us look at a simple elliptical objective function.

$$f(x_1, x_2) = 0.25\,x_1^2 + x_2^2 \qquad (3A.1)$$

This is shown in Figure 3.8 for various values of f.

Note that the minimum value f can attain is zero, but that it has no finite maximum value. The following is an example of a constrained optimization problem.

Minimize: $\qquad\qquad\qquad f(x_1, x_2) = 0.25\,x_1^2 + x_2^2$

Subject to the Constraint: $\quad \omega(x_1, x_2) = 0 \qquad\qquad\qquad (3A.2)$

Where: $\qquad\qquad\qquad\qquad \omega(x_1, x_2) = 5 - x_1 - x_2.$

This optimization problem can be pictured as in Figure 3.9.

We need to observe that the optimum pictured gives the minimum value for our objective function, f, while also meeting the constraint function, ω. This optimum

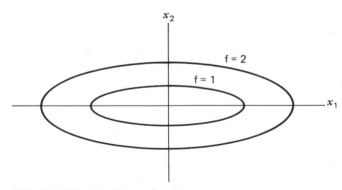

FIG. 3.8 Elliptical objective function.

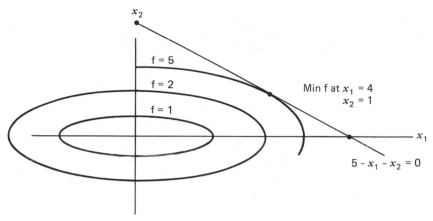

FIG. 3.9 Elliptical objective function with equality constraint.

point occurs where the function f is exactly tangent to the function ω. Indeed, this observation can be made more rigorous and will form the basis for our development of LaGrange multipliers.

First, redraw the function f for several values of f around the optimum point. At the point (x_1', x_2') calculate the gradient vector of f. This is pictured in Figure 3.10 as $\nabla f(x_1', x_2')$. Note that the gradient at (x_1', x_2') is perpendicular to f but not to ω and therefore has a nonzero component along ω. Similarly, at the point (x_1'', x_2'') the gradient of f has a nonzero component along ω. The nonzero component of the

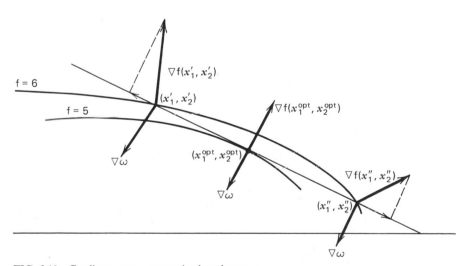

FIG. 3.10 Gradients near a constrained optimum.

gradient along ω tells us that a small move along ω in the direction of this component will increase the objective function. Therefore, to minimize f we should go along ω in the opposite direction to the component of the gradient projected onto ω. At the optimum point, the gradient of f is perpendicular (mathematicians say "normal"), to ω and therefore there can be no improvement in f by moving off this point. We can solve for this optimum point mathematically by using this "normal" property at the optimum. To guarantee that the gradient of f, (i.e. ∇f,) is normal to ω we simply require that ∇f and the gradient of ω, $\nabla\omega$, be linearly dependent vectors. Vectors that are linearly dependent must "line up" with each other (i.e., they point in exactly the same or exactly the opposite direction) although they may be different in magnitude. Mathematically, we can then set up the following equation.

$$\nabla f + \lambda \nabla \omega = 0 \tag{3A.3}$$

That is, the two gradients can be added together in such a way that they cancel each other as long as one of them is scaled. The scaling variable, λ, is called a *LaGrange multiplier*, and instead of using the gradients as shown in Eq. 3A.3, we will restate them as

$$\mathscr{L}(x_1, x_2, \lambda) = f(x_1, x_2) + \lambda\omega(x_1, x_2) \tag{3A.4}$$

This equation is called the *LaGrange equation* and consists of three variables, x_1, x_2, and λ. When we solve for the optimum values for x_1 and x_2, we will automatically calculate the correct value for λ. To meet the conditions set down in Eq. 3A.3, we simply require that the partial derivative of $\mathscr{L}$ with respect to each of the unknown variables, x_1, x_2, and λ, be equal to zero. That is,

At the Optimum

$$\frac{\partial \mathscr{L}}{\partial x_1} = 0$$

$$\frac{\partial \mathscr{L}}{\partial x_2} = 0 \tag{3A.5}$$

$$\frac{\partial \mathscr{L}}{\partial \lambda} = 0$$

To show how this works, solve for the optimum point for the sample problem using LaGrange's method.

$$\mathscr{L}(x_1, x_2, \lambda) = 0.25 x_1^2 + x_2^2 + \lambda(5 - x_1 - x_2)$$

$$\frac{\partial \mathscr{L}}{\partial x_1} = 0.5 x_1 - \lambda = 0$$

$$\frac{\partial \mathscr{L}}{\partial x_2} = 2 x_2 - \lambda = 0 \tag{3A.6}$$

$$\frac{\partial \mathscr{L}}{\partial \lambda} = 5 - x_1 - x_2 = 0$$

Note that the last equation in (3A.6) is simply the original constraint equation. The solution to Eq. 3A.6 is

$$x_1 = 4$$

$$x_2 = 1 \tag{3A.7}$$

$$\lambda = 2$$

When there is more than one constraints present in the problem, the optimum point can be found in a similar manner to that just used. Suppose there were three constraints to be met, then our problem would be as follows,

Minimize: $f(x_1, x_2)$

Subject to: $\omega_1(x_1, x_2) = 0$

$$\omega_2(x_1, x_2) = 0 \tag{3A.8}$$

$$\omega_3(x_1, x_2) = 0$$

The optimum point would possess the property that the gradient of f and the gradients of ω_1, ω_2, and ω_3 are linearly dependent. That is,

$$\nabla f + \lambda_1 \nabla \omega_1 + \lambda_2 \nabla \omega_2 + \lambda_3 \nabla \omega_3 = 0 \tag{3A.9}$$

Again, we can set up a LaGrangian equation as before.

$$\mathscr{L} = f(x_1, x_2) + \lambda_1 \omega_1(x_1, x_2) + \lambda_2 \omega_2(x_1, x_2) + \lambda_3 \omega_3(x_1, x_2) \tag{3A.10}$$

whose optimum occurs at

$$\frac{\partial \mathscr{L}}{\partial x_1} = 0 \qquad \frac{\partial \mathscr{L}}{\partial x_2} = 0$$

$$\frac{\partial \mathscr{L}}{\partial \lambda_1} = 0 \qquad \frac{\partial \mathscr{L}}{\partial \lambda_2} = 0 \qquad \frac{\partial \mathscr{L}}{\partial \lambda_3} = 0 \tag{3A.11}$$

Up until now we have assumed that all the constraints in the problem were equality constraints, that is, $\omega(x_1, x_2, \ldots) = 0$. In general, however, optimization problems involve inequality constraints, that is, $g(x_1, x_2, \ldots) \leq 0$ as well as equality constraints. The optimal solution to such problems will not necessarily require all the inequality constraints to be binding. Those that are binding will result in $g(x_1, x_2, \ldots) = 0$ at the optimum.

The fundmental rule that tells when the optimum has been reached is presented in a famous paper by Kuhn and Tucker (3). *The Kuhn-Tucker conditions*, as they are called, are presented here.

Minimize: $f(\mathbf{x})$ $= F(P_i)$

Subject to: $\omega_i(\mathbf{x}) = 0 \qquad i = 1, 2, \ldots, N\omega$ $\quad P_R - \sum_i P_i$

$g_i(\mathbf{x}) \leq 0 \qquad i = 1, 2, \ldots, Ng$ $\quad \begin{cases} P_i - P_{max} \leq 0 \\ P_{min} - P_i \leq 0 \end{cases}$

$\mathbf{x} = $ vector of real numbers, dimension $= N$

$$\mathcal{L} = F(P_i) + \lambda\left(P_r - \sum_{c=1}^{N} P_i\right) + \sum_{i=1}^{N} \mu_i\left(P_i - \ell_{max}\right) + \sum_{i=1}^{N} \mu_i^*\left(\ell_{min_i} - P_i\right)$$

Then forming the LaGrange function,

$$\mathcal{L}(\mathbf{x}, \lambda, \mu) = f(\mathbf{x}) + \sum_{i=1}^{N\omega} \lambda_i \omega_i(\mathbf{x}) + \sum_{i=1}^{Ng} \mu_i g_i(\mathbf{x})$$

The conditions for an optimum for the point $\mathbf{x}^0$, λ^0, μ^0 are

1. $\dfrac{\partial \mathcal{L}}{\partial x_i}(\mathbf{x}^0, \lambda^0, \mu^0) = 0$ for $i = 1 \cdots N$

2. $\omega_i(\mathbf{x}^0) = 0$ $\qquad$ for $i = 1 \cdots N\omega$

3. $g_i(\mathbf{x}^0) \leq 0$ $\qquad$ for $i = 1 \cdots Ng$

4. $\left.\begin{array}{l} \mu_i^0 g_i(\mathbf{x}^0) = 0 \\ \mu_i^0 \geq 0 \end{array}\right\}$ $\qquad$ for $i = 1 \cdots Ng$

The first condition is simply the familiar set of partial derivatives of the LaGrange function that must equal zero at the optimum. The second and third conditions are simply a restatement of the constraint conditions on the problem. The fourth condition, often referred to *as the complimentary slackness condition*, provides a concise mathematical way to handle the problem of binding and nonbinding constraints. Since the product $\mu_i^0 g_i(\mathbf{x}^0)$ equals zero, either μ_i^0 is equal to zero or $g_i(\mathbf{x}^0)$ is equal to zero, or both are equal to zero. If μ_i^0 is equal to zero, $g_i(\mathbf{x}^0)$ is free to be nonbinding; if μ_i^0 is positive, then $g_i(\mathbf{x}^0)$ must be zero. Thus we have a clear indication of whether the constraint is binding or not by looking at μ_i^0.

To illustrate how the Kuhn-Tucker equations are used, we will add an inequality constraint to the sample problem used earlier in this appendix. The problem we will solve is as follows.

Minimize: $\qquad\qquad f(x_1, x_2) = 0.25\,x_1^2 + x_2^2$

Subject to: $\qquad\qquad \omega(x_1, x_2) = 5 - x_1 - x_2 = 0$

$\qquad\qquad\qquad\qquad g(x_1, x_2) = x_1 + 0.2\,x_2 - 3 \leq 0$

which can be illustrated as in Figure 3.11.

First, set up the LaGrange equation for the problem.

$$\mathcal{L} = f(x_1, x_2) + \lambda[\omega(x_1, x_2)] + \mu[g(x_1, x_2)]$$
$$= 0.25\,x_1^2 + x_2^2 + \lambda(5 - x_1 - x_2) + \mu(x_1 + 0.2\,x_2 - 3)$$

The first condition gives

$$\frac{\partial \mathcal{L}}{\partial x_1} = 0.5\,x_1 - \lambda + \mu = 0$$

$$\frac{\partial \mathcal{L}}{\partial x_2} = 2\,x_2 - \lambda + 0.2\,\mu = 0$$

The second condition gives

$$5 - x_1 - x_2 = 0$$

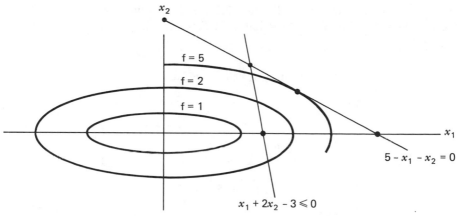

FIG. 3.11 Elliptical objective function with equality and inequality constraints.

The third condition gives

$$x_1 + 0.2\, x_2 - 3 \le 0$$

The fourth condition gives

$$\mu(x_1 + 0.2\, x_2 - 3) = 0$$

$$\mu \ge 0$$

At this point we are confronted with the fact that the Kuhn-Tucker conditions only give necessary conditions for a minimum, not a precise, procedure as to how that minimum is to be found. To solve the problem just presented we must literally experiment with various solutions until we can verify that one of the solutions meets all four conditions. First, let $\mu = 0$, which implies that $g(x_1, x_2)$ can be less than or equal to zero. However, if $\mu = 0$, we can see that the first and second conditions give the same solution as we had previously without the inequality constraint. But the previous solution violates our inequality constraint; and therefore the four Kuhn-Tucker conditions do not hold with $\mu = 0$. In summary

---

If $\mu = 0$, then by conditions 1 and 2

$$x_1 = 4$$

$$x_2 = 1$$

$$\lambda = 2$$

but

$$g(x_1, x_2)\big|_{\substack{x_1=4 \\ x_2=1}} = 4 + 0.2(1) - 3 = 1.2 \nleq 0$$

Now we will try a solution in which $\mu > 0$. In this case $g(x_1, x_2)$ must be exactly zero and our solution can be found by solving for the intersection of $g(x_1, x_2)$ and

$\omega(x_1, x_2)$, which occurs at $x_1 = 2.5$, $x_2 = 2.5$. Further, condition 1 gives $\lambda = 5.9375$ and $\mu = 4.6875$ and all four of the Kuhn-Tucker conditions are met. In summary

---

If $\mu > 0$, then by conditions 2 and 3

$$x_1 = 2.5$$

$$x_2 = 2.5$$

by condition 1

$$\lambda = 5.9375$$

$$\mu = 4.6875$$

and

$$g(x_1, x_2)|_{x_1 = x_2 = 2.5} = 2.5 + 0.2(2.5) - 3 = 0$$

All conditions are met.

Considerable insight can be gained into the characteristics of optimal solutions through use of the Kuhn-Tucker conditions. One important insight comes from formulating the optimization problem so that it reflects our standard power system economic dispatch problems. Specifically, we will assume that the objective function consists of a sum of individual cost functions each of which is a function of only one variable. For example,

$$f(x_1, x_2) = C_1(x_1) + C_2(x_2)$$

Further, we will restrict this problem to have one equality constraint of the form

$$\omega(x_1, x_2) = L - x_1 - x_2 = 0$$

and a set of inequality constraints that act to restrict the problem variables within an upper and lower limit. That is,

$$x_1^- \leq x_1 \leq x_1^+ \rightarrow \begin{cases} g_1(x_1) = x_1 - x_1^+ \leq 0 \\ g_2(x_1) = x_1^- - x_1 \leq 0 \end{cases}$$

$$x_2^- \leq x_2 \leq x_2^+ \rightarrow \begin{cases} g_3(x_2) = x_2 - x_2^+ \leq 0 \\ g_4(x_2) = x_2^- - x_2 \leq 0 \end{cases}$$

Then the LaGrange function becomes

$$= f(x_1, x_2) + \lambda\omega(x_1, x_2) + \mu_1 g_1(x_1) + \mu_2 g_2(x_1) + \mu_3 g_3(x_2) + \mu_4 g_4(x_2)$$

$$= C_1(x_1) + C_2(x_2) + \lambda(L - x_1 - x_2) + \mu_1(x_1 - x_1^+) + \mu_2(x_1^- - x_1)$$

$$+ \mu_3(x_2 - x_2^+) + \mu_4(x_2^- - x_2)$$

Condition 1 gives

$$C_1'(x_1) - \lambda + \mu_1 - \mu_2 = 0$$

$$C_2'(x_2) - \lambda + \mu_3 - \mu_4 = 0$$

Condition 2 gives

$$L - x_1 - x_2 = 0$$

Condition 3 gives

$$x_1 - x_1^+ \leq 0$$
$$x_1^- - x_1 \leq 0$$
$$x_2 - x_2^+ \leq 0$$
$$x_2^- - x_2 \leq 0$$

Condition 4 gives

$$\mu_1(x_1 - x_1^+) = 0 \qquad \mu_1 \geq 0$$
$$\mu_2(x_1^- - x_1) = 0 \qquad \mu_2 \geq 0$$
$$\mu_3(x_2 - x_2^+) = 0 \qquad \mu_3 \geq 0$$
$$\mu_4(x_2^- - x_2) = 0 \qquad \mu_4 \geq 0$$

Case 1

If the optimum solution occurs at values for x_1 and x_2 that are not at either an upper or a lower limit, then all μ's are equal to zero and

$$C_1'(x_1) = C_2'(x_2) = \lambda$$

That is, the incremental costs associated with each variable are equal and this value is exactly the λ we are interested in.

Case 2

Now suppose that the optimum solution requires that x_1 be at its upper limit (i.e., $x_1 - x_1^+ = 0$) and that x_2 is not at its upper or lower limit. Then

$$\mu_1 \geq 0$$

and μ_2, μ_3, and μ_4 will each equal zero. Then, from condition 1,

$$C_1'(x_1) = \lambda - \mu_1 \rightarrow C_1'(x_1) \leq \lambda$$
$$C_2'(x_2) = \lambda$$

Therefore, the incremental cost associated with the variable that is at its upper limit will always be less than or equal to λ, whereas the incremental cost associated with the variable that is not at limit will exactly equal λ.

Case 3

Now suppose the opposite of Case 2 obtains, that is, let the optimum solution require x_1 to be at its lower limit (i.e., $x_1^- - x_1 = 0$) and again assume that x_2 is not at its upper or lower limit. Then

$$\mu_2 \geq 0$$

and μ_1, μ_3, and μ_4 will each equal zero. Then from condition 1

$$C_1'(x_1) = \lambda + \mu_2 \Rightarrow C_1'(x_1) \geq \lambda$$
$$C_2'(x_2) = \lambda$$

Therefore, the incremental cost associated with a variable at its lower limit will be greater than or equal to λ whereas, again, the incremental cost associated with the variable that is not at limit will equal λ.

Case 4

If the optimum solution requires that both x_1, x_2 are at limit and the equality constraint can be met, then λ and the nonzero μ's are indeterminate. For example, suppose the optimum required that

$$x_1 - x_1^+ = 0$$

and

$$x_2 - x_2^+ = 0$$

Then

$$\mu_1 \geq 0 \qquad \mu_3 \geq 0 \qquad \mu_2 = \mu_4 = 0$$

Condition 1 would give

$$C_1'(x_1) = \lambda - \mu_1$$
$$C_2'(x_2) = \lambda - \mu_3$$

and the specific values for λ, μ_1, and μ_3 would be undetermined. In summary, for the general problem of N variables

Minimize: $\qquad\qquad C_1(x_1) + C_2(x_2) + \cdots C_N(x_1)$

Subject to: $\qquad\qquad L - x_1 - x_2 - \cdots - x_N = 0$

And: $\qquad\qquad \left.\begin{array}{l} x_i - x_i^+ \leq 0 \\ x_i^- - x_i \leq 0 \end{array}\right\} \text{ for } i = 1 \cdots N$

Let the optimum lie at $x_i = x_i^{\text{opt}}$ $i = 1 \cdots N$ and assume that at least one x_i is not at limit. Then

If $x_i^{\text{opt}} < x_i^+$ and $x_i^{\text{opt}} > x_i^-$, then $C_i(x_i^{\text{opt}}) = \lambda$

If $x_i^{\text{opt}} = x_i^+ \qquad C_i'(x_i^{\text{opt}}) \leq \lambda$

If $x_i^{\text{opt}} = x_i^- \qquad C_i'(x_i^{\text{opt}}) \geq \lambda$

Slack Variable Formulation

An alternate approach to the optimization problem with inequality constraints requires that all inequality constraints be made into equality constraints. This is done by adding slack variables in the following way.

If:
$$g(x_1) = x_1 - x_1^+ \leq 0$$

Then:
$$g(x_1, S_1) = x_1 - x_1^+ + S_1^2 = 0$$

We add S_1^2 rather than S_1 so that S_1 need not be limited in sign.

Making all inequality constraints into equality constraints eliminates the need for conditions 3 and 4 of the Kuhn-Tucker conditions. However, as we will see shortly, the result is essentially the same. Let us use our two variable problem again.

Minimize: $f(x_1, x_2) = C_1(x_1) + C_2(x_2)$

Subject to: $\omega(x_1, x_2) = L - x_1 - x_2 = 0$

And:

$$g_1(x_1) = x_1 - x_1^+ \leq 0 \qquad \text{or} \qquad g_1(x_1, S_1) = x_1 - x_1^+ + S_1^2 = 0$$

$$g_2(x_1) = x_1^- - x_1 \leq 0 \qquad\qquad\qquad g_1(x_1, S_2) = x_1^- - x_1 + S_2^2 = 0$$

$$g_3(x_2) = x_2 - x_2^+ \leq 0 \qquad\qquad\qquad g_3(x_2, S_3) = x_2 - x_2^+ + S_3^2 = 0$$

$$g_4(x_2) = x_2^- - x_2 \leq 0 \qquad\qquad\qquad g_4(x_2, S_4) = x_2^- - x_2 + S_4^2 = 0$$

The resulting LaGrange function is

$$\mathscr{L} = f(x_1, x_2) + \lambda_0 \omega(x_1, x_2) + \lambda_1 g_2(x_1, S_2)$$
$$+ \lambda_2 g_2(x_1, S_2) + \lambda_3 g_3(x_2, S_2) + \lambda_4 g_4(x_2, S_4)$$

Note that all constraints are now equality constraints, so we have used only λ's as LaGrange multipliers.

Condition 1 Gives:

$$\frac{\partial \mathscr{L}}{\partial x_1} = C_1'(x_1) - \lambda_0 + \lambda_1 - \lambda_2 = 0$$

$$\frac{\partial \mathscr{L}}{\partial x_2} = C_2'(x_2) - \lambda_0 + \lambda_3 - \lambda_4 = 0$$

$$\frac{\partial \mathscr{L}}{\partial S_1} = 2\lambda_1 S_1 = 0$$

$$\frac{\partial \mathscr{L}}{\partial S_2} = 2\lambda_2 S_2 = 0$$

$$\frac{\partial \mathscr{L}}{\partial S_3} = 2\lambda_3 S_3 = 0$$

$$\frac{\partial \mathscr{L}}{\partial S_4} = 2\lambda_4 S_4 = 0$$

Condition 2 Gives:
$$L - x_1 - x_2 = 0$$
$$(x_1 - x_1^+ + S_1^2) = 0$$
$$(x_1^- - x_1 + S_2^2) = 0$$
$$(x_2 - x_2^+ + S_3^2) = 0$$
$$(x_2^- - x_2 + S_4^2) = 0$$

We can see that the derivatives of the LaGrange function with respect to the slack variables provide us once again with a complimentary slackness rule. For example, if $2\lambda_1 S_1 = 0$, then either $\lambda_1 = 0$ and S_1 is free to be any value or $S_1 = 0$ and λ_1 is free (or λ_1 and S_1 can both be zero). Since there are as many problem variables whether one uses the slack variable form or the inequality constraint form, there is little advantage to either other than perhaps a conceptual advantage to the student.

PROBLEMS

3.1 Assume that the fuel inputs in Btu per hour for units 1 and 2, which are both on-line, are given by
$$H_1 = 8\,P_1 + 0.024\,P_1^2 + 80$$
$$H_2 = 6\,P_2 + 0.04\,P_2^2 + 120$$

where

H_n = fuel input to unit n in MBtu per hour (millions of Btu per hour)

P_n = unit output in megawatts

a. Plot the input-output characteristics for each unit expressing input in MBtu per hour and output in megawatts. Assume that the minimum loading of each unit is 20 MW and that the maximum loading is 100 MW.

b. Calculate the net heat rate in Btu/kWh, and plot against output in megawatts.

c. Assume that the cost of fuel is 1.5 Ɍ/MBtu. Calculate the incremental production cost in Ɍ/MWh of each unit, and plot against output in megawatts.

3.2 Dispatch with Three Segment Piecewise Linear Incremental Heat Rate Function

Given: Two generating units with incremental heat rate curves(IHR) specified as three- connected line segments (four points as shown in Figure 3.12).

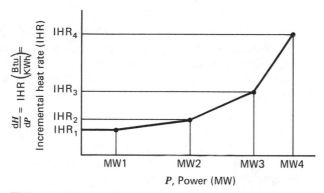

FIG. 3.12 Piecewise linear incremental heat rate curve for Problem 3.2.

Unit 1

Point	MW	IHR (Btu/kWh)
1	100	7000
2	200	8200
3	300	8900
4	400	11000

Fuel cost for unit 1 = 1.60 ₹/MBtu.

Unit 2

Point	MW	IHR (Btu/kWh)
1	150	7500
2	275	7700
3	390	8100
4	450	8500

Fuel cost for unit 2 = 2.10 ₹/MBtu.

Both units are running. Calculate the optimum schedule (i.e., the unit megawatt output for each unit) for various total megawatt values to be supplied by the units. Find the schedule for these total megawatt values.

300 MW

500 MW

700 MW

840 MW

Notes: Piecewise linear increment cost curves are quite common in digital computer executions of economic dispatch. The problem is best solved by using a "search" technique. In such a technique, the incremental cost is given a value and the units are scheduled to meet this incremental cost. The megawatt outputs for the units are added together and compared to the desired total. Depending on the difference, and whether the resulting total is above or below the desired total, a new value of incremental cost is "tried." This is repeated until the incremental cost is found that gives the correct desired value. The trick is to search in an efficient manner so that the number of iterations is minimized.

3.3 Assume the system load served by the two units of Problem 3.1 varies from 50 to 200 MW. For the data of Problem 3.1 plot the outputs of units 1 and 2 as a function of total system load when scheduling generation by equal incremental production costs. Assume that both units are operating.

3.4 As an exercise, obtain the optimum loading of the two generating units in Problem 3.1 using the following technique. The two units are to deliver 100 MW. Assume both units are on-line and delivering power. Plot the total fuel cost for 100 MW of delivered power as generation is shifted from one unit to the other. Find the minimum cost. The optimum schedule should check with the schedule obtained by equal incremental production costs.

3.5 This problem demonstrates the complexity involved when we must commit (turn on) generating units as well as dispatch them economically. This problem is known as the *unit commitment problem* and is the subject of Chapter 5.

Given the two generating units in Problem 3.1, assume that they are both off-line at the start. Also assume that load starts at 50 MW and increases to 200 MW. The most economic schedule to supply this varying load will require committing one unit first, followed by commitment of the second unit when the load reaches a higher level.

Determine which unit to commit first and at what load the remaining unit should be committed. Assume no "start-up" costs for either unit.

3.6 The system to be studied consists of two units described in Problem 3.1. Assume a daily load cycle as follows.

Time Band	Load (MW)
0000–0600	50
0600–1800	150
1800–0000	50

Also, assume that a cost of 180 R is incurred in taking either unit off line and returning it to service after 12 h. Consider the 24-h period from 0600 one morning to 0600 the next morning.

a. Would it be more economical to keep both units in service for this 24-h period or to remove one of the units from service for the 12-h period from 1800 one evening to 0600 the next morning?

b. What is the economic schedule for the period of time from 0600 to 1800 (load = 150 MW)?

c. What is the economic schedule for the period of time from 1800 to 0600 (load = 50 MW)?

3.7 Assume that all three of the thermal units described in the following table are running. Find the economic dispatch schedules as requested in each part. Use the method and starting conditions given

Unit Data	Minimum (MW)	Maximum (MW)	Fuel Cost (R/MBtu)
$H_1 = 225 + 8.4\,P_1 + 0.0025\,P_1^2$	45	350	0.80
$H_2 = 729 + 6.3\,P_2 + 0.0081\,P_2^2$	45	350	1.02
$H_3 = 400 + 7.5\,P_3 + 0.0025\,P_3^2$	47.5	450	0.90

a. Use the lambda-iteration method to find the economic dispatch for a total demand of 450 MW.

b. Use the base-point and participation factor method to find the economic schedule for a demand of 495 MW. Start from the solution to part a.

c. Use a gradient method to find the economic schedule for a total demand of 500 MW assuming the initial conditions (i.e., loadings) on the three units are

$$P_1 = P_3 = 100 \text{ MW} \quad \text{and} \quad P_2 = 300 \text{ MW}$$

Give the individual unit loadings and cost per hour as well as the total cost per hour to supply each load level. (MBtu = millions of BTU; H_j = heat input in MBtu/h; P_i = electric power output in MW; i = 1, 2, 3.)

3.8 Thermal Scheduling with Straight-Line Segments for Input-Output Curves

The following data apply to three thermal units. Compute and sketch the input-output characteristics and the incremental heat rate characteristics. Assume the unit input-output curves consist of straight-line segments between the given power points.

Unit No.	Power Output (MW)	Net Heat Rate (Btu/kWh)
1	45	13.512.5
	300	9,900.0
	350	9,918.0
2	45	22,764.5
	200	11,465.0
	300	11,060.0
	350	11,117.9
3	47.5	16,039.8
	200	10,000.0
	300	9,583.3
	450	9,513.9

Fuel costs are

Unit No.	Fuel Cost (R/MBtu)
1	0.61
2	0.75
3	0.75

Compute the economic schedule for system demands of 300, 460, 500, and 650 MW, assuming all three units are on-line. Give unit loadings and costs per hour as well as total costs in R per hour.

3.9 Environmental Dispatch

Recently there has been concern that optimum *economic* dispatch was not the best environmentally. The principles of economic dispatch can fairly easily be extended to handle this problem. Following is a problem based on a real situation that occurred in the midwestern United States in 1973. Other cases have arisen with "NOX" emission in Los Angeles.

Two steam units have input-output curves as follows.

$$H_1 = 400 + 5 P_1 + 0.01 P_1^2, \text{MBtu/h}, 20 \le P_1 \le 200 \text{ MW}$$

$$H_2 = 600 + 4 P_2 + 0.015 P_2^2, \text{MBtu/h}, 20 \le P_2 \le 200 \text{ MW}$$

The units each burn coal with a heat content of 11,500 Btu/lb that costs 13.50 R per ton (2000 lb). The combustion process in each unit results in 11.75% of the coal by weight going up the stack as fly ash.

a. Calculate the net heat rates of both units at 200 MW.

b. Calculate the incremental heat rates; schedule each unit for optimum *economy* to serve a total load of 250 MW with both on-line.

c. Calculate the cost of supplying that load.

d. Calculate the rate of emission of fly ash for that case in lb/h, assuming no fly ash removal devices are in service.

e. Unit 1 has a precipitator installed that removes 85% of the fly ash; unit 2's precipitator is 89.0% efficient. Reschedule the two units for the minimum *total fly ash emission rate* with both on-line to serve a 250 MW load.

f. Calculate the rate of emission of ash and the cost for this schedule to serve the 250 MW load. What is the cost penalty?

g. Where does all that fly ash go?

FURTHER READING

Since this chapter introduces several optimization concepts it would be useful to refer to some of the general works on optimization such as references 1 and 2. The importance of the Kuhn-Tucker theorem is given in their paper (3). A very thorough discussion of the Kuhn-Tucker theorem is found in Chapter 1 of reference 4.

For an overview of recent power system optimization practices see references 5 and 6. Several other applications of optimization have been presented. Reference 7 discusses the allocation of regulating margin while dispatching generator units. References 8–10 discuss how to formulate the dispatch problem as one that minimizes air pollution from power plants.

1. *Application of Optimization Methods in Power System Engineering*, IEEE Tutorial Course Text 76CH1107-2-PWR, IEEE, New York, 1976.

2. Wilde, P. J., Beightler, C. S., *Foundations of Optimization*, Prentice-Hall, Englewood Cliffs, N.J., 1967.

3. Kuhn, H. W., Tucker, A. W., "Nonlinear Programming", in *Second Berkeley Symposium on Mathematical Programming Statistics and Probability*, 1950, University of California Press, Berkeley, 1951.

4. Wismer, D. A., *Optimization Methods for Large-Scale Systems With Applications*, McGraw-Hill, New York, 1971.

5. IEEE Committee Report, "Present Practices in the Economic Operation of Power Systems," *IEEE Transactions on Power Apparatus and Systems*, Vol. PAS-90, July/August 1971, pp. 1768–1775.

6. Najaf-Zadeh, K., Nikolas, J. T., Anderson, S. W., "Optimal Power System Operation Analysis Techniques," *Proceedings American Power Conference*, 1977.

7. Stadlin, W. O., "Economic Allocation of Regulating Margin," *IEEE Transactions on Power Apparatus and Systems*, Vol. PAS-90, July/August 1971, pp. 1777–1781.

8. Gent, M. R., Lamont, J. W., "Minimum Emission Dispatch," *IEEE Transactions on Power Apparatus and Systems*, Vol. PAS-90, November/December 1971, pp. 2650–2260.

9. Sullivan, R. L., "Minimum Pollution Dispatching," IEEE Summer Power Meeting, Paper C-72-468, 1972.

10. Friedman, P. G., "Power Dispatch Strategies for Emission and Environmental Control," *Proceedings of the Instrument Society of America*, Vol. 16, 1973.

Transmission Losses

This chapter introduces the concept of the steady-state transmission network solution or load flow and the relationship of losses in the transmission network to economic dispatch. *Load flow* is the name given to a network solution that shows currents, voltages, and power flows at every bus in the system. It is normally assumed that the system is balanced and the common use of the term load flow implies a positive sequence solution only. Full three-phase load-flow solution techniques are available for special purpose calculations. As used here, we are only interested in balanced solutions. Load flow is not a single calculation such as $E = IR$ or $\mathbf{E} = [Z]\mathbf{I}$ involving linear circuit analysis. Such circuit analysis problems start with a given set of currents or voltages, and one must solve for the linearly dependent unknowns. In the load-flow problem we are given a nonlinear relationship between voltage and current at each bus and we must solve for all voltages and currents such that these nonlinear relationships are met. The nonlinear relationships involve, for example, the real and reactive power consumption at a bus, or the generated real power and scheduled voltage magnitude at a generator bus. As such, the load flow gives us the electrical response of the transmission system to a particular set of loads and generator unit outputs. Load flows are an important part of power system design procedures (system planning). Modern digital computer load-flow programs are routinely run for systems up to 5000 or more buses and also are used widely in power system control centers to study unique operating problems and to provide accurate calculations of bus penalty factors.

One of the quantities available from a load-flow solution is the electrical losses of the network. The electrical losses are important in economically dispatching generation. However, it is not simply the value of the losses but the derivatives of losses with respect to generator outputs that are important. Before the advent of large, fast digital computers, the losses were treated by various approximate methods. Present state-of-the-art system-control centers build the economic dispatch problem around the load flow itself.

4.1 THE LOAD-FLOW PROBLEM AND ITS SOLUTION

The load-flow problem consists of a given transmission network where all lines are represented by a Pi-equivalent circuit and transformers by an ideal voltage transformer in series with an impedance. Generators and loads represent the boundary conditions of the solution. Generator or load real and reactive power involves products of voltage and current. Mathematically, the load flow requires a solution of a system of simultaneous nonlinear equations.

4.1.1 The Load-Flow Problem on a Direct Current Network

The problems involved in solving a load flow can be illustrated by the use of direct current circuit examples. The circuit shown in Figure 4.1 has a resistance of 0.25 Ω tied to a constant voltage of 1.0 V (called the *reference voltage*). We wish to find the voltage at bus 2 that results in a net inflow of 1.2 W. *Buses* are electrical nodes. Power is said to be "injected" into a network; therefore, loads are simply negative injections.

The current from bus 2 to bus 1 is

$$I_{21} = (E_2 - 1.0) \times 4 \tag{4.1}$$

Power P_2 is

$$P_2 = 1.2 = E_2 I_{21} = E_2(E_2 - 1) \times 4 \tag{4.2}$$

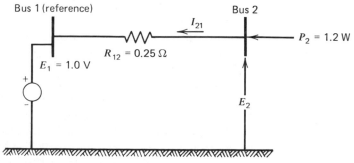

FIG. 4.1 Two-bus dc network.

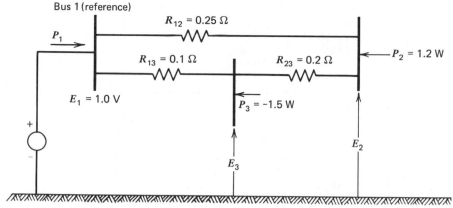

FIG. 4.2 Three-bus dc network.

or

$$4E_2^2 - 4E_2 - 1.2 = 0 \tag{4.3}$$

The solutions to this quadratic equation are $E_2 = 1.24162$ V and $E_2 = -0.24162$ V. Note that 1.2 W enter bus 2, producing a current of 0.96648 A ($E_2 = 1.24162$), which means that 0.96648 W enter reference bus and 0.23352 W are consumed in the 0.25 Ω resistor.

Let us complicate the problem by adding a third bus and two more lines (see Figure 4.2). The problem is more complicated because we cannot simply write out the solutions using a quadratic formula. The admittance equations are

$$\begin{bmatrix} I_1 \\ I_2 \\ I_3 \end{bmatrix} = \begin{bmatrix} 14 & -4 & -10 \\ -4 & 9 & -5 \\ -10 & -5 & 15 \end{bmatrix} \begin{bmatrix} E_1 \\ E_2 \\ E_3 \end{bmatrix} \tag{4.4}$$

In this case we know the power injected at buses 2 and 3 and we know the voltage at bus 1. To solve for the unknowns (E_2, E_3, and P_1), we write Eqs. 4.5, 4.6, and 4.7. The solution procedure is known as the *Gauss-Seidel procedure*, wherein a calculation for a new voltage at each bus is made based on the most recently calculated voltages at all neighboring buses.

Bus 2:
$$I_2 = \frac{P_2}{E_2} = -4(1.0) + 9E_2 - 5E_3$$

$$E_2^{\text{new}} = \frac{1}{9}\left(\frac{1.2}{E_2^{\text{old}}} + 4 + 5E_3^{\text{old}}\right) \tag{4.5}$$

where E_2^{old} and E_3^{old} are the initial values for E_2 and E_3, respectively.

Bus 3:
$$I_3 = \frac{P_3}{E_3} = -10(1.0) - E_2^{\text{new}} + 15E_3$$

$$E_3^{\text{new}} = \frac{1}{15}\left[\frac{-1.5}{E_3^{\text{old}}} + 10 + 5E_2^{\text{new}}\right] \tag{4.6}$$

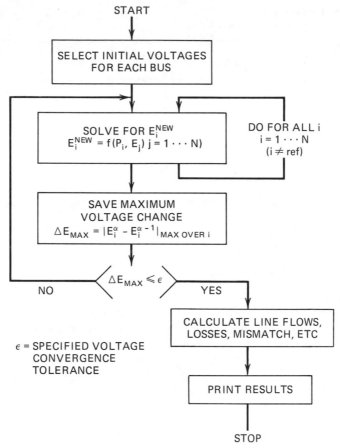

FIG. 4.3 Gauss-Seidel load-flow solution.

where E_2^{new} is the voltage found in solving Eq. 4.5 and E_3^{old} is the initial value of E_3.

Bus 1:
$$P_1 = E_1 I_1^{new} = 1.0\, I_1^{new} = 14 - 4\,E_2^{new} - 10\,E_3^{new} \tag{4.7}$$

The Gauss-Seidel method first assumes a set of voltages at buses 2 and 3 and then uses Eqs. 4.5 and 4.6 to solve for new voltages. The new voltages are compared to the voltage's most recent values, and the process continues until the change in voltage is very small. This is illustrated in the flowchart in Figure 4.3 and Eqs. 4.8 and 4.9.

First Iteration:
$$E_2^{(0)} = E_3^{(0)} = 1.0$$

$$E_0^{(1)} = \frac{1}{9}\left(\frac{1.2}{1.0} + 4 + 5\right) = 1.133$$

$$E_3^{(1)} = \frac{1}{15}\left[\frac{-1.5}{1.0} + 10 + 5(1.133)\right] = 0.944 \tag{4.8}$$

$$\Delta E_{max} = 0.133 \text{ too large}$$

NOTE: In calculating $E_3^{(1)}$ we used the new value of E_2 found in the first correction.

Second Iteration: $\quad E_2^{(2)} = \dfrac{1}{9}\left[\dfrac{1.2}{1.133} + 4 + 5(0.944) \right] = 1.087$

$$E_3^{(2)} = \frac{1}{15}\left[\frac{-1.5}{0.944} + 10 + 5(1.087) \right] = 0.923 \qquad (4.9)$$

$$\Delta E_{max} = 0.046$$

And so forth until $\Delta E_{max} < \varepsilon$.

4.1.2 The Formulation of the AC Load Flow

AC load flows involve several types of bus specifications, as shown in Figure 4.4. Note that $[P, \theta]$, $[Q, |E|]$, and $[Q, \theta]$ combinations are generally not used.

The transmission network consists of complex impedances between buses and from the buses to ground. An example is given in Figure 4.5. The equations are written in matrix form as

$$\begin{bmatrix} I_1 \\ I_2 \\ I_3 \\ I_4 \end{bmatrix} = \begin{bmatrix} y_{12} & -y_{12} & 0 & 0 \\ -y_{12} & (y_{12} + y_{2g} + y_{23}) & -y_{23} & 0 \\ 0 & -y_{23} & (y_{23} + y_{3g} + y_{34}) & -y_{34} \\ 0 & 0 & -y_{34} & (y_{34} + y_{4g}) \end{bmatrix} \begin{bmatrix} E_1 \\ E_2 \\ E_3 \\ E_4 \end{bmatrix} \qquad (4.10)$$

(All I's, E's, y's complex)

This matrix is called the *network Y matrix*, which is written as

$$\begin{bmatrix} I_1 \\ I_2 \\ I_3 \\ I_4 \end{bmatrix} = \begin{bmatrix} Y_{11} & Y_{12} & Y_{13} & Y_{14} \\ Y_{21} & Y_{22} & Y_{23} & Y_{24} \\ Y_{31} & Y_{32} & Y_{33} & Y_{34} \\ Y_{41} & Y_{42} & Y_{43} & Y_{44} \end{bmatrix} \begin{bmatrix} E_1 \\ E_2 \\ E_3 \\ E_4 \end{bmatrix} \qquad (4.11)$$

The rules for forming a Y matrix are

If a line exists from i to j

$$Y_{ij} = -y_{ij}$$

and

$$Y_{ii} = \sum_j y_{ij} + y_{ig}$$

j over all lines connected to i.

Bus Type	P	Q	$\lvert E\rvert$	θ	Comments
Load	✓	✓			Usual load representation
Voltage Controlled	✓		✓		Assume $\lvert E\rvert$ is held constant no matter what Q is
Generator or Synchronous Condenser	✓		✓ when $Q^- < Q_g < Q^+$		Generator or synchronous condenser ($P = 0$) has VAR limits
	✓	✓ when $Q_g < Q^-$ $Q_g > Q^+$			Q^- minimum VAR limit Q^+ maximum VAR limit $\lvert E\rvert$ is held as long as Q_g is within limit
Fixed Z to Ground					Only Z is given
Reference			✓	✓	"Swing bus" must adjust net power to hold voltage constant (essential for solution)

FIG. 4.4 Load-flow bus specifications (quantities checked are the bus boundary conditions).

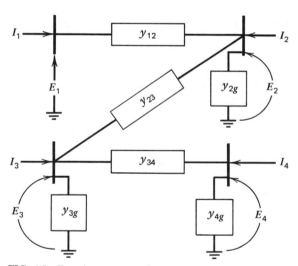

FIG. 4.5 Four-bus ac network.

The equation of net power injection at a bus is usually written as

$$\frac{P_k - jQ_k}{E_k^*} = \sum_{\substack{j=1 \\ j \neq k}}^{n} Y_{jk}E_j + Y_{kk}E_k \tag{4.12}$$

$$P + jQ = EI^*$$

4.1.2.1 The Gauss-Seidel Method

The voltages at each bus can be solved for by using the Gauss-Seidel method. The equation in this case is

$$E_k^{(\alpha)} = \frac{1}{Y_{kk}} \frac{(P_k - jQ_k)}{E_k^{(\alpha-1)*}} - \frac{1}{Y_{kk}} \left[\sum_{j<k} Y_{jk} E_j^{(\alpha)} + \sum_{j>k} Y_{jk} E_j^{(\alpha-1)} \right] \quad (4.13)$$

Voltage at
iteration α

The Gauss-Seidel method was the first ac load-flow method to be developed for solution on digital computers. This method is characteristically long in solving due to its slow convergence and often difficulty is experienced with unusual network conditions such as negative reactance branches. The solution procedure is the same as shown in Figure 4.3.

4.1.2.2 The Newton-Raphson Method

One of the disadvantages of the Gauss-Seidel method lies in the fact that each bus is treated independently. Each correction to one bus requires subsequent correction to all the buses to which it is connected. The Newton-Raphson method is based on the idea of calculating the corrections while taking account of all the interactions.

Newton's method involves the idea of an error in a function f(x) being driven to zero by making adjustments Δx to the independent variable associated with the function. Suppose we wish to solve

$$f(x) = K \quad (4.14)$$

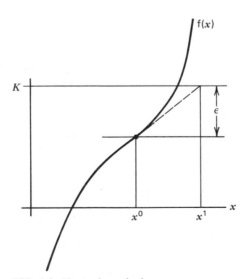

FIG. 4.6 Newton's method.

In Newton's method we pick a starting value of x and call it x^0. The error is the difference between K and $f(x^0)$. Call the error ε. This is shown in Figure 4.6 and given in Eq. 4.15.

$$f(x^0) + \varepsilon = K \tag{4.15}$$

To drive the error to zero, we use a Taylor expansion of the function about x^0.

$$f(x^0) + \frac{df(x^0)}{dx} \Delta x + \varepsilon = K \tag{4.16}$$

Setting the error to zero we calculate

$$\Delta x = \left(\frac{df(x^0)}{dx}\right)^{-1} [K - f(x^0)] \tag{4.17}$$

When we wish to solve a load flow we extend Newton's method to the multivariable case (the multivariable case is called the *Newton-Raphson method*). An equation is written for each bus "i."

$$P_i + jQ_i = E_i I_i^* \tag{4.18}$$

where

$$I_i = \sum_{k=1}^{N} Y_{ik} E_k$$

then

$$P_i + jQ_i = E_i \left[\sum_{k=1}^{N} Y_{ik} E_k\right]^*$$

$$= |E_i|^2 Y_{ii}^* + \sum_{\substack{k=1 \\ k \neq i}}^{N} Y_{ik}^* E_i E_k^*$$

As in the Gauss-Seidel method, a set of starting voltages is used to get things going. The $P + jQ$ calculated is subtracted from the scheduled $P + jQ$ at the bus and the resulting errors stored in a vector. As shown following, we will assume that the voltages are in polar coordinates and that we are going to adjust each voltage's magnitude and phase angle as separate independent variables. Each bus injection equation is differentiated with respect to all independent variables. Note that at this point two equations are written for each bus: one for real power and one for reactive power. For each bus

$$\Delta P_i = \sum_{k=1}^{N} \frac{\partial P_i}{\partial \theta_k} \Delta \theta_k + \sum_{k=1}^{N} \frac{\partial P_i}{\partial |E_k|} \Delta |E_k|$$

$$\Delta Q_i = \sum_{k=1}^{N} \frac{\partial Q_i}{\partial \theta_k} \Delta \theta_k + \sum_{k=1}^{N} \frac{\partial Q_i}{\partial |E_k|} \Delta |E_k| \tag{4.19}$$

All the terms are arranged in a matrix (the Jacobian matrix) as follows.

$$
\begin{bmatrix} \Delta P_1 \\ \Delta Q_1 \\ \Delta P_2 \\ \Delta Q_2 \\ \vdots \end{bmatrix} = \underbrace{\begin{bmatrix} \dfrac{\partial P_1}{\partial \theta_1} & \dfrac{\partial P_1}{\partial |E_1|} & \cdots \\[2ex] \dfrac{\partial Q_1}{\partial \theta_1} & \dfrac{\partial Q_1}{\partial |E_1|} & \cdots \\[2ex] \vdots & \vdots \end{bmatrix}}_{\text{Jacobian matrix}} \begin{bmatrix} \Delta \theta_1 \\ \Delta |E_1| \end{bmatrix}
\qquad (4.20)
$$

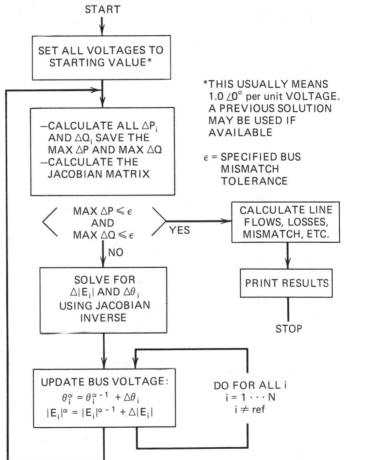

FIG. 4.7 Newton-Raphson load-flow solution.

The solution to the Newton-Raphson load flow runs according to the flowchart in Figure 4.7. Note that solving for $\Delta\theta$ and $\Delta|E|$ requires the solution of a set of linear equations whose coefficients make up the Jacobian matrix. (See Appendix C for details of the Jacobian matrix.) The Jacobian matrix generally has only a

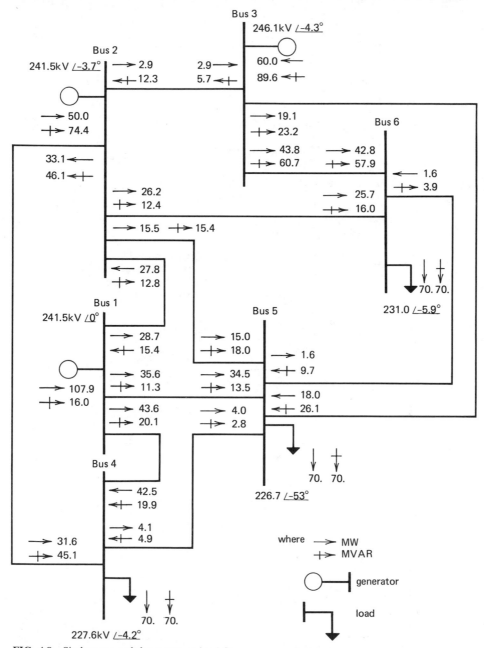

FIG. 4.8 Six-bus network base case ac load flow.

few percent of its entries that are nonzero. Programs that solve an ac load flow using the Newton-Raphson method are successful because they take advantage of the Jacobian's "sparsity." The solution procedure uses Gaussian elimination on the Jacobian matrix and does not calculate J^{-1} explicitly. (See reference 3 for introduction to "sparsity" techniques.)

EXAMPLE 4A

The six-bus network shown in Figure 4.8 will be used to demonstrate several aspects of load flows and transmission loss factors. The voltages and flows shown are for the "base case" of 210 MW total load. The impedance values and other data for this system may be found in Appendix A.

4.1.3 The "DC" Load Flow

A very widely used approximation of the AC load flow is the "linearized" or "DC" *load flow*, which converts the AC solution into a simple linear circuit analysis problem. Assume we are given a transmission line Pi-equivalent circuit as shown in Figure 4.9. The equation for power flowing through the line calculated at bus i can be written as

$$
\begin{aligned}
P_{ij} + jQ_{ij} &= E_i[(E_i - E_j)y_{ij}]^* + E_i[E_i y_{\text{cap}_{ij}}]^* \\
&= |E_i|e^{j\theta_i}[(|E_i|e^{j\theta_i} - |E_j|e^{j\theta_j})(G_{ij} + jB_{ij})]^* - j|E_i|^2 B_{\text{cap}_{ij}} \\
&= (|E_i|^2 - |E_i||E_j|\cos(\theta_i - \theta_j) - |E_i||E_j|j\sin(\theta_i - \theta_j)) \\
&\quad \times (G_{ij} + jB_{ij})^* - j|E_i|^2 B_{\text{cap}_{ij}}
\end{aligned}
\tag{4.21}
$$

Then

$$
P_{ij} = G_{ij}|E_i|^2 - G_{ij}|E_i||E_j|\cos(\theta_i - \theta_j) - B_{ij}|E_i||E_j|\sin(\theta_i - \theta_j) \tag{4.22}
$$

Assuming

1. $|E_i| = |E_j| = 1$

2. $x_{ij} \gg r_{ij} \rightarrow G_{ij} = \dfrac{r_{ij}}{r_{ij}^2 + x_{ij}^2} \simeq 0$

$$
B_{ij} = \dfrac{-x_{ij}}{r_{ij}^2 + x_{ij}^2} = \dfrac{-1}{x_{ij}}
$$

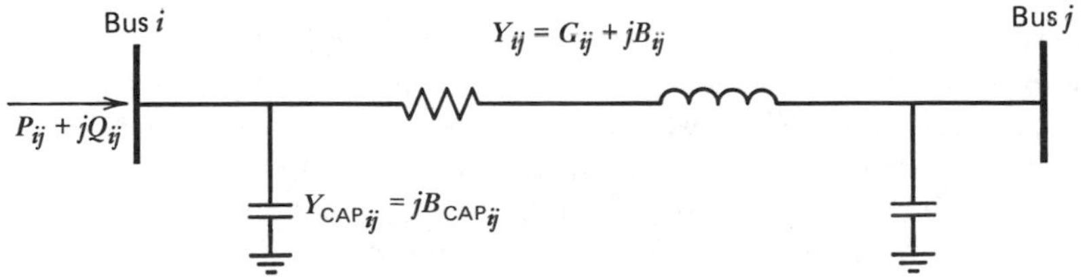

FIG. 4.9 Transmission line Pi-equivalent circuit.

3. $(\theta_i - \theta_j)$ is quite small.

$$\cos(\theta_i - \theta_j) \cong 1$$

$$\sin(\theta_i - \theta_j) \cong (\theta_i - \theta_j)$$

Then

$$P_{ij} = \frac{1}{x_{ij}}(\theta_i - \theta_j) \tag{4.23}$$

This linearized form may be used to calculate all bus phase angles in a network.

$$P_i = \sum_j P_{ij} = \sum_j \frac{1}{x_{ij}}(\theta_i - \theta_j) \tag{4.24}$$

j over all
nodes connected
directly to i

Then in matrix form,

$$\begin{bmatrix} P_1 \\ P_2 \\ \vdots \end{bmatrix} = [B_x] \begin{bmatrix} \theta_1 \\ \theta_2 \\ \vdots \end{bmatrix} \tag{4.25a}$$

or

$$\begin{bmatrix} \theta_1 \\ \theta_2 \\ \vdots \end{bmatrix} = [X] \begin{bmatrix} P_1 \\ P_2 \\ \vdots \end{bmatrix} \tag{4.25b}$$

where $\quad B_{x_{ii}} = \sum_j \dfrac{1}{x_{ij}} \qquad$ for $i \neq$ ref

j over all lines
connected to i

$B_{x_{ii}} = 0.0 \qquad$ for $i =$ ref

$B_{x_{ij}} = \dfrac{-1}{x_{ij}} \qquad$ for $i \neq$ ref and $j \neq$ ref

$B_{x_{ij}} = 0.0 \qquad$ for $i =$ ref or $j =$ ref

Strictly speaking, $[B_x]$ has no inverse since the row and column corresponding to the reference bus contain all zeros. If there are N buses, we only have $N - 1$ linearly independent equations. Thus, we have an $(N - 1) \times (N - 1)$ submatrix of $[B_x]$, which can be inverted. Therefore, when we refer to the matrix $[X]$, we will mean that matrix with the inverse of the submatrix of $[B_x]$ plus a zero row and column corresponding to the reference bus. See Example 4B. We have written $[B_x]$ and $[X]$ in this way so that the vectors θ and $\mathbf{P}$ can contain all buses. Note that we must

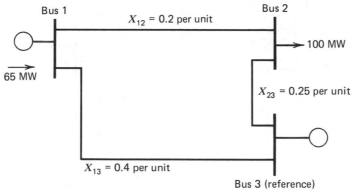

Bus 1

$X_{12} = 0.2$ per unit

Bus 2

100 MW

65 MW

$X_{23} = 0.25$ per unit

$X_{13} = 0.4$ per unit

Bus 3 (reference)

FIG. 4.10   Three-bus network.

always assume a value for the phase angle at the reference bus. Usually we assign this value to be zero radians.

EXAMPLE 4B

The megawatt flows on the network in Figure 4.10 will be solved using the DC load flow. The B_x matrix equation is

$$\theta_3 = 0$$

$$\begin{bmatrix} 7.5 & -5.0 & 0 \\ -5.0 & 9.0 & 0 \\ 0 & 0 & 0 \end{bmatrix} \begin{bmatrix} \theta_1 \\ \theta_2 \\ 0 \end{bmatrix} = \begin{bmatrix} P_1 \\ P_2 \\ P_3 \end{bmatrix}$$

Note that all megawatt quantities and network quantities are expressed in pu (per unit, 100 MVA base). All phase angles will then be in radians.

The solution to the preceding matrix equation is

$$\begin{bmatrix} \theta_1 \\ \theta_2 \\ 0 \end{bmatrix} = \begin{bmatrix} 0.2118 & 0.1177 & 0 \\ 0.1177 & 0.1765 & 0 \\ 0 & 0 & 0 \end{bmatrix} \begin{bmatrix} 0.65 \\ -1.00 \\ 0.35 \end{bmatrix} = \begin{bmatrix} 0.02 \\ -0.1 \\ 0 \end{bmatrix}$$

The resulting flows are shown in Figure 4.11 and calculated using Eq. 4.24. Note that all flows in Figure 4.11 were converted to actual megawatt values.

EXAMPLE 4C

The network of Example 4A was solved using the dc load flow with resulting power flows as shown in Figure 4.12. The dc load flow is useful for rapid calculations of real power flows, and, as will be shown later, it can be used to calculate incremental losses.

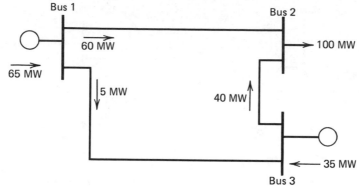

FIG. 4.11 Three-bus network showing flows calculated by dc load flow.

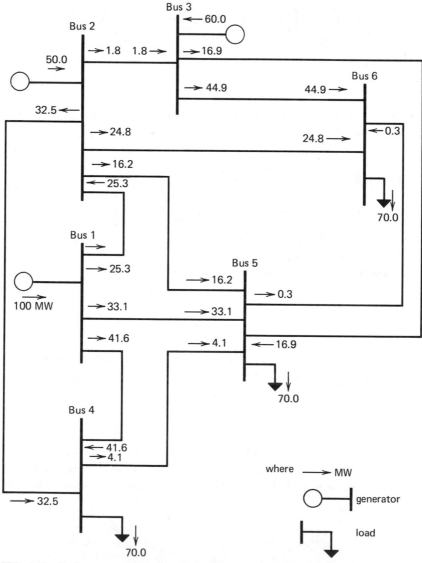

FIG. 4.12 Six-bus network base case dc load flow for Example 4C.

4.2 TRANSMISSION LOSSES

4.2.1 A Two-Generator System

We are given the power system in Figure 4.13. The losses on the transmission line are proportional to the square of the power flow. The generating units are identical, and the production cost modeled using the quadratic curve shown. If both units were loaded to 250 MW, we would fall short of the 500 MW load value by 12.5 MW lost on the transmission line, as shown in Figure 4.14.

Where should the extra 12.5 MW be generated? Solve the LaGrange equation that was given in Chapter 3, Eqs. 3.7, 3.8, and 3.9.

$$\mathcal{L} = F_1(P_1) + F_2(P_2) + \lambda(500 + P_{\text{loss}} - P_1 - P_2) \qquad (4.26)$$

where

$$P_{\text{loss}} = 0.0002 \, P_1^2$$

Then

$$\frac{\partial \mathcal{L}}{\partial P_1} = \frac{dF_1(P_1)}{dP_1} - \lambda\left(1 - \frac{\partial P_{\text{loss}}}{\partial P_1}\right) = 0$$

$$\frac{\partial \mathcal{L}}{\partial P_2} = \frac{dF_2(P_2)}{dP_2} - \lambda\left(1 - \frac{\partial P_{\text{loss}}}{\partial P_2}\right) = 0 \qquad (4.27)$$

$$P_1 + P_2 - 500 - P_{\text{loss}} = 0$$

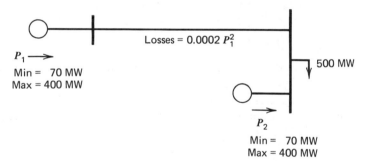

$$P_1 \longrightarrow$$
Min = 70 MW
Max = 400 MW

Losses = 0.0002 P_1^2

500 MW

$$P_2$$
Min = 70 MW
Max = 400 MW

$$F_1(P) = F_2(P) = 400 + 7.0 \, P + 0.002 \, P^2$$

FIG. 4.13   Two-generator system.

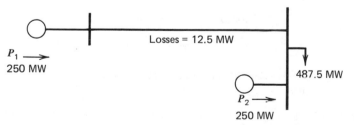

$$P_1 \longrightarrow$$
250 MW

Losses = 12.5 MW

487.5 MW

$$P_2 \longrightarrow$$
250 MW

FIG. 4.14   Two-generator system with both generators at 250 MW output.

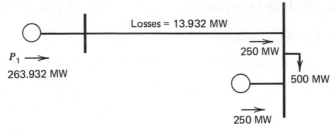

FIG. 4.15 Two-generator system with generator 1 supplying all losses.

Substituting into Eq. 4.27,

$$7.0 + 0.004\,P_1 - \lambda(1 - 0.0004\,P_1) = 0$$

$$7.0 + 0.004\,P_2 - \lambda = 0$$

$$P_1 + P_2 - 500 - 0.0002\,P_1^2 = 0$$

Solution: $P_1 = 178.882$

$P_2 = 327.496$

Production cost: $F_1(P_1) + F_2(P_2) = 4623.15\text{R/h}$

Losses: 6.378 MW

Suppose we had decided simply to ignore the economic influence of losses and ran unit 1 up until it supplied all the losses. It would need to be run at 263.932 MW as shown in Figure 4.15. In this case, the total production cost would be

$$F_1(263.932) + F_2(250) = 4661.84 \; \text{R/h}$$

Note that the optimum dispatch tends toward supplying the losses from the unit close to the load, and it also resulted in a lower value of losses. Also note that best economics are not necessarily attained at minimum losses. The minimum loss solution for this case would simply run unit 1 down and unit 2 up as far as possible. The result is unit 2 on high limit.

$$P_1 = 102.084 \; \text{MW}$$

$$P_2 = 400.00 \; \text{MW (high limit)}$$

The minimum loss production cost would be

$$F_1(102.084) + F_2(400) = 4655.43 \; \text{R/h}$$

$$\text{Min losses} = 2.084 \; \text{MW}$$

4.2.2 Coordination Equations, Incremental Losses, and Penalty Factors

The classic LaGrange multiplier solution to the economic dispatch problem was given in Chapter 3, Eqs. 3.7, 3.8, and 3.9. These are repeated here and expanded.

Minimize: $\mathcal{L} = F_T + \lambda\phi$

Where: $F_T = \sum\limits_{i=1}^{N} F_i(P_i)$

$$\phi = P_R + P_L(P_1, P_2 \cdots P_N) - \sum\limits_{i=1}^{N} P_i$$

Load   Losses   Generation

Solution: $\dfrac{\partial\mathcal{L}}{\partial P_i} = 0$ for all $P_{i\min} \leq P_i \leq P_{i\max}$

Then

$$\frac{\partial\mathcal{L}}{\partial P_i} = \frac{dF_i}{dP_i} - \lambda\left(1 - \frac{\partial P_L}{\partial P_i}\right) = 0$$

The equations are rearranged

$$\left(\frac{1}{1 - \dfrac{\partial P_L}{\partial P_i}}\right)\frac{dF_i(P_i)}{dP_i} = \lambda \tag{4.28}$$

where

$$\frac{\partial P_L}{\partial P_i}$$

is called the *incremental loss* for bus i, and

$$Pf_i = \left(\frac{1}{1 - \dfrac{\partial P_L}{\partial P_i}}\right)$$

is called the *penalty factor* for bus i. Note that if the losses increase for an increase in power from bus i, the incremental loss is positive and the penalty factor is greater than unity.

When we did not take account of transmission losses, the economic dispatch problem was solved by making the incremental cost at each unit the same. We can still use this concept by observing that the penalty factor, Pf_i, will have the following effect. For

$$Pf_i > 1$$

(Positive increase in P_i results in increase in losses.)

$$Pf_i \frac{dF_i(P_i)}{dP_i}$$

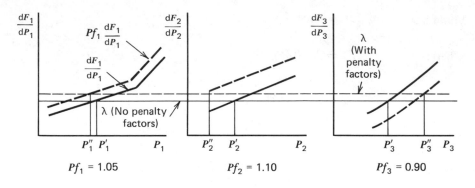

P'_i = Dispatch ignoring losses
P''_i = Dispatch with penalty factors

FIG. 4.16 Economic dispatch with and without penalty factors.

acts as if

$$\frac{dF_i(P_i)}{dP_i}$$

had been slightly increased (moved up). For $Pf_i < 1$ (positive increase in P_i results in decrease in losses).

$$Pf_i \frac{dF_i(P_i)}{dP_i}$$

acts as if

$$\frac{dF_i(P_i)}{dP_i}$$

had been slightly decreased (moved down). The resulting set of equations look like

$$Pf_i \frac{dF_i(P_i)}{dP_i} = \lambda \qquad \text{for all } P_{i\min} \leq P_i \leq P_{i\max} \qquad (4.29)$$

and are called *coordination equations*. The P_i values that result when penalty factors are used will be somewhat different from the dispatch, which ignores the losses (depending on the Pf_i and $dF_i(P_i)/dP_i$ values). This is illustrated in Figure 4.16.

4.2.3 The B Matrix Loss Formula

The B matrix loss formula was originally introduced in the early 1950s as a practical method for loss and incremental loss calculations. At the time, automatic dispatching was performed by analog computers and the loss formula was "stored" in the analog computers by setting precision potentiometers. The equation for the B matrix loss formula is as follows.

where $\mathbf{P}$ = vector of all generator bus net MW
        $[B]$ = square matrix of the same dimension as $\mathbf{P}$
        $\mathbf{B}_0$ = vector of the same length as $\mathbf{P}$
        B_{00} = constant

$$P_L = \mathbf{P}^T[\mathbf{B}]\mathbf{P} + \mathbf{P}^T\mathbf{B}_0 + B_{00} \tag{4.30}$$

This can be written:

$$P_L = \sum_i \sum_j P_i B_{ij} P_j + \sum_i B_{i0} P_i + B_{00} \tag{4.31}$$

Before we discuss the calculation of the B coefficients, we will discuss how the coefficients are used in an economic dispatch calculation. Substitute Eq. 4.31 into Eqs. 3.7, 3.8, and 3.9.

$$\phi = \sum_{i=1}^N P_i - P_R - \left(\sum_i \sum_j P_i B_{ij} P_j + \sum_i B_{i0} P_i + B_{00} \right) \tag{4.32}$$

Then

$$\frac{\partial \mathscr{L}}{\partial P_i} = \frac{dF_i P_i}{dP_i} - \lambda \left(1 - 2 \sum_j B_{ij} P_j - B_{i0} \right) \tag{4.33}$$

Note that the presence of the incremental losses has coupled the coordination equations; this makes solution somewhat more difficult. A method of solution that is often used is shown in Figure 4.17.

EXAMPLE 4D

The B matrix loss formula for the network in Example 4A is given here. (Note that all P_i's must be per unit on 100 MVA base, which results in P_{loss} in per unit on 100 MVA base.)

$$P_{loss} = [P_1 \quad P_2 \quad P_3] \begin{bmatrix} 0.0676 & 0.00953 & -0.00507 \\ 0.00953 & 0.0521 & 0.00901 \\ -0.00507 & 0.00901 & 0.0294 \end{bmatrix} \begin{bmatrix} P_1 \\ P_2 \\ P_3 \end{bmatrix}$$

$$+ [-0.0766 \quad -0.00342 \quad 0.0189] \begin{bmatrix} P_1 \\ P_2 \\ P_3 \end{bmatrix} + 0.040357$$

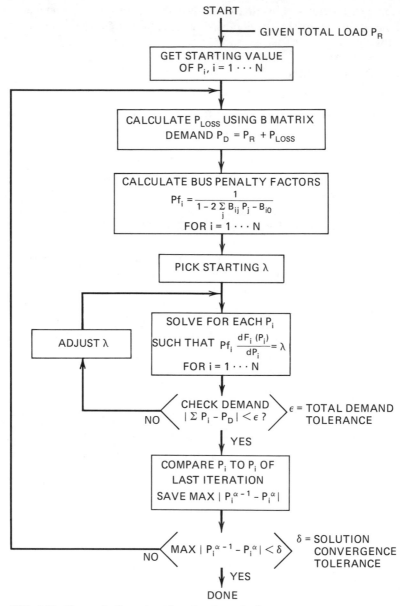

FIG. 4.17 Economic dispatch with updated penalty factors.

From the base case load flow we have

$$P_1 = 107.9 \text{ MW}$$

$$P_2 = 50.0 \text{ MW}$$

$$P_3 = 60.0 \text{ MW}$$

$$P_{\text{loss}} = 7.9 \text{ MW (as calculated by the load flow)}$$

With these generation values placed in the B matrix, we see a very close agreement with the load flow calculation.

$$P_{loss} = [1.079 \quad 0.50 \quad 0.60] \begin{bmatrix} 0.0676 & 0.00953 & -0.00507 \\ 0.00953 & 0.0521 & 0.00901 \\ -0.00507 & 0.00901 & 0.0294 \end{bmatrix} \begin{bmatrix} 1.079 \\ 0.50 \\ 0.60 \end{bmatrix}$$

$$+ [-0.0766 \quad -0.00342 \quad 0.0189] \begin{bmatrix} 1.079 \\ 0.50 \\ 0.60 \end{bmatrix} + 0.040357$$

$$= 0.07877 \text{ pu (or 7.877 MW) loss}$$

EXAMPLE 4E

Let the fuel cost curves for the three units in the six-bus network of Example 4A be given as

$$F_1(P_1) = 213.1 + 11.669\,P_1 + 0.00533\,P_1^2 \; \text{R/h}$$

$$F_2(P_2) = 200.0 + 10.333\,P_2 + 0.00889\,P_2^2 \; \text{R/h}$$

$$F_3(P_3) = 240.0 + 10.833\,P_3 + 0.00741\,P_3^2 \; \text{R/h}$$

With unit dispatch limits

$$50.0 \text{ MW} \le P_1 \le 200 \text{ MW}$$

$$37.5 \text{ MW} \le P_2 \le 150 \text{ MW}$$

$$45.0 \text{ MW} \le P_3 \le 180 \text{ MW}$$

A computer program using the method of Figure 4.17 was run using the following starting conditions from the load-flow solution.

$$P_1 = 107.9 \text{ MW}$$

$$P_2 = 50.0 \text{ MW}$$

$$P_3 = 60.0 \text{ MW}$$

and

$$P_R \text{ (total load to be supplied)} = 210 \text{ MW}$$

The resulting iterations (Table 4.1) show how the program must redispatch again and again to account for the changes in losses and penalty factors.

Note that the flowchart of Figure 4.17 shows a "two-loop" procedure. The "inner" loop adjusts λ until total demand is met; then the outer loop recalculates the penalty factors. (Under some circumstances the penalty factors are quite sensitive to changes in dispatch. If the incremental costs are relatively "flat," this procedure may be unstable and special precautions may need to be employed to ensure convergence.)

TABLE 4.1 Iterations for Example 4E

Iteration	λ	P_{loss}	P_D	P_1	P_2	P_3
1	11.9626	7.8771	217.8771	50.0000	91.6545	76.2225
2	12.6500	10.8536	220.8535	50.0000	48.2477	122.6055
3	13.2807	10.5843	220.5843	60.5845	79.6274	80.3727
4	13.2772	9.9045	219.9044	60.2852	79.4562	80.1625
5	13.2771	9.8672	219.8672	60.2686	79.4467	80.1509
6	13.2770	9.8651	219.8651	60.2677	79.4462	80.1503
7	13.2770	9.8650	219.8650	60.2677	79.4462	80.1503
8	13.2770	9.8650	219.8650	60.2677	79.4462	80.1503

4.2.3.1 Derivation of the B Matrix Loss Formula

The derivation of the B matrix loss formula involves the concept of network reference-frame transformations. A *reference frame* is nothing more than a set of voltages and currents that completely describe the electrical network. At any bus i in an electric power network, power "injected" into the network is defined as

$$P_i + jQ_i = E_i I_i^* \qquad (4.34)$$

Injected
power

Let the charging capacitance and all other impedances to ground be removed. Then whatever power goes into the network equals the power taken out of the network plus the losses in the network. If we count power put into the network as a positive injection and power taken from the network as a negative injection, we can express losses as

$$P + jQ = \sum_{\substack{\text{All network} \\ \text{buses}}} P_i + jQ_i \qquad (4.35)$$

Losses

If we express the injected power as a function of the voltage and injected current at each bus we have

$$P + jQ = \sum_{\substack{\text{All network} \\ \text{buses}}} E_i I_i^* \qquad (4.36)$$

Losses

Equation 4.36 can be written in terms of the voltage vector and current vector as follows.

$$\mathbf{E} = \begin{bmatrix} E_1 \\ \vdots \\ E_N \end{bmatrix} \qquad \mathbf{I} = \begin{bmatrix} I_1 \\ \vdots \\ I_N \end{bmatrix}$$

Then

$$P + jQ = \mathbf{E}^T\mathbf{I}*$$
Losses

where

$\mathbf{E}^T$ is the transpose of $\mathbf{E}$

$\mathbf{I}*$ is the conjugate of $\mathbf{I}$

so

$$\mathbf{E}^T = \begin{bmatrix} E_1 & \cdots & E_N \end{bmatrix}$$

$$\mathbf{I}* = \begin{bmatrix} I_1^* \\ \vdots \\ I_N^* \end{bmatrix} \tag{4.37}$$

We are going to transform the voltages and currents so that the network losses remain constant. First we define a matrix $[C]$ that contains both real and complex numbers. We transform using the following matrix multiplications.

$$\mathbf{E}_{new} = [C]^{T*}\mathbf{E}_{old}$$
$$\mathbf{I}_{old} = [C]\mathbf{I}_{new} \tag{4.38}$$

where $[C]^{T*}$ is the transpose conjugate of $[C]$. Let $\mathbf{E}_{old}$ and $\mathbf{I}_{old}$ be the original voltage and current vectors. From Eq. 4.37

$$P + jQ = \mathbf{E}^T\mathbf{I}* = \mathbf{E}_{old}^T\mathbf{I}_{old}^* \tag{4.39}$$
Losses

what we want is

$$P + jQ = \mathbf{E}_{new}^T\mathbf{I}_{new}^* \tag{4.40}$$
Losses

substituting into Eq. 4.40,

$$P + jQ = ([C]^{T*}\mathbf{E}_{old})^T\mathbf{I}_{new}^*$$
Losses

$$= \mathbf{E}_{old}^T[C]*\mathbf{I}_{new}^*$$
$$= \mathbf{E}_{old}^T([C]\mathbf{I}_{new})*$$
$$= \mathbf{E}_{old}^T\mathbf{I}_{old}^* \tag{4.41}$$

REMEMBER: (1) The transpose of a product of matrices is the product of the transpose of each written in reverse order. (2) The conjugate of the product of matrices is the product of their conjugates in the same order.

Furthermore, we can express the impedance matrix as a new matrix referenced to the new voltage and currents. First,

$$\mathbf{E}_{old} = [Z_{old}]\mathbf{I}_{old} \tag{4.42}$$

Then

$$\mathbf{E}_{new} = [C]^{T*}\mathbf{E}_{old} = [C]^{T*}[Z_{old}]\mathbf{I}_{old}$$
$$= \underbrace{[C]^{T*}[Z_{old}][C]}_{[Z_{new}]}\mathbf{I}_{new}$$

or

$$[Z_{new}] = [C]^{T*}[Z_{old}][C]$$

and

$$\mathbf{E}_{new} = [Z_{new}]\mathbf{I}_{new} \tag{4.43}$$

EXAMPLE 4F

A simple DC circuit example will be used to show how the transforms are used. The circuit is given in Figure 4.18. Note that one bus is the reference and is grounded. Then the Y and Z matrix equations are:

$$\mathbf{I} = [Y]\mathbf{E}$$

$$\begin{bmatrix} I_1 \\ I_2 \\ I_3 \end{bmatrix} = \begin{bmatrix} 5 & -5 & 0 \\ -5 & 7 & -2 \\ 0 & -2 & 3 \end{bmatrix} \begin{bmatrix} E_1 \\ E_2 \\ E_3 \end{bmatrix} \tag{4.44}$$

$$\mathbf{E} = [Z]\mathbf{I}$$

$$\begin{bmatrix} E_1 \\ E_2 \\ E_3 \end{bmatrix} = \begin{bmatrix} 1.7 & 1.5 & 1 \\ 1.5 & 1.5 & 1 \\ 1 & 1 & 1 \end{bmatrix} \begin{bmatrix} I_1 \\ I_2 \\ I_3 \end{bmatrix} \tag{4.45}$$

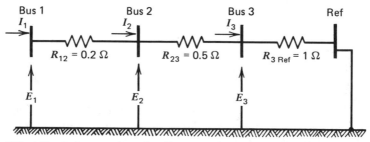

FIG. 4.18   A dc network used to illustrate reference transformations.

Let the currents be given as

$$
\begin{bmatrix} I_1 \\ I_2 \\ I_3 \end{bmatrix} = \begin{bmatrix} +10 \\ -4 \\ -6 \end{bmatrix}
$$

The voltages are (using Eq. 4.45)

$$
\begin{bmatrix} E_1 \\ E_2 \\ E_3 \end{bmatrix} = \begin{bmatrix} 5 \\ 3 \\ 0 \end{bmatrix}
$$

The net losses are (we will write the conjugate but ignore it since all quantities here are real numbers):

$$
P_{\text{loss}} = \mathbf{E}^T \mathbf{I}^* = \begin{bmatrix} 5 & 3 & 0 \end{bmatrix} \begin{bmatrix} 10 \\ -4 \\ -6 \end{bmatrix} = 38 \text{ W} \tag{4.46}
$$

Think of I_2 and I_3 as load currents that are always in the same proportion of an equivalent total load current I_{Leq}. (Such loads in a power system are said to be "conforming.") Similarly, let I_1 be thought of as a generator current. Then for our example here let

$$
\begin{aligned}
I_2 &= 0.4\, I_{Leq} \\
I_3 &= 0.6\, I_{Leq}
\end{aligned} \tag{4.47}
$$

The vector relationship between $\mathbf{I}_{\text{old}}$ and $\mathbf{I}_{\text{new}}$ becomes

$$
\mathbf{I}_{\text{old}} = [C]\mathbf{I}_{\text{new}}
$$

$$
\begin{bmatrix} I_1 \\ I_2 \\ I_3 \end{bmatrix} = \begin{bmatrix} 1 & 0 \\ 0 & 0.4 \\ 0 & 0.6 \end{bmatrix} \begin{bmatrix} I_1 \\ I_{Leq} \end{bmatrix} \tag{4.48}
$$

Then by definition

$$
\mathbf{E}_{\text{new}} = [C]^{T*}\mathbf{E}_{\text{old}}
$$

$$
\begin{bmatrix} E_1 \\ E_{Leq} \end{bmatrix} = \begin{bmatrix} 1 & 0 & 0 \\ 0 & 0.4 & 0.6 \end{bmatrix} \begin{bmatrix} E_1 \\ E_2 \\ E_3 \end{bmatrix} \tag{4.49}
$$

and $[Z_{\text{new}}]$ is

$$
[Z_{\text{new}}] = \begin{bmatrix} 1 & 0 & 0 \\ 0 & 0.4 & 0.6 \end{bmatrix} \begin{bmatrix} 1.7 & 1.5 & 1 \\ 1.5 & 1.5 & 1 \\ 1 & 1 & 1 \end{bmatrix} \begin{bmatrix} 1 & 0 \\ 0 & 0.4 \\ 0 & 0.6 \end{bmatrix}
$$

$$
= \begin{bmatrix} 1.7 & 1.2 \\ 1.2 & 1.08 \end{bmatrix} \tag{4.50}
$$

Let us calculate the losses for a total equivalent load current of -10 A. (Note that $I_{L_{eq}} = -10$ results in $I_2 = -4$, $I_3 = -6$, as before.) Then

$$\begin{bmatrix} E_1 \\ E_{L_{eq}} \end{bmatrix} = \begin{bmatrix} 1.7 & 1.2 \\ 1.2 & 1.08 \end{bmatrix}\begin{bmatrix} I_1 \\ I_{L_{eq}} \end{bmatrix} = \begin{bmatrix} 1.7 & 1.2 \\ 1.2 & 1.08 \end{bmatrix}\begin{bmatrix} 10 \\ -10 \end{bmatrix} = \begin{bmatrix} 5 \\ 1.2 \end{bmatrix} \tag{4.51}$$

and the losses are

$$\mathbf{E}_{new}^T \mathbf{I}_{new}^* = \begin{bmatrix} 5 & 1.2 \end{bmatrix}\begin{bmatrix} 10 \\ -10 \end{bmatrix} = 38 \text{ W} \tag{4.52}$$

If we further make the assumption that $I_L = -I_1$, that is, the equivalent load current will always be the negative of the generator current. [*Note:* The next new reference frame is designed with a prime ($\mathbf{I}'$, $\mathbf{E}'$, etc.).] Then

$$\mathbf{I}_{new} = [C]\mathbf{I}'_{new}$$

$$\begin{bmatrix} I_1 \\ I_{L_{eq}} \end{bmatrix} = \begin{bmatrix} 1 \\ -1 \end{bmatrix}I_1 \tag{4.53}$$

and

$$\mathbf{E}'_{new} = \begin{bmatrix} 1 & -1 \end{bmatrix}\begin{bmatrix} E_1 \\ E_{L_{eq}} \end{bmatrix}$$

and

$$[\mathbf{Z}'_{new}] = \begin{bmatrix} 1 & -1 \end{bmatrix}\begin{bmatrix} 1.7 & 1.2 \\ 1.2 & 1.08 \end{bmatrix}\begin{bmatrix} 1 \\ -1 \end{bmatrix} = 0.38$$

if as before $I_1 = 10$ A. Then

$$E'_{new} = 3.8 \text{ V}$$

and

$$E'_{new}I'^*_{new} = (3.8)(10) = 38 \text{ W} \tag{4.54}$$

By following this procedure we have reduced the effort to compute losses. As long as the load currents "conform" as in the derivation, the losses can be calculated using only the generator input current itself. This concept is extended to an electric power system to allow losses to be calculated using only the megawatt input for each of the system's generators.

4.2.3.2 Loss-Matrix Calculation

There are different methods for calculating a loss matrix. One of these methods is shown in detail in Appendix B to this chapter. This discussion will simply highlight some of the types of transformations and calculations used in loss-matrix algorithms. The first assumption usually made is that the load at each bus conforms to the total load. That is,

$$P_{\text{load}_i} = \ell_i P_{\text{total load}} + P^0_{\text{load}_i} \tag{4.55}$$

Since the load on the network buses is represented by complex load currents, we can more properly write

$$I^L_i = \ell_i I_L + I^0_i \tag{4.56}$$

where I^L_i = load current at bus i

   ℓ_i = complex load distribution factor for bus i

   I_L = total system load current

   I^0_i = base or constant load current at bus i

Similarly, the generator real and reactive powers are assumed to be related to each other by a constant, S_i.

$$Q^G_i = Q^{G^0}_i + S_i P^G_i \tag{4.57}$$

The S_i, $Q^{G^0}_i$, ℓ_i, and I^0_i quantities are derived from two load-flow calculations, one called the "base" the other called the "off-base." These load-flow calculations are run at loading conditions typically found on the network. The base-load flow may, for example, correspond to the system's peak load conditions and the off-base to its low load conditions.

The ℓ_i factors are used in a series of transformations on the network impedance matrix to eliminate the individual load currents (see Example 4F). The S_i factors together with the voltage magnitudes and phase angles at the generator buses are used to convert generator bus currents to generator bus real powers. Thus, the basic system voltage-current relationship

$$\mathbf{V} = [Z]\mathbf{I} \tag{4.58}$$

is used to set up an equation for losses.

$$\begin{aligned} P_{\text{loss}} &= R_e\{P_{\text{loss}} - jQ_{\text{loss}}\} \\ &= R_e\{\mathbf{I}^{*^T}[Z]\mathbf{I}\} \end{aligned} \tag{4.59}$$

and the transformations just described give a final expression:

$$P_{\text{loss}} = \mathbf{P}^{G^T}[B]\mathbf{P}^G + \mathbf{B}_0\mathbf{P}^G + B_{00} \tag{4.60}$$

where $\mathbf{P}^G$ is a vector of generator real powers and B, $\mathbf{B}_0$, B_{00} are the loss-matrix terms that are a function of the $[Z]$ matrix elements, the base-case load flow, voltages and phase angles, as well as the ℓ_i's, and S_i's.

The calculations are not simple, as even a casual reading of Appendix B will show. Further, it must be remembered that the assumptions of conforming load and conforming generator reactive power are never seen exactly on a real system; therefore, the loss matrix will always remain an approximation. Nonetheless, loss matrices are widely used because of their convenience and the speed with which losses and incremental losses can be calculated.

4.2.4 Other Methods of Calculating Penalty Factors

4.2.4.1 A Discussion of Reference Bus versus Load Center Penalty Factors

The B matrix assumes that all load currents conform to an equivalent total load current and that the equivalent load current is the negative of the sum of all generator currents. When incremental losses are calculated, something is implied.

$$\text{Total loss} = \mathbf{P}^T[B]\mathbf{P} + \mathbf{B}_0\mathbf{P} + B_{00}$$

$$\text{Incremental loss at generator bus } i = \frac{\partial P_{\text{loss}}}{\partial P_i}$$

The incremental loss is the change in losses when an increment is made in generation output. As just derived, the incremental loss for bus i assumed that all the other generators remained fixed. By the original assumption, however, the load currents all conform to each other and always balance with the generation, then the implication in using a B matrix is that an *incremental increase in generator output is matched by an equivalent increment in load.*

An alternative approach to economic dispatch is to use a reference bus that always moves when an increment in generation is made. Figure 4.19 shows a power system with several generator buses and a reference-generator bus. Suppose we change the generation on bus i by ΔP_i,

$$P_i^{\text{new}} = P_i^{\text{old}} + \Delta P_i \tag{4.61}$$

Furthermore, we will assume that *load stays constant* and that to compensate for the increase in ΔP_i the reference bus just drops off by ΔP_{ref}.

$$P_{\text{ref}}^{\text{new}} = P_{\text{ref}}^{\text{old}} + \Delta P_{\text{ref}} \tag{4.62}$$

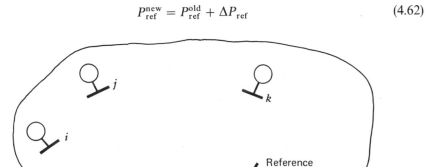

FIG. 4.19 Power system with reference generator.

If nothing else changed, ΔP_{ref} would be the negative of ΔP_i; however, the flows on the system can change as a result of the two generation adjustments. The change in flow is apt to cause a change in losses so that ΔP_{ref} is not necessarily equal to ΔP_i. That is,

$$\Delta P_{\text{ref}} = -\Delta P_i + \Delta P_{\text{loss}} \tag{4.63}$$

Next, we can define β_i as the ratio of the negative change in the reference-bus power to the change ΔP_i.

$$\beta_i = \frac{-\Delta P_{\text{ref}}}{\Delta P_i} = \frac{(\Delta P_i - \Delta P_{\text{loss}})}{\Delta P_i} \tag{4.64}$$

or

$$\beta_i = 1 - \frac{\partial P_{\text{loss}}}{\partial P_i} \tag{4.65}$$

We can define economic dispatch as follows.

All generators are in economic dispatch when a shift of ΔP MW from any generator to the reference bus results in no change in net production cost. Where ΔP is arbitrarily small.

That is,

$$\text{if total production cost} = \sum F_i(P_i)$$

then the change in production cost with a shift ΔP_i from plant i is

$$\Delta \text{Production cost} = \frac{dF_i(P_i)}{dP_i}\,\Delta P_i + \frac{dF_{\text{ref}}(P_{\text{ref}})}{dP_{\text{ref}}}\,\Delta P_{\text{ref}} \tag{4.66}$$

But

$$\Delta P_{\text{ref}} = -\beta_i\,\Delta P_i$$

Then

$$\Delta \text{Production cost} = \frac{dF_i(P_i)}{dP_i}\,\Delta P_i - \beta_i \frac{dF_{\text{ref}}(P_{\text{ref}})}{dP_{\text{ref}}}\,\Delta P_i \tag{4.67}$$

To satisfy the economic conditions,

$$\Delta \text{Production cost} = 0$$

or

$$\frac{dF_i(P_i)}{dP_i} = \beta_i \frac{dF_{\text{ref}}(P_{\text{ref}})}{dP_{\text{ref}}} \tag{4.68}$$

which could be written as

$$\frac{1}{\beta_i}\frac{dF_i(P_i)}{dP_i} = \frac{dF_{ref}(P_{ref})}{dP_{ref}} \tag{4.69}$$

This is very similar to Eq. 4.28. To obtain an economic dispatch solution, pick a value of generation on the reference bus and then set all other generators according to Eq. 4.69 and check for total demand and readjust reference generation as needed until a solution is reached.

Note further that this method is exactly the first-order gradient method (Section 3.4) with losses. Where (see Eq. 3.14 and substitute ref for x)

$$\Delta F_T = \sum_{i \neq ref} \left[\frac{dF_i}{dP_i} - \beta_i \frac{dF_{ref}}{dP_{ref}} \right] \Delta P_i \tag{4.70}$$

4.2.4.2 Reference-Bus Penalty Factors Direct from the AC Load Flow

The reference-bus penalty factors may be derived using the Newton-Raphson load flow. What we wish to know is the ratio of change in power on the reference bus when a change ΔP_i is made.

Where P_{ref} is a function of the voltage magnitude and phase angle on the network

When a change in ΔP_i is made, all phase angles and voltages in the network will change

Then

$$\begin{aligned}\Delta P_{ref} &= \sum_i \frac{\partial P_{ref}}{\partial \theta_i}\Delta \theta_i + \sum_i \frac{\partial P_{ref}}{\partial |E_i|}\Delta |E_i| \\ &= \sum_i \frac{\partial P_{ref}}{\partial \theta_i}\frac{\partial \theta_i}{\partial P_i}\Delta P_i + \sum_i \frac{\partial P_{ref}}{\partial |E_i|}\frac{\partial |E_i|}{\partial P_i}\Delta P_i\end{aligned} \tag{4.71}$$

To carry out the matrix manipulations, we will also need the following.

$$\begin{aligned}\Delta P_{ref} &= \sum_i \frac{\partial P_{ref}}{\partial \theta_i}\Delta \theta_i + \sum_i \frac{\partial P_{ref}}{\partial |E_i|}\Delta |E_i| \\ &= \sum_i \frac{\partial P_{ref}}{\partial \theta_i}\frac{\partial \theta_i}{\partial Q_i}\Delta Q_i + \sum_i \frac{\partial P_{ref}}{\partial |E_i|}\frac{\partial |E_i|}{\partial Q_i}\Delta Q_i\end{aligned} \tag{4.72}$$

The terms $\partial P_{ref}/\partial \theta_i$ are derived by differentiating Eq. 4.18 for the reference bus. The terms $\partial \theta_i/\partial P_i$ and $\partial |E_i|/\partial P_i$ are from the inverse Jacobian matrix (see Eq. 4.20). We can write Eqs. 4.71 and 4.72 for every bus i in the network. The resulting equation is

$$\begin{aligned}\left[\frac{\partial P_{ref}}{\partial P_1}\frac{\partial P_{ref}}{\partial Q_1}\frac{\partial P_{ref}}{\partial P_2}\frac{\partial P_{ref}}{\partial Q_2} \cdots \frac{\partial P_{ref}}{\partial P_N}\frac{\partial P_{ref}}{\partial Q_N} \right] \\ = \left[\frac{\partial P_{ref}}{\partial \theta_1}\frac{\partial P_{ref}}{\partial |E_1|}\frac{\partial P_{ref}}{\partial \theta_2}\frac{\partial P_{ref}}{\partial |E_2|} \cdots \frac{\partial P_{ref}}{\partial \theta_N}\frac{\partial P_{ref}}{\partial |E_N|} \right][J^{-1}]\end{aligned} \tag{4.73}$$

By transposing we get

$$
\begin{bmatrix}
\dfrac{\partial P_{\mathrm{ref}}}{\partial P_1} \\[8pt]
\dfrac{\partial P_{\mathrm{ref}}}{\partial Q_1} \\[8pt]
\dfrac{\partial P_{\mathrm{ref}}}{\partial P_2} \\[8pt]
\dfrac{\partial P_{\mathrm{ref}}}{\partial Q_2} \\[8pt]
\vdots \\[8pt]
\dfrac{\partial P_{\mathrm{ref}}}{\partial P_N} \\[8pt]
\dfrac{\partial P_{\mathrm{ref}}}{\partial Q_N}
\end{bmatrix}
= [J^{T-1}]
\begin{bmatrix}
\dfrac{\partial P_{\mathrm{ref}}}{\partial \theta_1} \\[8pt]
\dfrac{\partial P_{\mathrm{ref}}}{\partial |E_1|} \\[8pt]
\dfrac{\partial P_{\mathrm{ref}}}{\partial \theta_2} \\[8pt]
\dfrac{\partial P_{\mathrm{ref}}}{\partial |E_2|} \\[8pt]
\vdots \\[8pt]
\dfrac{\partial P_{\mathrm{ref}}}{\partial \theta_N} \\[8pt]
\dfrac{\partial P_{\mathrm{ref}}}{\partial |E_N|}
\end{bmatrix}
$$

In practice, instead of calculating J^{T-1} explicitly, we use Gaussian elimination on J^T in the same way we operate on J in the Newton load-flow solution.

4.2.4.3 Reference-Bus Penalty Factors Using the DC Load Flow

The real (MW) losses on a transmission line (where $R_e(\ )=$ real part of) are

$$
R_e(\mathrm{loss}_{ij}) = [|E_i|^2 + |E_j|^2 - 2|E_i||E_j|\cos(\theta_i - \theta_j)]R_e(Y_{ij}) \tag{4.74}
$$

Taking the derivatives with respect to θ_i and θ_j,

$$
\frac{\partial P_{\mathrm{loss}_{ij}}}{\partial \theta_i} = 2|E_i||E_j|\sin(\theta_i - \theta_j)R_e(Y_{ij})
$$

$$
\frac{\partial P_{\mathrm{loss}_{ij}}}{\partial \theta_j} = -2|E_i||E_j|\sin(\theta_i - \theta_j)R_e(Y_{ij})
$$

$$\tag{4.75}$$

We can calculate incremental losses for the entire system by first obtaining the change in phase angle at every bus with a change in bus power injection and then multiplying these derivatives times the derivative in Eq. 4.75 and adding the contribution for all lines. That is,

$$
\Delta P_{\mathrm{loss}} = \sum_{\substack{\text{All} \\ \text{lines } \ell}} \frac{\partial P_{\mathrm{loss}} \text{ on line } \ell}{\partial \theta_j} \frac{\partial \theta_j}{\partial P_i} \Delta P_i \tag{4.76}
$$

To facilitate matrix operations, we will sum over all buses and then over all lines attached to each bus.

$$\frac{\Delta P_{\text{loss}}}{\Delta P_i} = \sum_{\substack{\text{All} \\ \text{buses} \\ j}} \left[\sum_{\substack{\text{All lines } \ell \\ \text{attached to} \\ \text{bus } j}} \frac{\partial P_{\text{loss}} \text{ line } \ell}{\partial \theta_j} \right] \frac{\partial \theta_j}{\partial P_i} \tag{4.77}$$

The terms $\partial \theta_j / \partial P_i$ are in the columns of the X matrix given in Eq. 4.25b. That is,

$$\begin{bmatrix} \Delta \theta_1 \\ \Delta \theta_2 \\ \vdots \end{bmatrix} = \begin{bmatrix} \partial \theta_1 / \partial P_1 & \partial \theta_1 / \partial P_2 & \cdots \\ \vdots & \vdots & \end{bmatrix} \begin{bmatrix} \Delta P_1 \\ \Delta P_2 \\ \vdots \end{bmatrix} \tag{4.78}$$

Then Eq. 4.77 can be written

$$\begin{bmatrix} \dfrac{\partial P_{\text{loss}}}{\partial P_1} \\[2ex] \dfrac{\partial P_{\text{loss}}}{\partial P_2} \\[2ex] \vdots \end{bmatrix} = \begin{bmatrix} \dfrac{\partial \theta_1}{\partial P_1} & \dfrac{\partial \theta_2}{\partial P_1} & \cdots \\[2ex] \vdots & & \end{bmatrix}^{\text{T}} \begin{bmatrix} \dfrac{\partial P_{\text{loss}}}{\partial \theta_1} \\[2ex] \dfrac{\partial P_{\text{loss}}}{\partial \theta_2} \\[2ex] \vdots \end{bmatrix} \tag{4.79}$$

where the right-hand side vector terms come from the bracketed terms in Eq. 4.77. Where

$$\frac{\partial P_{\text{loss}}}{\partial \theta_j} = \sum_{\substack{\text{All lines } \ell \\ \text{attached to} \\ \text{bus } j}} \frac{\partial P_{\text{loss}} \text{ line } \ell}{\partial \theta_j} \tag{4.80}$$

Also note that the matrix in Eq. 4.79 is the transpose of the $[X]$ matrix. Incremental losses derived by Eq. 4.79 are very nearly the same as those derived by Eq. 4.73, but the solution of Eq. 4.79 is much faster.

4.2.4.4 Mathematical Relationship of Loss-Matrix Penalty Factors and and Reference-Bus Penalty Factors

Suppose we are given a loss matrix from which we can calculate incremental losses $\partial P_{\text{loss}} / \partial P_i$ for each generator bus in a network. Assume that for a change ΔP_i in generation at bus i a compensating change in load, ΔP_{load_i}, will take place at an "equivalent load center" bus. We are going to look at such a shift for two separate generator buses with each shift taken independently. We will call the first generator bus, bus i, and the second, bus r. For the first shift, from bus i, see Figure 4.20, where

$$\Delta P_i = \Delta P_{\text{loss}_i} + \Delta P_{\text{load}_i} \tag{4.81}$$

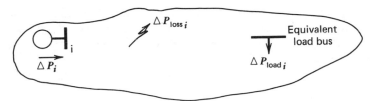

FIG. 4.20   Change at equivalent load bus and bus i.

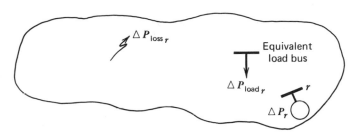

FIG. 4.21   Change at equivalent load bus and reference bus.

If we approximate $\Delta P_{\text{loss}}/\Delta P_i$ using $\partial P_{\text{loss}}/\partial P_i$ from the loss matrix, we can write Eq. 4.81 as

$$\Delta P_{\text{load}_i} = \left(1 - \frac{\partial P_{\text{loss}}^*}{\partial P_i}\right)\Delta P_i \qquad (4.82)$$

where the asterisk (*) indicates an incremental loss calculated from a loss matrix. Similarly, we can do the same thing, independently, for bus r (see Figure 4.21), where

$$\Delta P_r = \Delta P_{\text{loss}_r} + \Delta P_{\text{load}_r} \qquad (4.83)$$

and again we approximate $\Delta P_{\text{loss}}/\Delta P_i$ as $\partial P_{\text{loss}}/\partial P_r$ from the loss matrix. Then,

$$\Delta P_{\text{load}_r} = \left(1 - \frac{\partial P_{\text{loss}}^*}{\partial P_r}\right)\Delta P_r \qquad (4.84)$$

Now suppose we carry out these two steps in sequence as follows. First, we carry out the step shown in Eq. 4.82 so that ΔP_{load_i} exactly compensates for the change ΔP_i and the change in losses. Then we adjust ΔP_r so that ΔP_{load_r} is precisely the *negative* of ΔP_{load_i}. That is,

$$\Delta P_{\text{load}_r} = -\Delta P_{\text{load}_i} \qquad (4.85)$$

Then,

$$\left(1 - \frac{\partial P_{\text{loss}}^*}{\partial P_r}\right)\Delta P_r = -\left(1 - \frac{\partial P_{\text{loss}}^*}{\partial P_r}\right)\Delta P_i \qquad (4.86)$$

By performing this sequence, we have artificially created the actions that consitute the basis for reference-bus penalty factors. That is, we have increased generation at bus i by ΔP_i and compensated with a change ΔP_r at a reference bus while holding load constant. By definition (see Eq. 4.64),

$$\beta_i = \frac{-\Delta P_r}{\Delta P_i} \tag{4.87}$$

but by manipulating Eq. 4.86,

$$\beta_i = \frac{-\Delta P_r}{\Delta P_i} = \frac{(1 - \partial P^*_{\text{loss}}/\partial P_i)}{(1 - \partial P^*_{\text{loss}}/\partial P_r)} \tag{4.88}$$

If we define penalty factors as

$$Pf_i^{\text{loss matrix}} = \frac{1}{1 - \dfrac{\partial P^*_{\text{loss}}}{\partial P_i}} \tag{4.89}$$

and

$$Pf_i^{\text{ref bus}} = \frac{1}{\beta_i} \tag{4.90}$$

Then,

$$Pf_i^{\text{ref bus}} = \left(1 - \frac{\partial P^*_{\text{loss}}}{\partial P_r}\right)\left(\frac{1}{1 - \dfrac{\partial P^*_{\text{loss}}}{\partial P_i}}\right)$$

$$= \left(1 - \frac{\partial P^*_{\text{loss}}}{\partial P_r}\right) Pf_i^* \tag{4.91}$$

Thus we can see that the reference-bus penalty factors are simply a constant times the loss-matrix penalty factors. Furthermore, it should be realized that this means that an economic dispatch calculation performed with either reference-bus penalty factors or loss-matrix penalty factors will produce the same generation dispatch and only differ in the value of λ.

APPENDIX A
Load Flow Input Data

Figure 4.22 lists the input data to the six-bus sample power system used in the examples in Chapter 4.

LINE DATA

FROM	TO	R	X	BCAP*
1	2	0.1000	0.2000	0.0200
1	4	0.0500	0.2000	0.0200
1	5	0.0800	0.3000	0.0300
2	3	0.0500	0.2500	0.0300
2	4	0.0500	0.1000	0.0100
2	5	0.1000	0.3000	0.0200
2	6	0.0700	0.2000	0.0250
3	5	0.1200	0.2600	0.0250
3	6	0.0200	0.1000	0.0100
4	5	0.2000	0.4000	0.0400
5	6	0.1000	0.3000	0.0300

BUS DATA

BUS NO.		GEN (PU MW)	VOLTAGE (PU KV)	P LOAD (PU MW)	Q LOAD (PU MVAR)
SWING	1	0.00	1.050	0.00	0.00
	2	0.50	1.050	0.00	0.00
	3	0.60	1.070	0.00	0.00
	4	0.00	1.000	0.70	0.70
	5	0.00	1.000	0.70	0.70
	6	0.00	1.000	0.70	0.70

* BCAP = 1/2 Total line charging

FIG. 4.22 Input data to a six-bus sample power system.

APPENDIX B
Derivation of the B Matrix Loss Formula

The B matrix loss formula is developed using a series of transformations on the full-impedance matrix of the transmission system network. The derivation that follows is taken from the paper by Meyer (7). A computer program based on this technique was used to derive the loss formula shown in Example 4D. The derivation in Meyer's paper is related to much earlier pioneering work by Kron (4) and Early and Watson (6), which should be consulted to appreciate Meyer's method fully.

To start, assume that the N transmission system buses are divided up into M generator buses and $N - M$ load buses. The generator buses are ordered $1 \cdots M$, and the load buses $M + 1 \cdots N$. The network is shown in Figure 4.23. The basic relationships between the voltages and currents will be stated throughout the derivation using the impedance (Z) matrix. Meyer's paper should be consulted for specific computational details involving the Z matrix terms.

Our first relationship is then

$$\mathbf{V} = [Z]\mathbf{I} \qquad (4B.1)$$

where $\mathbf{V}$ and $\mathbf{I}$ are $N \times 1$ vectors and $[Z]$ is an $N \times N$ matrix. The total real and reactive losses in the network may be expressed as

$$P_{\text{loss}} + jQ_{\text{loss}} = \mathbf{I}^{*T}\mathbf{V} = \mathbf{I}^{*T}[Z]\mathbf{I} \qquad (4B.2)$$

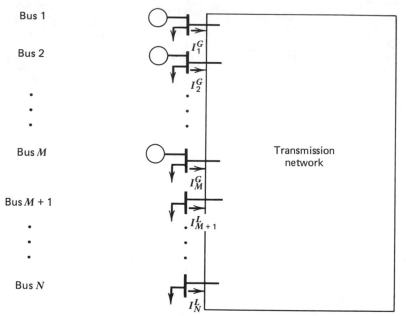

FIG. 4.23 Representation of loads and generators attached to a transmission network.

Since we are only interested in the real power losses, we will use

$$P_{\text{loss}} = R_e\{\mathbf{I}^{*T}[Z]\mathbf{I}\} \tag{4B.3}$$

Each bus-current injection consists of two components, the generation component and the load component designated I_i^G and I_i^L, respectively. Then

$$I_i = I_i^G + I_i^L \qquad \text{for } i = 1 \cdots N \tag{4B.4}$$

where by our ordering of buses $I_i^G = 0$ for $i = M + 1 \cdots N$.

Further, the load currents will be modeled as a linear function of the total load current I_L. This is similar to the "conforming load" assumption used in Example 4F. Then

$$I_i^L = \ell_i I_L + I_i^0 \tag{4B.5}$$

where I_i^L = load current at bus i

ℓ_i = load distribution factor for bus i (ℓ is complex)

I_L = total system load current

I_i^0 = constant or base-load current at bus i

Now we can begin to use transformations as follow.

$$\mathbf{I} = [C_1]\mathbf{I}_2 + \mathbf{I}^0 \tag{4B.6}$$

or

$$
\begin{bmatrix} I_1 \\ \vdots \\ I_M \\ \hline I_{M+1} \\ \vdots \\ I_N \end{bmatrix} = \left[\begin{array}{ccc|c} 1 & & & \ell_1 \\ & \ddots & & \vdots \\ & & 1 & \ell_M \\ \hline & & & \ell_{M+1} \\ & 0 & & \vdots \\ & & & \ell_N \end{array} \right] \begin{bmatrix} I_1^G \\ \vdots \\ I_N^G \\ \hline I_L \end{bmatrix} + \begin{bmatrix} I_1^0 \\ \vdots \\ I_M^0 \\ \hline I_{M+1}^0 \\ \vdots \\ I_N^0 \end{bmatrix}
$$

where $\mathbf{I}_2$ is a vector as shown with the first N elements equal to the N generator currents and the $N + 1$ element equal to the total load current I_L. Substituting (4B.6) into (4B.3), gives

$$ P_{\text{loss}} = R_e\{(\mathbf{I}_2^{*T}[C_1]^{*T} + \mathbf{I}^{0*T})[Z]([C_1]\mathbf{I}_2 + \mathbf{I}^0)\} \tag{4B.7} $$

which on being expanded has four terms.

$$ P_{\text{loss}} = R_e\{H_1 + H_2 + H_3 + H_4\} \tag{4B.8} $$

where $H_1 = \mathbf{I}_2^{*T}[C_1]^{*T}[Z][C_1]\mathbf{I}_2$
$H_2 = \mathbf{I}^{0*T}[Z]\mathbf{I}^0$
$H_3 = \mathbf{I}^{0*T}[Z][C]\mathbf{I}_2$
$H_4 = \mathbf{I}_2^{*T}[C]^{*T}[Z]\mathbf{I}^0$

It should be observed here that the H_1 term contributes only to the quadratic part (B_{ij}) of the loss formula, H_2 contributes to the constant part (B_{00}), and H_3 and H_4 contribute to the linear part (B_{i0}).

Before proceeding, we will observe two assumptions that are built into this loss-formula derivation. The first assumes that each generator's reactive power is a linear function of its real power output. That is,

$$ Q_i^G = Q_i^{G0} + S_i P_i^G \tag{4B.9} $$

The value of S_i is quite important and must be obtained carefully. The usual procedure is to run base and off-base load-flow calculations that differ by the amount of load on the network. Each generator's S_i value and each load's ℓ_i and I_i^0 values are derived from the two load flows.

We use the S_i factors together with the bus voltage magnitudes and phase angles from the base-load flow as follows.

$$ I_i^G = \left(\frac{d_1}{|V_i|} + j\, \frac{d_2}{|V_i|} \right) P_i^G \tag{4B.10} $$

Note that we have ignored the Q_i^{G0} by simply lumping it into the load bus reactive power. The d_1 and d_2 terms are

$$ \begin{aligned} d_1 &= \cos \theta_i + S_i \sin \theta_i \\ d_2 &= \sin \theta_i + S_i \cos \theta_i \end{aligned} \tag{4B.11} $$

The second assumption built into the calculation concerns the reference-bus voltage. This voltage is assumed known and constant as loading varies on the system. Let $i = r$ be the reference bus for the network. It is normal practice to have r be one of the generator buses. Then

$$V_r = \sum_{i=1}^{M} z_{ri} I_i^G + I_L \sum_{i=1}^{N} z_{ri} \ell_i + \sum_{i=1}^{N} z_{ri} I_i^0 \tag{4B.12}$$

We can use 4B.12 to eliminate I_L from our equations.

$$I_L = - \sum_{i=1}^{M} T_i I_i^G + T'' \tilde{V}_r \tag{4B.13}$$

where $\quad T_i = \dfrac{z_{ri}}{\displaystyle\sum_{j=1}^{N} z_{rj} \ell_j}$

$$T'' = \dfrac{1}{\displaystyle\sum_{j=1}^{N} z_{rj} \ell_j}$$

$$\tilde{V}_r = V_r - \sum_{i=1}^{N} z_{ri} I_i^0$$

Using 4B.13, we can build another transformation on I_2.

$$I_2 = [C_2] I_2' \tag{4B.14}$$

or

$$\begin{bmatrix} I_1^G \\ \vdots \\ I_M^G \\ \hline I_L \end{bmatrix} = \left[\begin{array}{ccc|c} & 1 & & 0 \\ & & & \vdots \\ & & 1 & 0 \\ \hline -T_1 & \cdots & -T_M & T'' \end{array} \right] \begin{bmatrix} I_1^G \\ \vdots \\ I_M^G \\ \hline V_R \end{bmatrix}$$

Now we can start to reduce H_1 through H_4. First, H_1; using 4B.14, H_1 becomes

$$H_1 = I_2'^{*T} [C_2]^{*T} [C_1]^{*T} [Z] [C_1][C_2] I_2'$$
$$= I_2'^{*T} [Z^3] I_2' \tag{4B.15}$$

where

$$[Z^3] = [C_2]^{*T} [C_1]^{*T} [Z] [C_1][C_2]$$

We can reduce H_2 to

$$H_2 = I^{0*T} V^A \tag{4B.16}$$

where

$$V^A = [Z] I^0$$

Then

$$R_e\{H_2\} = \sum_{i=1}^{N} R_e\{I_i^{0*} V_i^A\} \tag{4B.17}$$

H_3 and H_4 will be treated together. First, since H_3 and H_4 are both scalers,

$$\begin{aligned}
R_e\{H_3 + H_4\} &= R_e\{H_3^T + H_4^*\} \\
&= R_e\{\mathbf{I}_2^T[C_1]^T[Z]^T\mathbf{I}^{0*} + \mathbf{I}_2^T[C_1]^T[Z]^*\mathbf{I}^{0*}\} \\
&= R_e\{\mathbf{I}_2^T\underbrace{([C_1]^T[Z]^T\mathbf{I}^{0*} + [C_1]^T[Z]^*\mathbf{I}^{0*})}_{V^D}\}
\end{aligned} \tag{4B.18}$$

where

$$V^D = [C_1]^T(\underbrace{Z^T\underline{I}^{0*}}_{\mathbf{V}^B} + \underbrace{Z^*\underline{I}^{0*}}_{\mathbf{V}^A}) \tag{4B.19}$$

Then

$$\begin{aligned}
R_e\{H_3 + H_4\} &= R_e\{\mathbf{I}_2^T\mathbf{V}^D\} \\
&= R_e\left\{\sum_{i=1}^{M} I_i^G V_i^D + V_{M+1}^D I_2\right\}
\end{aligned} \tag{4B.20}$$

and using 4B.13,

$$R_e\{H_3 + H_4\} = R_e\left\{\sum_{i=1}^{M} (V_i^D - V_{M+1}^D T_j)I_i^G + V_{M+1}^D T^n\tilde{V}_r\right\} \tag{4B.21}$$

We can now write out the terms in the loss matrix.

$$\begin{aligned}
B_{ij} = &\frac{(R_{ij}^3 + R_{ji}^3)}{2}\left[\frac{(1 + S_iS_j)\cos(\theta_i - \theta_j) + (S_i - S_j)\sin(\theta_i - \theta_j)}{|V_i||V_j|}\right] \\
&+ \frac{(X_{ij}^3 - X_{ji}^3)}{2}\left[\frac{(1 + S_iS_j)\sin(\theta_i - \theta_j) + (S_i - S_j)\cos(\theta_i - \theta_j)}{|V_i||V_j|}\right]
\end{aligned} \tag{4B.22}$$

$$\begin{aligned}
B_{i0} = &\frac{1}{|V_i|}[(R_e\{\tilde{V}_r\}d_8 - I_m\{\tilde{V}_r\}d_9 + R_e\{\zeta\})d_1 \\
&+ R_e\{\tilde{V}_r\}d_9 + I_m\{\tilde{V}_r\}d_8 - I_m\{\zeta\})d_2]
\end{aligned}$$

$$B_{00} = |\tilde{V}_r|^2 R_e\{\Psi\} + \sum_{i=1}^{N} R_e\{I_i^{0*}V_i^A\} + R_e\{V_{M+1}^D T^n\tilde{V}_r\}$$

where R^3, X^3 are elements from $[Z^3]$ [see Eq. 4B.15].

S_i, S_j are as defined in Eq. 4B.9.

$\tilde{V}_r$ is defined in Eq. 4B.13.

$|V_i|$, $|V_j|$, θ_i, θ_j are from the base-load flow.

$$d_8 = R_e\{T^n(a_i - T_i^*\omega) + T^{n*}(b_j - T_j\omega)\}$$
$$d_9 = I_m\{T^n(a_i - T_i^*\omega) + T^{n*}(b_j - T_j\omega)\}$$

where $a_i = \sum\limits_{j=1}^{N} z_{ij}\ell_j$

$b_j = \sum\limits_{i=1}^{N} Z_{ij}\ell_i^*$

$\omega = \sum\limits_{i=1}^{N} \ell_i^* a_i$

$\zeta = V_i^D - T_i V_{M+1}^D$ where V^D is defined in Eq. 4B.19

$\Psi = T^n T^{n*} \omega$

and V^A is defined in Eq. 4B.16

APPENDIX C
Jacobian Matrix in the Newton Power Flow

The Jacobian matrix in Eq. 4.20 and the transpose Jacobian shown in Eq. 4.73 both start with the equation for the real and reactive power at each bus. This equation, shown in the text as Eq. 4.18, is repeated here.

$$P_i + jQ_i = E_i \sum_{k=1}^{N} Y_{ik}^* E_k^* \qquad (4C.1)$$

This can be expanded as

$$P_i + jQ_i = \sum_{k=1}^{N} |Y_{ik}||E_i||E_k|\varepsilon^{j(\theta_i - \theta_k - \theta_{ik})} \qquad (4C.2)$$

where θ_i, θ_k = phase angle at buses i and k, respectively

$\theta_{ik} = \arctan \dfrac{B_{ik}}{G_{ik}}$

$|E_i|, |E_k|$ = magnitude of E_i and E_k, respectively
$|Y_{ik}| = \sqrt{G_{ik}^2 + B_{ik}^2}$

General practice in solving load flows by the Newton method has been to use $\Delta|E_i|/|E_i|$ instead of simply $\Delta|E_i|$. This simplifies writing the equations. The derivatives are

For $i \neq k$:

$$\dfrac{\partial P_i}{\partial \theta_k} = |Y_{ik}||E_i||E_k|\sin(\theta_i - \theta_k - \theta_{ik})$$

$$= (a_k f_i - b_k e_i)$$

$$\dfrac{\dfrac{\partial P_i}{\partial |E_k|}}{|E_k|} = |Y_{ik}||E_i||E_k|\cos(\theta_i - \theta_k - \theta_{ik}) \qquad (4C.3)$$

$$= (a_k e_i + b_k f_i)$$

where $a_k + jb_k = (e_k + jf_k)(G_{ik} + jB_{ik})$
$$e_k = R_e(E_k)$$
$$f_k = I_m(E_k)$$

and

$$\frac{\partial Q_i}{\partial \theta_k} = -\frac{\partial P_i}{\dfrac{\partial |E_k|}{|E_k|}}$$

$$\frac{\partial Q_i}{\dfrac{\partial |E_k|}{|E_k|}} = \frac{\partial P_i}{\partial \theta_k}$$

For $i = k$:

$$\frac{\partial P_i}{\partial \theta_i} = -Q_i - B_{ii}E_i^2$$

$$\frac{\partial P_i}{\dfrac{\partial |E_i|}{|E_i|}} = P_i + G_{ii}E_i^2$$

$$\frac{\partial Q_i}{\partial \theta_i} = P_i - G_{ii}E_i^2$$

$$\frac{\partial Q_i}{\dfrac{\partial |E_i|}{|E_i|}} = Q_i - B_{ii}E_i^2$$

Equation 4.20 now becomes

$$\begin{bmatrix} \Delta P_1 \\ \Delta Q_1 \\ \Delta P_2 \\ \Delta Q_2 \\ \vdots \end{bmatrix} = [J] \begin{bmatrix} \Delta\theta_1 \\ \dfrac{\Delta|E_1|}{|E_1|} \\ \Delta\theta_2 \\ \dfrac{\Delta|E_2|}{|E_2|} \\ \vdots \end{bmatrix}$$

Similarly for Eq. 4.73.

PROBLEMS

4.1 The circuit elements in the 138 kV circuit in Figure 4.24 are in per unit on a 100 MVA base with the nominal 138 kV voltage as base. The $P + jQ$ load is scheduled to be 170 MW and 50 MVAR.

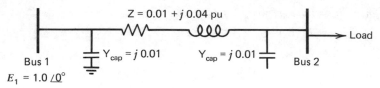

Bus 1
$E_1 = 1.0 \angle 0°$

FIG. 4.24 Two-bus ac system for Problem 4.1.

a. Write the Y matrix for this two-bus system.

b. Assume bus 1 as the reference bus and set up the Gauss-Seidel correction equation for bus 2. (Use $1.0 \angle 0°$ as the initial voltage on bus 2.) Carry out two or three iterations and show that you are converging.

c. Apply the "DC" load flow conventions to this circuit and solve for the phase angle at bus 2 for the same load real power of 1.7 per unit.

4.2 Given the network in Figure 4.25 (base = 100 MVA):

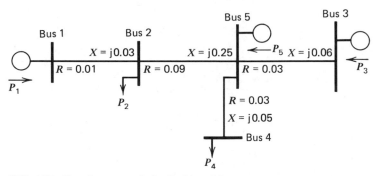

FIG. 4.25 Four-bus network for Problem 4.2.

a. Develop the $[B_x]$ matrix for this system.

$$\begin{bmatrix} P_1 \\ P_2 \\ P_3 \\ P_4 \\ P_5 \end{bmatrix} = [B_x] \begin{bmatrix} \theta_1 \\ \theta_2 \\ \theta_3 \\ \theta_4 \\ \theta_5 \end{bmatrix} \quad \begin{array}{l} P \text{ in per unit MW} \\ \theta \text{ in radians} \end{array}$$

b. Assume bus 5 as the reference bus. To carry out a "DC" load flow, we will set $\theta_5 = 0$ rad. Row 5 and column 5 will be zeroed.

$$\begin{bmatrix} P_1 \\ P_2 \\ P_3 \\ P_4 \\ P_5 \end{bmatrix} = \begin{bmatrix} & & & & 0 \\ & \text{Remainder} & & 0 \\ & \text{of } B_x & & 0 \\ & & & & 0 \\ 0 & 0 & 0 & 0 & 0 \end{bmatrix} \begin{bmatrix} \theta_1 \\ \theta_2 \\ \theta_3 \\ \theta_4 \\ \theta_5 \end{bmatrix}$$

Solve for the X matrix.

$$\begin{bmatrix} \theta_1 \\ \theta_2 \\ \theta_3 \\ \theta_4 \\ \theta_5 \end{bmatrix} = [X] \begin{bmatrix} P_1 \\ P_2 \\ P_3 \\ P_4 \\ P_5 \end{bmatrix}$$

c. Calculate the phase angles for the set of power injections.

$$P_1 = 100 \text{ MW generation}$$

$$P_2 = 120 \text{ MW load}$$

$$P_3 = 150 \text{ MW generation}$$

$$P_4 = 200 \text{ MW load}$$

d. Calculate P_5 according to the "DC" load flow.

e. Calculate all power flows on the system using the phase angles found in part c.

f. (Optional) Calculate the reference-bus penalty factors for buses 1, 2, 3, and 4. Assume all bus voltage magnitudes are 1.0 per unit.

4.3 Given the following loss formula (use P's in MW):

$$B_{ij} = \begin{matrix} & 1 & 2 & 3 \\ 1 & \\ 2 & \\ 3 & \end{matrix} \begin{bmatrix} 1.36255 \cdot 10^{-4} & 1.753 \cdot 10^{-5} & 1.8394 \cdot 10^{-4} \\ 1.753 \cdot 10^{-5} & 1.5448 \cdot 10^{-4} & 2.82765 \cdot 10^{-4} \\ 1.8394 \cdot 10^{-4} & 2.82765 \cdot 10^{-4} & 1.6147 \cdot 10^{-3} \end{bmatrix}$$

B_{i0} and B_{00} are neglected.
Assume three units are on line and have the following characteristics:

Unit 1: $H_1 = 312.5 + 8.25 P_1 + 0.005 P_1^2$, MBtu/h

$$50 \leq P_1 \leq 250 \text{ MW}$$

Fuel cost = 1.05 ₹/MBtu

Unit 2: $H_2 = 112.5 + 8.25 P_2 + 0.005 P_2^2$, MBtu/h

$$5 \leq P_2 \leq 150 \text{ MW}$$

Fuel cost = 1.217 ₹/MBtu

Unit 3: $H_3 = 50 + 8.25 P_3 + 0.005 P_3^2$, MBtu/h

$$15 \leq P_3 \leq 100 \text{ MW}$$

Fuel cost = 1.1831 ₹/MBtu

a. No Losses Used in Scheduling

   i. Calculate the optimum dispatch and total cost neglecting losses for $P_D = 190$ MW.*

   ii. Using this dispatch and the loss formula, calculate the system losses.

b. Losses Included in Scheduling

   i. Find the optimum dispatch for a total generation of $P_D = 109$ MW* using the coordination equations and the loss formula.

   ii. Calculate the cost rate.

   iii. Calculate the total losses using the loss formula.

   iv. Calculate the resulting load supplied.

4.4 Given the network in Figure 4.26:

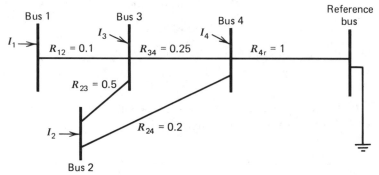

FIG. 4.26 Network for Problem 4.4.

  a. Develop the Y matrix and Z matrix equations for this network.

  b. Assume I_3 and I_4 are load-bus currents. Transform E's and I's and $[Z]$ to reflect a conforming load assumption.

$$I_3 = 0.35 I_{L_{eq}}$$
$$I_4 = 0.65 I_{L_{eq}}$$

  c. Transform to eliminate $I_{L_{eq}}$. That is, $I_{L_{eq}} = -(I_1 + I_2)$

  d. Calculate network losses for each "frame," assuming the following initial currents.

$$I_1 = 10$$
$$I_2 = 15$$
$$I_3 = -8.75$$
$$I_4 = -16.25$$

---

* $P_{demand} = P_1 + P_2 + P_3 = P_D$.

  P_L = power loss.

  $P_R = P_D - P_L$ = net load.

---

e. Calculate the incremental losses for a change in current ΔI_1 at bus 1 with $I_2 = 15$ A.

Compare the incremental loss just calculated with an approximation obtained by setting $I_1 = 11$ A (leave $I_2 = 15$ A and assuming the load currents compensate this 1-A adjustment as described in the text).

4.5 This problem requires the use of a load-flow program. Set up and solve the load-flow base case shown in Appendix A and Example 4A. You should be able to match the solution shown in Figure 4.8.

a. Resolve the load flow with each load set to 71 MW, 71 MVAR. Require that bus 1 be the swing bus so that any change in generation will appear on bus 1 (i.e., keep generator 2 at 50 MW and generator 3 at 60 MW).

b. Resolve the load flow with loads set as in (a). Require that bus 2 be the swing bus so that all generation change appears on bus 2 (i.e., keep generator 1 at 107.9 MW and generator 3 at 60 MW).

c. Set all loads back to their base-case conditions of 70 MW, 70 MVAR. Increase the generation on bus 2 to 55 MW. Require bus 1 to be the swing bus and resolve the load flow (bus 3 remains at 60 MW).

In each of these three load-flow runs, you perturbed the network in some way. For each solution, you should notice a difference in the network losses as compared to the base case.

From the data of part (a) calculate

$$\frac{\Delta P_{losses}}{\Delta P_1}$$

From the data of part (b) calculate

$$\frac{\Delta P_{losses}}{\Delta P_2}$$

All loads were perturbed equally in parts (a) and (b) to simulate a slight increase in total load. Compare the incremental losses as calculated from the load flow to the incremental losses calculated from the B matrix shown in Example 4D.

The perturbation in part (c) is to demonstrate the change in losses on the network for a shift of generation such as proposed in Section 4.2.4.4. Calculate B_i and Pf_i for bus 2 from the data of part (c).

FURTHER READING

The basic papers on solution of the load flow can be found in references 1–3. The development of the loss-matrix equations is based on the work of Kron (4), who developed the reference-frame transformation theory. Other developments of the transmission-loss formula are seen in references 5 and 6. Meyer's paper (7) is representative of recent adaptation of sparsity programming methods to calculation of the loss matrix.

The development of the reference-bus penalty factor method can be seen in references 8 and 9. Reference 10 gives an excellent derivation of the reference-bus penalty factors derived from the Newton load-flow equations. Reference 10 provides an excellent summary of recent developments in power system dispatch.

1. Ward, J. B., Hale, H. W., "Digital Computer Solution of Power-Flow Problems," *AIEE Transactions, Part III Power Apparatus and Systems*, Vol. 75, June 1956, pp. 398–404.

2. VanNess, J. E., "Iteration Methods for Digital Load Flow Studies," *AIEE Transactions on Power Apparatus and Systems*, Vol. 78A, August 1959, pp. 583–588.

3. Tinney, W. F., Hart, C. E., "Power Flow Solution by Newton's Method," *IEEE Transactions on Power Apparatus and Systems*, Vol. PAS-86, November 1967, pp. 1449–1460.

4. Kron, G., "Tensorial Analysis of Integrated Transmission Systems—Part I: The Six Basic Reference Frames," *AIEE Transactions*, Vol. 70, Part I, 1951, pp. 1239–1248.

5. Kirchmayer, L. K., Stagg, G. W., "Analysis of Total and Incremental Losses in Transmission Systems," *AIEE Transactions*, Vol. 70, Part I, 1951, pp. 1197–1205.

6. Early, E. D., Watson, R. E., "A New Method of Determining Constants for the General Transmission Loss Equation," *AIEE Transactions on Power Apparatus and Systems*, Vol. PAS-74, February 1956, pp. 1417–1423.

7. Meyer, W. S., "Efficient Computer Solution for Kron and Kron-Early Loss Formulas," *Proc 1973 PICA Conference*, IEEE 73 CHO 740-1, PWR, pp. 428–432.

8. Shipley, R. B., Hochdorf, M., "Exact Economic Dispatch—Digital Computer Solution," *AIEE Transactions on Power Apparatus and Systems*, Vol. PAS-75, November 1956, pp. 1147–1152.

9. Dommel, H. W., Tinney, W. F., "Optimal Power Flow Solutions," *IEEE Transactions on Power Apparatus and Systems*, Vol. PAS 87, October 1968, pp. 1866–1876.

10. Happ, H. H., "Optimal Power Dispatch," *IEEE Transactions on Power Apparatus and Systems*, Vol. PAS-93, May/June 1974, pp. 820–830.

chapter 5

Unit
Commitment

5.1 INTRODUCTION

Because human activity follows cycles, most systems
supplying services to a large population will experience cycles. This includes transportation systems,
communication systems, as well as electric power
systems. In the case of an electric power system,
the total load on the system will generally be higher
during the daytime and early evening when industrial loads are high, lights are on, and so forth
and lower during the late evening and early morning
when most of the population is alseep. In addition,
the use of electric power has weekly cycles, the load
being lower over weekend days than weekdays. But
why is this a problem in the operation of an electric
power system? Why not just simply commit enough
units to cover the maximum system load and leave
them running? Note that to "commit" a generating
unit is to "turn it on," that is, to bring the unit
up to speed, synchronize it to the system, and
connect it so it can deliver power to the network.
The problem with "commit enough units and leave
them on line" is one of economics. As will be shown
in Example 5A, it is quite expensive to run too many
generating units. A great deal of money can be saved
by turning units off (decommiting them) when they
are not needed.

EXAMPLE 5A

Suppose one had the three units given here.

Unit 1:   Min = 150 MW

Max = 600 MW

$H_1 = 510.0 + 7.2 P_1 + 0.00142 P_1^2$ MBtu/h

Unit 2:   Min = 100 MW

Max = 400 MW

$H_2 = 310.0 + 7.85 P_2 + 0.00194 P_2^2$ MBtu/h

Unit 3:   Min = 50 MW

Max = 200 MW

$H_3 = 78.0 + 7.97 P_3 + 0.00482 P_3^2$ MBtu/h

with fuel costs:

$$F_1 = 1.1 \text{ R/MBtu}$$
$$F_2 = 1.0 \text{ R/MBtu}$$
$$F_3 = 1.2 \text{ R/MBtu}$$

If we are to supply a load of 550 MW, what unit or combination of units should be used to supply this load most economically? To solve this problem, simply try all combinations of the three units. Some combinations will be infeasible if the sum of all maximum MW for the units committed is less than the load or if the sum of all minimum MW for the units committed is greater than the load. For each feasible combination, the units will be dispatched using the techniques of Chapter 3. The results are presented in Table 5.1.

TABLE 5.1   Unit Combinations and Dispatch for 550 MW Load Example 5A

Unit 1	Unit 2	Unit 3	Max Generation	Min Generation	P_1	P_2	P_3	F_1	F_2	F_3	Total Generation Cost $F_1 + F_2 + F_3$
OFF	OFF	OFF	0	0			Infeasible				
OFF	OFF	ON	200	50			Infeasible				
OFF	ON	OFF	400	100			Infeasible				
OFF	ON	ON	600	150	0	400	150	0	3760	1658	5418
ON	OFF	OFF	600	150	550	0	0	5389	0	0	5389
ON	OFF	ON	800	200	500	0	50	4911	0	586	5497
ON	ON	OFF	1000	250	295	255	0	3030	2440	0	5471
ON	ON	ON	1200	300	267	233	50	2787	2244	586	5617

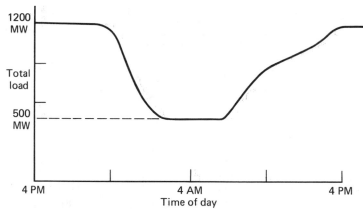

FIG. 5.1a Simple "peak-valley" load pattern.

Note that the least expensive way to supply the generation is not with all three units running, or even any combination involving two units. Rather, the optimum commitment is to only run unit 1, the most economic unit. By only running the most economic unit, the load can be supplied by that unit operating closer to its best efficiency. If another unit is committed, both unit 1 and the other unit will be loaded further from the best efficiency point such that the net cost is greater than unit 1 alone.

Suppose the load follows a simple "peak-valley" pattern as shown in Figure 5.1a. If the operation of the system is to be optimized, units must be shut down as the load goes down and then recommitted as it goes back up. We would like to know which units to drop and when. As we will show later, this problem is far from trivial when real generating units are considered. One approach to this solution is demonstrated in Example 5B, where a simple priority list scheme is developed.

EXAMPLE 5B

Suppose we wish to know which units to drop as a function of system load. Let the units and fuel costs be the same as in Example 5A with the load varying from a peak of 1200 MW to a valley of 500 MW. To obtain a "shut-down rule," simply use a brute-force technique wherein all combinations of units will be tried (as in Example 5A) for each load value taken in steps of 50 MW from 1200 to 500. The results of applying this brute-force technique are given in Table 5.2. Our shut-down rule is quite simple.

When load is above 1000 MW, run all three units; between 1000 MW and 600 MW, run units 1 and 2; below 600 MW, run only unit 1.

TABLE 5.2 "Shut-down Rule" Derivation for Example 5B

Load	Unit 1	Unit 2	Unit 3
	\multicolumn{3}{c}{Optimum combination}		
1200	ON	ON	ON
1150	ON	ON	ON
1100	ON	ON	ON
1050	ON	ON	ON
1000	ON	ON	OFF
950	ON	ON	OFF
900	ON	ON	OFF
850	ON	ON	OFF
800	ON	ON	OFF
750	ON	ON	OFF
700	ON	ON	OFF
650	ON	ON	OFF
600	ON	OFF	OFF
550	ON	OFF	OFF
500	ON	OFF	OFF

Figure 5.1b shows the unit commitment schedule derived from this shut-down rule as applied to the load curve of Figure 5.1a.

So far, we have only obeyed one simple constraint: *Enough units will be committed to supply the load.* If this were all that was involved in the unit commitment problem, that is, just meeting the load, we could stop here and state that the problem was "solved." Unfortunately, other constraints and other phenomena must be taken into account in order to claim an optimum solution. These constraints will be discussed in the next section followed by a description of some of the presently used methods of solution.

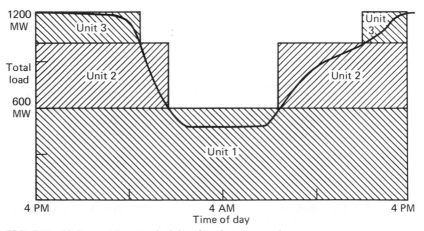

FIG. 5.1b Unit commitment schedule using shut-down rule.

5.1.1 Constraints in Unit Commitment: Introduction

Many constraints can be placed on the unit commitment problem. The list presented here is by no means exhaustive. Each individual power system, power pool, reliability council, and so forth may impose different rules on the scheduling of units, depending on the generation makeup, load-curve characteristics, and such.

5.1.2 Spinning Reserve

Spinning reserve is the term used to describe the total amount of generation available from all units synchronized (i.e., spinning) on the system minus the present load plus losses being supplied. Spinning reserve must be carried so that the loss of one or more units does not cause too far a drop in system frequency (see Chapter 9). Quite simply, if one unit is lost, there must be ample reserve on the other units to make up for the loss in a specified time period.

Spinning reserve must be allocated to obey certain rules, usually set by regional reliability councils (in the United States) that specify how the reserve is to be allocated to various units. Typical rules specify that reserve must be a given percentage of forecasted peak demand, or that reserve must be capable of making up the loss of the most heavily loaded unit in a given period of time, or such. Others calculate reserve requirements as a function of the probability of not having sufficient generation to meet the load.

Not only must the reserve be sufficient to make up for a generation unit failure, but the reserves must be allocated among fast-responding units and slow-responding units. This allows the automatic generation control system (see Chapter 9) to restore frequency and interchange quickly in the event of a generating-unit outage.

Beyond spinning reserve, the unit commitment problem may involve various classes of "scheduled reserves" or "off-line" reserves. These include quick-start diesel or gas-turbine units as well as most hydro and pumped-storage hydro units that can be brought on line, synchronized, and brought up to full capacity quickly. As such, these units can be "counted" in the overall reserve assessment as long as their time to come up to full capacity is taken into account.

Reserves, finally, must be spread around the power system to avoid transmission system limitations (often called "bottling" of reserves) and to allow various parts of the system to run as "islands," should they become electrically disconnected.

EXAMPLE 5C

Suppose a power system consisted of two isolated regions: a western region and an eastern region. Five units have been committed to supply 3090 MW as shown in Figure 5.2. The two regions are separated by transmission tie lines that can together transfer a maximum of 550 MW in either direction. This is shown in Figure 5.2. What can we say about the allocation of spinning reserve in this system?

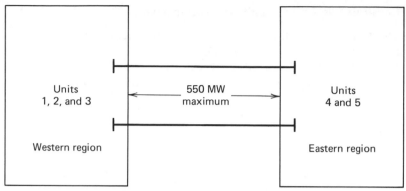

FIG. 5.2 Two-region system.

The data for the system in Figure 5.2 are given in Table 5.3. With the exception of unit 4, the loss of any unit on this system can be covered by the spinning reserve on the remaining units. Unit 4 presents a problem, however. If unit 4 were to be lost and unit 5 were to be run to its maximum of 600 MW, the eastern region would still need 590 MW to cover the load in the eastern region. The 590 MW would have to be transmitted over the tie lines from the western region, which can easily supply 590 MW from its reserves. However, the tie capacity of only 550 MW limits the transfer. Therefore, the loss of unit 4 cannot be covered even though the entire system has ample reserves. The only solution to this problem is to commit more units to operate in the eastern region.

5.1.3 Thermal Unit Constraints

Thermal units require a crew to operate them, especially when turned on and turned off. A thermal unit can undergo only gradual temperature changes, and this translates into a time period of some hours required to bring the unit on line. As a result of such restrictions in the operation of a thermal plant, various constraints

TABLE 5.3 Data for the System in Figure 5.2

Region	Unit	Unit capacity (MW)	Unit output (MW)	Regional generation (MW)	Spinning reserve	Regional load (MW)	Interchange (MW)
Western	1	1000	900 ⎫		100		
	2	800	420 ⎬	1740	380	1900	160 out
	3	800	420 ⎭		380		
Eastern	4	1200	1040 ⎫	1350	160 ⎫	1190	160 in
	5	600	310 ⎭		290 ⎭		
Total		4400	3090	3090	1310	3090	

the SR of the largest unit shouldn't be considered in total. SR.

arise such as

- **Minimum Up Time:** Once the unit is running, it should not be turned off immediately.

- **Minimum Down Time:** Once the unit is decommitted, there is a minimum time before it can be recommitted.

- **Crew Constraints:** If a plant consists of two or more units, they cannot both be turned on at the same time.

In addition, because the temperature and pressure of the thermal unit must be moved slowly, a certain amount of energy must be expended to bring the unit on line. This energy does not result in any MW generation from the unit and is brought into the unit commitment problem as a *start-up cost*.

The start-up cost can vary from a maximum "cold-start" value to a much smaller value if the unit was only turned off recently and is still relatively close to normal temperature. There are two approaches to treating a thermal unit during its down period. The first allows the unit's boiler to cool down and then heat back up to operating temperature in time for a scheduled turn on. The second (called *banking*) requires that sufficient energy be input to the boiler to just maintain operating temperature. The costs for the two can be compared so that, if possible, the best approach (cooling or banking) can be chosen.

$$\text{Start-up cost when cooling} = C_c(1 - \varepsilon^{-t/\alpha}) \cdot F + C_f$$

where C_c = cold-start cost (MBtu)

$\quad$ F = fuel cost

$\quad$ C_f = fixed cost (includes crew expense, maintenance expenses) (in R)

$\quad$ α = thermal time constant for the unit

$\quad$ t = time in hours the unit was cooled

$$\text{Start-up cost when banking} = C_t \cdot t \cdot F + C_f$$

where C_t = cost (MBtu/h) of maintaining unit at operating temperature

Up to a certain number of hours the cost of banking will be less than the cost of cooling, as is illustrated in Figure 5.3.

Finally, the capacity limits of thermal units may change frequently due to maintenance or unscheduled outages of various equipment in the plant; this must also be taken into account in unit commitment.

5.1.4 Other Constraints

5.1.4.1 Hydro Constraints

Unit commitment cannot be completely separated from the scheduling of hydro units. In this text we will assume that the hydrothermal scheduling (or "coordination") problem can be separated from the unit commitment problem. We,

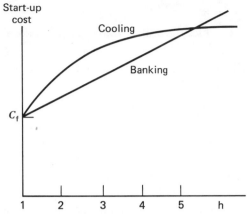

FIG. 5.3   Time-dependent Start-up costs.

of course, cannot assert flatly that our treatment in this fashion will always result in an optimal solution.

5.1.4.2 Must Run

Some units are given a must-run status during certain times of the year for reason of voltage support on the transmission network or for such purposes as supply of steam for uses outside the steam plant itself.

5.1.4.3 Fuel Constraints

We will treat the "fuel scheduling" problem briefly in Chapter 6. A system in which some units have limited fuel, or else have constraints that require them to burn a specified amount of fuel in a given time, presents a most challenging unit commitment problem for which no completely general solution (that works on a practical utility system) exists.

5.2 UNIT COMMITMENT SOLUTION METHODS

The commitment problem can be very difficult. As a theoretical exercise, let us postulate the following situation.

- We must establish a loading pattern for M periods.
- We have N units to commit and dispatch.
- The M load levels and operating limits on the N units are such that any one unit can supply the individual loads and that any combination of units can also supply the loads.

Next, assume we are going to establish the commitment by enumeration (brute force). The total number of combinations we need to try each hour is

$$C(N, 1) + C(N, 2) + \cdots + C(N, N - 1) + C(N, N) = 2^N - 1$$

where $C(N, j)$ is the combination of N items taken j at a time. That is,

$$C(N, j) = \left(\frac{N!}{(N - j)! \, j!} \right)$$

$$j! = 1 \times 2 \times 3 \times \cdots \times j$$

For the total period of M intervals, the maximum number of possible combinations is $(2^N - 1)^M$, which can become a horrid number to think about.

For example, take a 24-h period (e.g., 24 one-hour intervals) and consider systems with 5, 10, 20, and 40 units. The value of $(2^N - 1)^{24}$ becomes the following.

N	$(2^N - 1)^{24}$
5	6.2×10^{35}
10	1.73×10^{72}
20	3.12×10^{144}
40	(Too big)

These very large numbers are the upper bounds for the number of enumerations required. Fortunately, the constraints on the units and the load-capacity relationships of typical utility systems are such that we do not approach these large numbers. Nevertheless, the real practical barrier in the optimized unit commitment problem is the high dimensionality of the possible solution space.

The most talked about techniques for the solution of the unit commitment problem are

- Priority-list schemes.
- Dynamic programming (DP).
- Mixed integer-linear programming (MILP).

Of these, priority-list schemes are the most popular. DP algorithms are the only ones that approach an optimum solution for large systems. MILP algorithms are just beginning to be researched and are not widely used on large-system problems.

5.2.1 Priority-List Methods

The simplest unit commitment solution methods consists of creating a priority list of units. As we saw in Example 5B, a simple shut-down rule or priority-list scheme could be obtained after an exhaustive enumeration of all unit combinations at each load level. The priority list of Example 5B could be obtained in a much simpler manner by noting the full-load average production cost of each unit, where the full-load average production cost is simply the net heat rate at full load times the fuel cost.

EXAMPLE 5D

Construct a priority list for the units of Example 5A. (Use the same fuel costs as in Example 5A.) First, the full load average production cost will be calculated

Unit	Full load average production cost (R/MWh)
1	9.79
2	9.40
3	11.188

A strict priority order for these units based on the average production cost would order them as follows.

Unit	R/MWh	Min MW	Max MW
2	9.40	100	400
1	9.79	150	600
3	11.40	50	200

and the commitment scheme would (ignoring min up/down time, start-up costs, etc.) simply use only the following combinations.

Combination	Min MW from combination	Max MW from combination
2 + 1 + 3	300	1200
2 + 1	250	1000
2	150	400

Note that such a scheme would not completely parallel the shut-down sequence described in Example 5B where unit 2 was shut down at 600 MW leaving unit 1. With the priority-list scheme, both units would be held on until load reached 400 MW, then unit 1 would be dropped.

Most priority-list schemes are built around a simple shut-down algorithm that might operate as follows.

- At each hour when load is dropping, determine whether dropping the next unit on the priority list will leave sufficient generation to supply the load plus spinning-reserve requirements. If not, continue operating as is, if yes, go on to the next step.

- Determine the number of hours, H, before the unit will be needed again. That is, assuming that the load is dropping and will then go back up some hours later.

- If H is less than the minimum shut-down time for the unit, keep commitment as is. If not, go to next step.

- Calculate two costs. The first is the sum of the hourly production costs from the next H hours with the unit up. Then recalculate the same sum for the unit down and add in the start-up cost for either cooling the unit or banking it, whichever is less expensive. If there is sufficient savings from shutting down the unit, it should be shut down, otherwise keep it on.

- Repeat this entire procedure for the next unit on the priority list. If it is also dropped, go to the next and so forth.

Various enhancements to the priority-list scheme can be made by grouping of units to ensure that various constraints are met. We will note later that dynamic-programming methods usually create the same type of priority list for use in the DP search.

5.2.2 Dynamic-Programming Solution

5.2.2.1 Introduction

Dynamic programming has many advantages over the enumeration scheme, the chief advantage being a reduction in the dimensionality of the problem. Suppose we have four units on a system and any combination of them could serve the (single) load. There would be a maximum of $2^4 - 1 = 15$ combinations to test. However, if a strict priority order is imposed, there are only four combinations to try:

Priority 1 Unit
Priority 1 Unit + Priority 2 Unit
Priority 1 Unit + Priority 2 Unit + Priority 3 Unit
Priority 1 Unit + Priority 2 Unit + Priority 3 Unit + Priority 4 Unit

The imposition of a priority list arranged in order of the full-load average-cost rate would result in a theoretically correct dispatch and commitment only if

1. No load costs are zero.
2. Unit input-output characteristics are linear between zero output and full load.
3. There are no other restrictions.
4. Start-up costs are a fixed amount.

In the dynamic programming approach that follows, we assume that

1. A *state* consists of an array of units with specified units operating and the rest off-line.
2. The start-up cost of a unit is independent of the time it has been off-line (i.e., it is a fixed amount).
3. There are no costs for shutting down a unit.
4. There is a strict priority order, and in each interval a specified minimum amount of capacity must be operating.

A feasible state is one in which the committed units can supply the required load and that meets the minimum amount of capacity each period.

5.2.2.2 Backward DP Approach

The first dynamic-programming approach uses a backward (in time) approach in which the solution starts at the last interval and proceeds back to the initial point. There are M intervals in the period to be considered. The dynamic-programming equations for the computation of the minimum total fuel cost during a time period, K, are given here.

$$F_{cost}(K, I) = \text{Min}_{\{J\}} [P_{cost}(K, I) + S_{cost}(I, K:J, K + 1) + F_{cost}(K + 1, J)]$$

$$F_{cost}(M, I) = P_{cost}(M, I)$$

(5.1)

where

$F_{cost}(K, I)$ = minimum total fuel cost from state I in interval K to the last interval M

and

$P_{cost}(K, I)$ = minimum generation cost in supplying the load during interval K given state I

$S_{cost}(I, K:J, K + 1)$ = incremental start-up cost going from state I in the Kth interval to state J in the $(K + 1)^{th}$ interval

$\{J\}$ = set of feasible states in interval $K + 1$

The production cost $P_{cost}(K, I)$ is obtained by economically dispatching the units on-line in state I.

A *path* is a schedule starting from a state in an interval K to the final interval M. An *optimal path* is one for which the total fuel cost is minimum. Equation 5.1 asserts that, given the optimal paths starting from all the individual states in the $(K + 1)^{st}$ interval, the optimal path originating from any state in the Kth interval can be found. This is the main advantage of the dynamic-programming method. The procedure for determining the optimal schedule and the minimum total fuel cost is shown by the flowchart in Figure 5.4.

5.2.2.3 Forward DP Approach

The backward dynamic-programming approach does not cover many practical situations. For example, if the start-up cost of a unit is a function of the time it has been off-line (i.e., it's temperature), then a forward dynamic program approach is more suitable since the previous history of the unit can be computed at each stage. There are other practical reasons for going forward. The initial conditions are easily specified and the computations can go forward in time as long as required and as long as computer storage is available. A forward dynamic-programming algorithm similar to the previously given backward approach is shown by the flowchart in Figure 5.5.

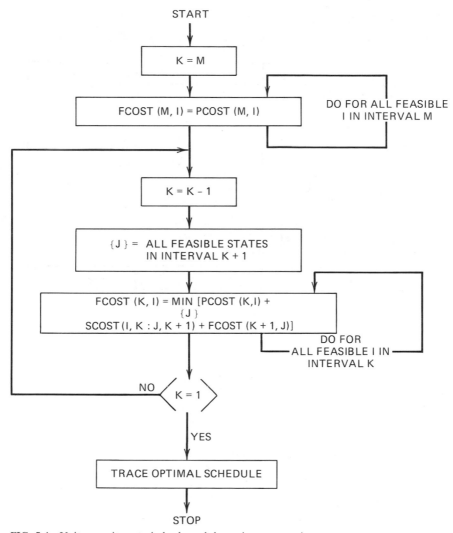

FIG. 5.4 Unit commitment via backward dynamic programming.

The recursive algorithm to compute the minimum cost in hour K with combination I is

$$F_{\text{cost}}(K, I) = \min_{\{L\}} \left[P_{\text{cost}}(K, I) + S_{\text{cost}}(K - 1, L: K, I) + F_{\text{cost}}(K - 1, L) \right]$$

where
$F_{\text{cost}}(K, I) = $ least total cost to arrive at state (K, I)

$P_{\text{cost}}(K, I) = $ production cost for state (K, I)

$S_{\text{cost}}(K - 1, L: K, I) = $ transition cost from state $(K - 1, \ L)$ to state (K, I)

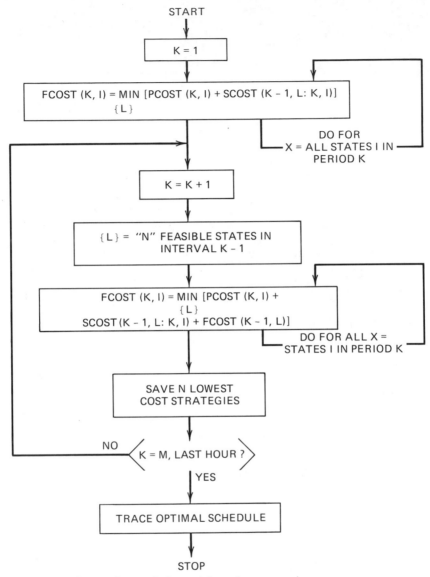

FIG. 5.5 Unit commitment via forward dynamic programming.

where state (K, I) is the Ith combination in hour K. For the forward dynamic-programming approach, we define a *strategy* as the transition, or path, from one state at a given hour to a state at the next hour.

Note that two new variables, X and N, have been introduced in Figure 5.5.

$$X = \text{Number of states to search each period}$$

$$N = \text{Number of strategies, or paths, to save at each step}$$

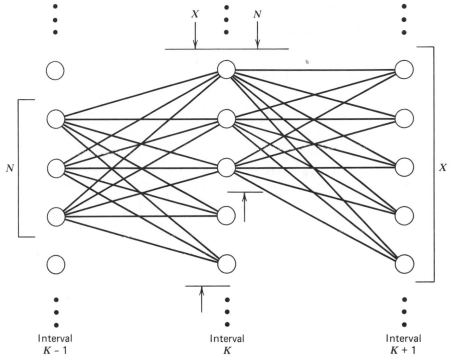

FIG. 5.6   Restricted search paths in DP algorithm with $N = 3$ and $X = 5$.

These variables allow control of the computational effort (see Figure 5.6). For complete enumeration, the maximum number of the value of X or N is $2^n - 1$.

For example, with a strict priority-list ordering, the upper bound on X is n, the number of units. Reducing the number N means that we are discarding the highest cost schedules at each time interval and saving only the lowest N paths or strategies. There is no assurance that the the theoretical optimal schedule will be found using a reduced number of strategies and search range (the X value): only experimentation with a particular program will indicate the potential error associated with limiting the values of X and N below their upper bounds.

EXAMPLE 5E

For this example, the complete search range will be used and three cases will be studied. The first is a priority-list schedule, the second is the same example with complete enumeration. Both of the first two cases ignore hot-start costs and minimum up and down times. The third case includes the hot-start costs as well as the minimum up and down times. Four units are to be committed to serve an 8-h load pattern. Data on the units and the load pattern are contained in Table 5.4.

In order to make the required computations more efficiently, a simplified model of the unit characteristics is used. In practical applications, two or three section

TABLE 5.4 Unit Characteristics, Load Pattern, and Initial Status for the Cases in Example 5E

Unit	Max (MW)	Min (MW)	Incremental heat rate (Btu/kWh)	No-load[a] cost (R/h)	Full-load ave. cost (R/mWh)	Minimum Times (h) Up	Down
1	80	25	10,440	213.00	23.54	4	2
2	250	60	9,000	585.62	20.34	5	3
3	300	75	8,730	684.74	19.74	5	4
4	60	20	11,900	252.00	28.00	1	1

	Initial conditions	Startup costs		
Unit	Hours off-line (−) or on-line (+)	Hot (R)	Cold (R)	Cold start (h)
1	−5	150	350	4
2	8	170	400	5
3	8	500	1,100	5
4	−6	0	.02	0

Load pattern	
Hour	Load (MW)
1	450
2	530
3	600
4	540
5	400
6	280
7	290
8	500

[a] This is the cost when the f(P) function (in this case a straight line) is extended to P = 0 MW. Note that we do not allow the unit to operate at zero output. That is, if the unit is on-line, it must be loaded between its min and max. If it is off-line, it must have zero output and its operating cost will be zero R/h. Fuel costs are 2.00 R/MBtu.

stepped incremental curves might be used, as shown in Figure 5.7. For our example, only a single step between minimum and the maximum power points is used. Startup costs for the first two cases are taken as the cold-start costs. The priority order for the four units in the example is: Unit 3, Unit 2, Unit 1, Unit 4. For the first two cases the minimum up and down times are taken as 1 h for all units.

In all three cases we will refer to the capacity ordering of the units. This is shown in Table 5.5 where the unit combinations or states are ordered by maximum net capacity for each combination.

Case 1

In Case 1 the units are scheduled according to a strict priority order. That is, units are committed in order until the load is satisfied. The total cost for the interval is the sum of the eight dispatch costs plus the transitional costs for starting any units. In this first case, a maximum of 24 dispatches must be considered.

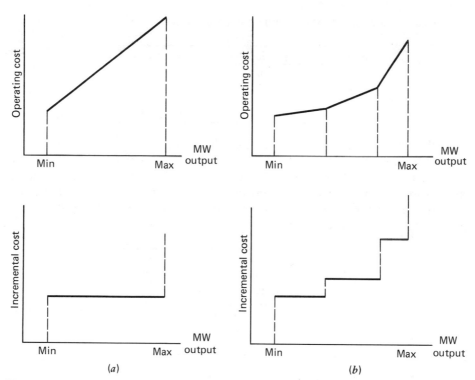

FIG. 5.7 (a) Single-step incremental cost curve and (b) multiple-step incremental cost curve.

TABLE 5.5 Capacity Ordering of the Units

State	Unit combination[a]	Maximum net capacity for combination
15	1 1 1 1	690
14	1 1 1 0	630
13	0 1 1 1	610
12	0 1 1 0	550
11	1 0 1 1	440
10	1 1 0 1	390
9	1 0 1 0	380
8	0 0 1 1	360
7	1 1 0 0	330
6	0 1 0 1	310
5	0 0 1 0	300
4	0 1 0 0	250
3	1 0 0 1	140
2	1 0 0 0	80
1	0 0 0 1	60
0	0 0 0 0	0
	Unit 1 2 3 4	

[a] 1 = Committed (unit operating).
0 = Uncommitted (unit shut down).

For case 1 the only states examined each hour consist of

State no.	Unit status	Capacity (MW)
5	0 0 1 0	300
12	0 1 1 0	550
14	1 1 1 0	630
15	1 1 1 1	690

Note that this is the priority order, that is, state 5 = unit 3, state 12 = units $3 + 2$, state 14 = unit $3 + 2 + 1$, and state 15 = units $3 + 2 + 1 + 4$. For the first 4 h only the last three states are of interest. The sample calculations illustrate the technique. All possible commitments start from state 12 since this was given as the initial condition. For hour 1 the minimum cost is state 12 and so on. The results for the priority-ordered case are as follows.

Hour	State with min total cost	Pointer for previous hour
1	12 (9208)	12
2	12 (19857)	12
3	14 (32472)	12
4	12 (43300)	14
⋮	⋮	⋮

Note that State 13 is not reachable in this strict priority ordering.

Sample Calculations for Case 1

$$F_{cost}(J, K) = \min_{\{L\}} [P_{cost}(J, K) + S_{cost}(J - 1, L; J, K) + F_{cost}(J - 1, L)]$$

Allowable states are

$$\{ \ \} = \{0010, 0110, 1110, 1111\} = \{5, 12, 14. 15\}$$

In hour 0 $\{L\} = \{12\}$, initial condition

$J = 1$:   1st hour

$$\frac{K}{15} \qquad F_{cost}(1, 15) = P_{cost}(1, 15) + S_{cost}(0, 12; 1,15)$$
$$= 9861 + 350 = 10211$$

$\quad$ 14 $\qquad F_{cost}(1, 14) = 9493 + 350 = 9843$

$\quad$ 12 $\qquad F_{cost}(1, 12) = 9208 + 0 = 9208$

$J = 2$:   2nd hour
Feasible states are $\{12, 14, 15\} = \{K\}$
So $X = 3$
Suppose two strategies are saved at each stage,

So $N = 2$ and

$$\{L\} = \{12, 14\}$$

$$\frac{K}{15} \quad F_{cost}(2, 15) = \min_{\{12, 14\}} \left[P_{cost}(2, 15) + S_{cost}(1, L; 2, 15) + F_{cost}(1, L) \right]$$

$$= 11301 + \min \begin{bmatrix} (350 + 9208) \\ (0 + 9843) \end{bmatrix} = 20860$$

and so on.

Case 2

In case 2 complete enumeration is tried with a limit of $(2^4 - 1) = 15$ dispatches each of the 8 hours, so that there is a theoretical maximum of $15^8 = 2.56 \times 10^9$ possibilities. Fortunately, most of these are not feasible because they do not supply sufficient capacity and can be discarded with little analysis required.

Figure 5.8 illustrates the computational process for the first 4 h for case 2. On the figure itself the circles denote states each hour. The numbers within the circles are the "pointers." That is, they denote the state number in the previous hour that provides the path to that particular state in the current hour. For example, in hour 2, the minimum costs for states 12, 13, 14, and 15 all result from transitions from state 12 in hour 1. Costs shown on the connections are the start-up costs. At each state the figures shown are the hourly cost/total cost.

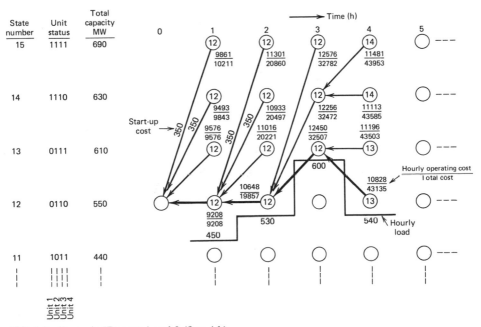

FIG. 5.8 Example 5D cases 1 and 2 (first 4 h).

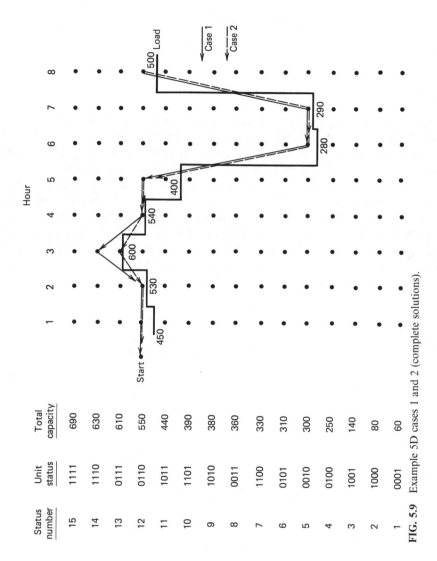

Status number	Unit status	Total capacity
15	1111	690
14	1110	630
13	0111	610
12	0110	550
11	1011	440
10	1101	390
9	1010	380
8	0011	360
7	1100	330
6	0101	310
5	0010	300
4	0100	250
3	1001	140
2	1000	80
1	0001	60

FIG. 5.9 Example 5D cases 1 and 2 (complete solutions).

In case 2 the true optimal commitment is found. That is, it is less expensive to turn on the less efficient peaking unit, number 4, for hour 3 than to start up the more efficient unit 1 for that period. By hour 3 the difference in total cost is ₹165, or ₹0.104/MWh. This is not an insignificant amount when compared with the fuel cost per MWh for an average thermal unit with a net heat rate of 10,000 Btu/kWh and a fuel cost of ₹2.00 MBtu. A savings of ₹165 every 3 h is equivalent to ₹481,800/yr.

The total 8-h trajectories for cases 1 and 2 are shown in Figure 5.9. The neglecting of start-up and shutdown restrictions in these two cases permits the shutting down of all but unit 3 in hours 6 and 7. The only difference in the two trajectories occurs in hour 3 as discussed in the previous paragraph.

Case 3

In case 3 the original unit data are used so that the minimum shutdown and operating times are observed. The forward dynamic-programming algorithm was repeated for the same 8-h period. Complete enumeration was used. That is, the upper bound on X shown in the flowchart was 15. Three different values for N, the number of strategies saved at each stage, were taken as 4, 8, and 10. The same trajectory was found for values of 8 and 10. This trajectory is shown in Figure 5.10. However, when only four strategies were saved, the procedure flounders (i.e., fails to find a feasible path) in hour 8 because the lowest cost strategies in hour 7 have shutdown units that cannot be restarted in hour 8 because of minimum unit downtime rules.

The practical remedy for this deficiency in the method shown in Figure 5.5 is to return to a period prior to the low-load hours and temporarily keep more (i.e., higher cost) strategies. This will permit keeping a nominal number of strategies at each stage. The other alternative is, of course, the method used here, run the entire period with more strategies saved.

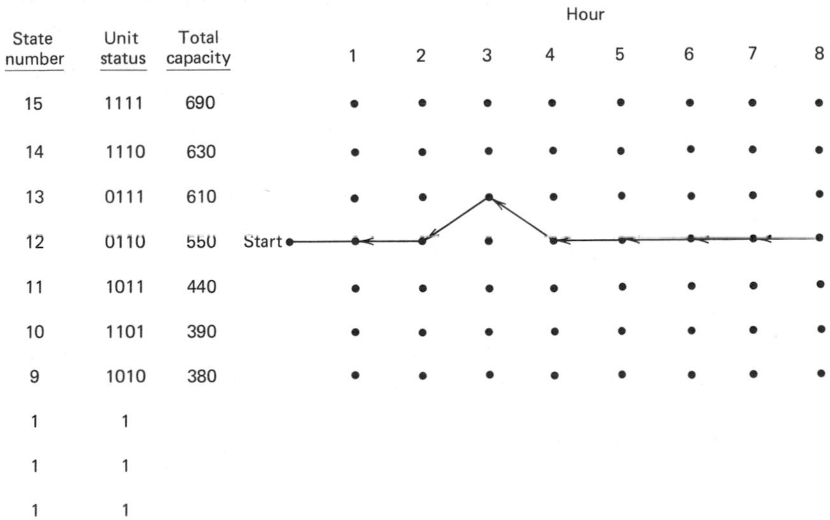

State number	Unit status	Total capacity	Hour 1	2	3	4	5	6	7	8
15	1111	690	•	•	•	•	•	•	•	•
14	1110	630	•	•	•	•	•	•	•	•
13	0111	610	•	•	•	•	•	•	•	•
12	0110	550	Start •	•	•	•	•	•	•	•
11	1011	440	•	•	•	•	•	•	•	•
10	1101	390	•	•	•	•	•	•	•	•
9	1010	380	•	•	•	•	•	•	•	•
1	1									
1	1									
1	1									

FIG. 5.10 Example 5D case 3.

TABLE 5.6 Summary of Cases 1–3

Case	Conditions	Total cost (R)
1	Priority order. Up and down times neglected	73,439
2	Enumeration ($X \leq 15$) with 4 strategies (N) saved. Up and down times neglected	73,274
3	$X \leq 15$. Up and down times observed	
	$N = 4$ strategies	No solution
	$N = 8$ strategies	74,110
	$N = 10$ strategies	74,110

These cases can be summarized in terms of the total costs found for the 8-h period, as shown in Table 5.6. These cases illustrate the forward dynamic-programming method and also point out the problems involved in the practical application of the method.

5.2.3.4 Restricted Search Ranges, Strategies, and Time-Dependent Problems

We can make use of the simple four-unit 8-h problem in Example 5E to illustrate the use of the hueristic methods to reduce the number of computations. We saw from the first two cases that the strict priority-list computation is apt to miss the optimal commitment. This is even more true when time-dependent effects are included. The most frequent of these are the inclusion of start-up costs, which are a function of the time a unit has been down, the inclusion of minimum downtimes, and the restriction on the maximum number of unit starts through the imposition of maximum start-ups per interval per plant and/or the imposition of minimum running-time requirements.

The heuristic approach, which restricts the number of strategies saved and the number of feasible states at each stage, is intermediate between the strict-priority ordering and a complete enumeration. The two facets of this approach are restrictions on the search range, X, and the number of strategies, N, preserved at each step. Let us consider the effect of restricting the number of strategies by considering one of the examples used in the appendix to this chapter.

Suppose on the first example in the appendix, the transportation problem, we had kept only a limited number of strategies at each stage. Now suppose we had attempted to solve this dynamic-programming problem in a backward fashion starting at the terminal node N and saving only the least costly path at each stage as the search proceeded toward the starting node A. Figure 5.11 illustrates the path we would have found. With this limited dynamic-programming approach, the minimum cost path found would have been ADGKMN, with a path cost of 21 units.

Next consider that two strategies are saved at each stage. Figure 5.12 shows the two paths. The minimum cost path found is the true optimal path, ACEILN, with a cost of 19.

The reduction in computational effort for this problem is, of course, trivial. (Would it be if there were 40 nodes at each stage?) However, we managed to find the

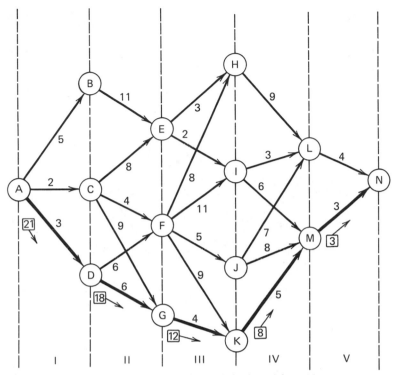

FIG. 5.11 Saving least-cost path each interval (backward from N).

optimal path with only two strategies at each stage. If this problem were being programmed on a computer, in order to ensure finding the optimal path, the maximum number of paths saved would have to be equal to the maximum number of nodes in any stage. For this problem, that would be four paths. (As an exercise, try this particular problem in a forward direction to establish the number of strategies needed in order to find the optimal path.)

As an observation it might be pointed out that the cost savings for finding the true optimal path is 10% of the total. This is usually not true for most scheduling problems. Unfortunately, the difference (savings in costs, in profit, and so forth) between the true optimum and a satisfactory near-optimal schedule is more apt to be less than 0.5 to 0.1% of the total. Since this is a heuristic method (i.e., one we have every reason to believe will work), it is necessary to try it out numerically in order to evaluate the benefit in added computational effort for saving more strategies.

The other heuristic technique that has been tried successfully in the unit commitment problem is that of using a limited "search range" in place of a strict priority order or complete enumeration. This is accomplished by making a unit selection list based on the priority-list concept modified by engineering judgment. There are always some units that are must-run, base-load units. Either these are so much more economical than the others that they always run or else they are constrained from shutting down. Some other units may be of the other extreme

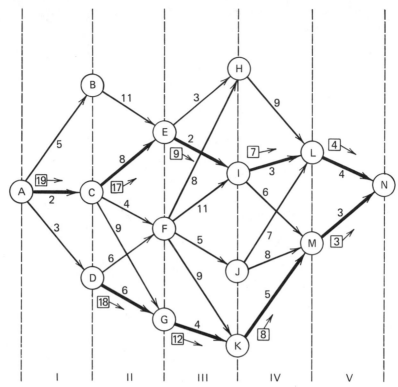

FIG. 5.12 Saving two strategies.

nature and are "must-not-run" units. Perhaps they are on maintenance or are so uneconomic that they are run only during emergencies. When these restrictions are considered along with the priority-list concept and the requirement that schedules must be feasible (i.e., they must meet the load plus spinning reserve and not violate any other constraint), the possible number of states is quite reduced.

The idea of a search range is illustrated in Figure 5.13. If complete enumeration was used, you would have to search over nine states at each interval. Suppose, however, that the load plus spinning-reserve requirement in a given interval could be met by commiting the first eight units in priority order. We could then consider searching over states involving units 7 through 10 for our optimal commitment procedure. This will greatly reduce the computational effort.

Take as another example the four-unit, 8-h case of Example 5E. Priority-list ordering gave us 24 dispatches, whereas the complete enumeration gave us 2.56×10^9. Suppose we were to cut the search range down from four to two each hour. The upper bound to the number of dispatches would now become $(2^2 - 1)^8 = 6561$, which is six orders of magnitude less than $(2^4 - 1)^8$.

This method of unit commitment works well with small and moderate sized systems. For large systems, further heuristic techniques must be applied to reduce the computational effort.

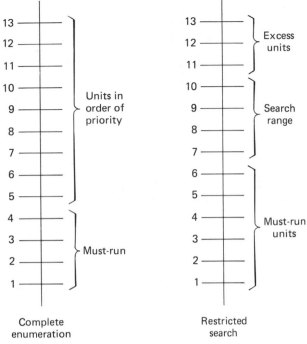

FIG. 5.13 Example of a unit selection list.

EXAMPLE 5F

Figures 5.14 and 5.15 show computer output from a typical unit commitment program. This program has the ability to include maintenance costs as a separate cost from generating cost and start-up costs. The unit statuses shown at the bottom of Figure 5.14 are

1 = OFF-FOUT: Unit is off-line due to a forced outage
2 = OFF-MAIN: Unit is off-line for maintenance work
3 = OFF-TIME:
4 = OFF-AVAL: Unit is off-line but available for start-up
5 = ON-MUST: Unit is on-line and must be running
6 = ON-TIME:
7 = ON-CBSD: Unit is on-line and can be shut down

Unit commitment programs, such as shown in Figures 5.14 and 5.15, are used for many different kinds of economic studies such as

1. **Costing a generator unit outage strategy.** In this type of study the user wishes to know if there is a significant difference in total generating cost between taking one unit off-line for maintenance versus another unit. A similar, but different, study would answer the question "Is it cheaper to take a given unit off for maintenance today or tomorrow?"

```
                                          UNIT COMMITMENT OUTPUT

********************************************  UNIT STARTUP AND SHUTDOWN SCHEDULE  ********************************************

                        ***** SAT 08-DEC-79 *****  ***** SUN 09-DEC-79 *****  ***** MON 10-DEC-79 *****  ***** TUE 11-DEC-79 *****
    TOTAL COST $ =             397152.                    339660.                    387307.                    387122.
    FUEL  COST $ =             379243.                    326991.                    370328.                    370265.
    MAINT COST $ =              16799.                     12669.                     15869.                     15747.
    START COST $ =               1110.                         0.                      1110.                      1110.

---UNIT ID--   STATUS            111 111111122222           111 111111122222           111 111111122222           111 111111122222
NO. --NAME--    IN FN   123456789012 345678901234  123456789012 345678901234  123456789012 345678901234  123456789012 345678901234

  1  LAKE #1      5 5   ************ ***********U  U*********** ************  ************ ************  ************ ************
  2  LAKE #2      5 5   *******UUUUU UUUUUUUUUUUU  UU********** ************  ************ ************  ************ ***U********
  3  LAKE #3      5 5   ************ ************  ************ ************  ************ ************  ************ ************
  4  LAKE #4      5 5   UUUUUUUUUUUU UUUUUUUUUUUU  UUUUUUU*UUUU UUUUUUUUUUUU  UUUUUUUUUUUU UUUUUUUUUU*U  ************ ************
  5  LAKE PKR     4 4
  6  ZENITH 1     4 4
  7  ZENITH 2     7 3   **LLLL****** ************  **LLLLLLLLLL LLLLL***LLL   LLLLLLL***** ***********L  LLLLLLL***** **********
  8  ZENITH 5     4 3                     L**LLL*  L                                 **** LLLLL***             **** LLLLL***
  9  ZENITH 6     4 3          L**** **LLL*****L*  **                               ***** **********          ***** ***********
 10  ZENITH T     4 4
 11  RI PKR 1     4 4
 12  RI PKR 2     4 4
 13  RI PKR 3     4 4
 14  BR PKR 1     4 4
 15  BR PKR 6     4 4
 16  STRK PKR     4 4
 17  JOINT M2     5 5   ************ ************  ************ ************  ************ ************  ************ ************
305  DIS TRAN     1 5                                                         LLLLLL*UUUUU UUUUUUUUUUU*  LLLLLL*UUUUU UUUUUUUUUUU*
308  DIS TRAN     5 5   *LLLLLLLLL** LLLLLLLLLLLL  LLLLLLLLLLLL LLLLLLLLLLLL  LLLLLLLLLLLL LLLLLLL*LLLL  LLLLLLLLLLLL LLLLLLLLLLLL
FIX  NET PUR      1 5                              FFFFFFFFFFFF FFFFFFFFFFFF  FFFFFFFFFFFF FFFFFFFFFFFF  FFFFFFFFFFFF FFFFFFFFFFFF
FIX  NET SALE     0 0

      STUDY PERIOD TOTALS:                        UNIT STATUS:                                             LEGEND:  F = FIXED MW
            TOTAL COST $ =  1511240.              (INIT/FINAL)  1 = OFF-FOUT    5 = ON-MUST                         * = DISPATCHED
            FUEL  COST $ =  1446826.                            2 = OFF-MAIN    6 = ON-TIME                         L = ON @ LOW
            MAINT COST $ =    61084.                            3 = OFF-TIME    7 = ON-CBSD                         U = ON @ HIGH
            START COST $ =     3330.                            4 = OFF-AVAL                                    BLANK = OFF
```

FIG. 5.14 Sample output from a unit commitment program. (Courtesy of Power Technologies, Inc.)

UNIT COMMITMENT OUTPUT

*** HOURLY MW SCHEDULES FOR SAT 08-DEC-79 ***

NAME	01	02	03	04	05	06	07	08	09	10	11	12	13	14	15	16	17	18	19	20	21	22	23	24
LAKE #1	129	129	129	129	129	129	129	131	129	129	129	129	129	130	131	131	131	133	132	133	134	134	135	136
LAKE #2	130	131	131	132	132	132	131	135	135	135	135	135	135	135	135	135	135	135	135	135	135	135	135	135
LAKE #3	138	138	138	138	138	138	138	138	138	138	138	138	138	138	138	138	138	138	138	138	138	138	138	139
LAKE #4	209	209	209	209	209	209	209	209	209	209	209	209	209	209	209	209	209	209	209	209	209	209	209	209
LAKE PKR	0	0	0	0	0	0	0	0	0	0	0	0	0	0	0	0	0	0	0	0	0	0	0	0
ZENITH 1	60	0	0	0	0	0	0	0	0	0	0	0	0	0	0	0	0	0	0	0	0	0	0	0
ZENITH 2	0	41	25	25	25	25	28	42	60	60	60	60	60	60	36	30	0	60	60	60	60	60	37	60
ZENITH 5	0	0	0	0	0	0	0	0	0	0	0	0	0	0	0	0	40	14	42	19	14	14	14	54
ZENITH 6	0	0	0	0	0	0	0	22	51	77	77	77	49	28	22	22	22	72	77	77	62	40	22	77
ZENITH T	0	0	0	0	0	0	0	0	0	0	0	0	0	0	0	0	0	0	0	0	0	0	0	0
RI PKR 1	0	0	0	0	0	0	0	0	0	0	0	0	0	0	0	0	0	0	0	0	0	0	0	0
RI PKR 2	0	0	0	0	0	0	0	0	0	0	0	0	0	0	0	0	0	0	0	0	0	0	0	0
RI PKR 3	0	0	0	0	0	0	0	0	0	0	0	0	0	0	0	0	0	0	0	0	0	0	0	0
BR PKR 1	0	0	0	0	0	0	0	0	0	0	0	0	0	0	0	0	0	0	0	0	0	0	0	0
BR PKR 6	0	0	0	0	0	0	0	0	0	0	0	0	0	0	0	0	0	0	0	0	0	0	0	0
STRK PKR	0	0	0	0	0	0	0	0	0	0	0	0	0	0	0	0	0	0	0	0	0	0	0	0
JOINT M2	132	132	124	109	107	117	132	132	132	132	132	132	132	132	132	132	132	132	132	132	132	132	132	132
305 TRAN	12	0	0	0	0	0	0	0	0	0	3	1	0	0	0	0	0	0	0	0	0	0	0	0
308 TRAN	0	0	0	0	0	0	0	0	0	0	0	0	0	0	0	0	0	0	0	0	0	0	0	0
FIX PUR	0	0	0	0	0	0	0	0	0	0	0	0	0	0	0	0	0	0	0	0	0	0	0	0
FIX SALE	0	0	0	0	0	0	0	0	0	0	0	0	0	0	0	0	0	0	0	0	0	0	0	-166
TOT GEN	798	780	756	742	740	750	767	809	854	880	880	880	852	832	803	797	807	893	925	903	884	862	822	942
NET INT	12	0	0	0	0	0	0	0	0	0	3	1	0	0	0	0	0	0	0	0	0	0	0	-166
TOT LOAD	808	778	754	740	738	748	765	807	852	878	881	879	850	830	801	745	805	891	923	901	882	860	820	774
OPER RES	105	104	104	103	103	103	104	105	107	107	107	107	107	106	105	105	105	108	108	108	107	107	106	104
SPIN RES	28	27	27	26	26	26	27	28	30	30	30	30	30	29	28	28	28	31	32	31	30	30	29	27

FIG. 5.15 Sample output from a unit commitment program. (Courtesy of Power Technologies, Inc.)

2. **Evaluating interchange power (see Chapter 10).** Here the user of the unit commitment program wishes to study the change in operating cost incurred if a block of power were sold or purchased over tie lines to a neighboring system. Often, these sales or purchases require a different unit commitment to give optimum economic operation.

Of course, both types of studies are in addition to using such a unit commitment simply to work out the commit/decommit schedules for the system operators to use in the daily operation of the power system.

APPENDIX
Dynamic Programming Applications

The application of digital methods to solve a wide variety of control and dynamic optimization problems in the late 1950s led Dr. Richard Bellman and his associates to the development of dynamic programming. These techniques are useful in solving a variety of problems and can greatly reduce the computational effort in finding optimal trajectories or control policies.

The theoretical mathematical background, based on the calculus of variations, is somewhat difficult. The applications are not, however, since they depend on a willingness to express the particular optimization problem in terms appropriate for a dynamic programming (DP) formulation.

In the scheduling of power generation systems, DP techniques have been developed for

- The economic dispatch of thermal systems.
- The solution of hydrothermal economic-scheduling problems.
- The practical solution of the unit commitment problem.

This text will touch on all three areas.

First, however, it will be well to introduce some of the notions of DP by means of some one-dimensional examples. Figure 5.16 represents the cost of transporting a unit shipment from Node A to Node N. The values on the arcs are the costs, or values, of shipping the unit from the originating to the terminating node of the arc. The problem is to find the minimum cost route from A to N. The method to be illustrated is that of dynamic programming. The first two examples are from reference 12 and are used by permission.

Starting at A, the minimum cost path to N is ACEILN.

Starting at C, the least cost path to N is   CEILN.

Starting at E, the least cost path to N is   EILN

Starting at I, the least cost path to N is   ILN.

Starting at L, the least cost path to N is   LN.

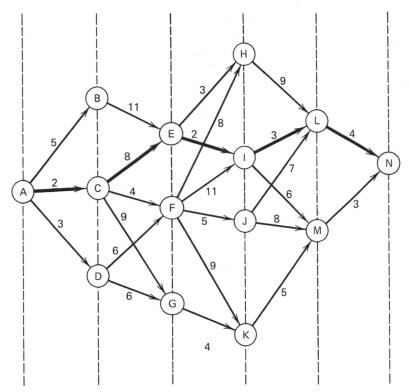

FIG. 5.16 Dynamic-programming example.

The same type of statements could be made about the maximum cost path from A to N (ABEHLN). That is, the maximum cost to N, starting from any node on the original maximal path, is contained in that original path.

The choice of route is made in sequence. There are various stages traversed. The optimum sequence is called the *optimal policy*. Any subsequence is a *subpolicy*. From this it may be seen that the optimal policy (i.e., the minimum cost route) contains only optimal subpolicies. This is the *Theorem of Optimality*.

> An optimal policy must contain only optimal subpolicies.

In Reference 15, Bellman and Dreyfus call it the "Principle of Optimality" and state it as

> A policy is optimal if, at a stated stage, whatever the preceding decisions may have been, the decisions still to be taken constitute an optimal policy when the result of the previous decisions is included.

We continue with the same example, only now let us find the minimum cost path. Figure 5.17 identifies the stages (I, II, III, IV, V). At the terminus of each stage there is a set of choices of nodes $\{X_i\}$ to be chosen $[\{X_3\} = \{H, I, J, K\},]$. The symbol $V_a(X_i, X_i + 1)$ represents the "cost" of traversing stage a $(= I, \ldots, V)$ and depends on the variables selected from the sets $\{X_i\}$ and $\{X_i + 1\}$. That is, the cost, V_a, depends on the starting and terminating nodes. Finally, $f_a(X_i)$ is the minimum cost for stages I through a to arrive at some particular node X_i at the end of that stage, starting from A. The numbers in the node circles in Figure 5.17 represent this minimum cost.

$\{X_0\}$: A $\{X_2\}$: E, F, G $\{X_4\}$: L, M
$\{X_1\}$: B, C, D $\{X_3\}$: H, I, J, K $\{X_5\}$: N

$f_I(X_1)$: Minimum cost for the first stage is obvious:

$$f_I(B) = V_I(A, B) = 5$$

$$f_I(C) = V_I(A, C) = 2$$

$$f_I(D) = V_I(A, D) = 3$$

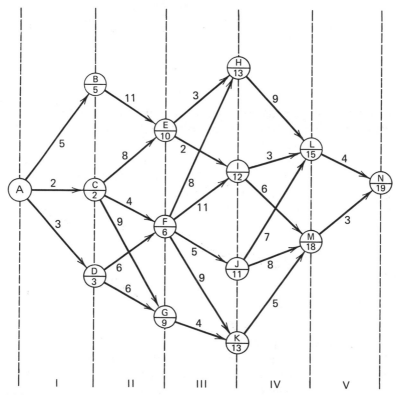

FIG. 5.17 Dynamic-programming example showing minimum cost at each node.

$f_{II}(X_2)$: Minimum cost for stages I and II as a function of X_2:

$$f_{II}(E) = \min_{\{X_1\}} [f_I(X_1) + V_{II}(X_1, E)]$$

$$= \min [5 + 11, \quad 2 + 8, \quad 3 + \infty] = 10$$
$$X_1 = B \quad = C \quad = D \quad X_1 = C$$

The cost is infinite for node D since there is no path from D to E:

$$f_{II}(F) = \min_{\{X_1\}} [f_I(X_1) + V_{II}(X_1, F)] = \min[\infty, 6, 9] = 6, \, X_1 = C$$

$$f_{II}(G) = \min_{\{X_1\}} [f_I(X_1) + V_{II}(X_1, G)] = \min[\infty, 11, 9] = 9, \, X_1 = D$$

Thus, at each stage we should record the minimum cost and the termination starting the stage to achieve the minimum cost path for each of the nodes terminating the current stage.

(X_2)	E	F	G
$f_{II}(X_2)$	10	6	9
Path X_0X_1	AC	AC	AD

$f_{III}(X_3)$: Minimum cost of stages I, II, and III as a function of X_3:

$$f_{III}(H) = \min_{\{X_2\}} (f_{II}(X_2) + V_{III}(X_2, H)) = \min(13, 14, \infty) = 13 \quad \text{with } X_2 = E$$

In general,

$$f_{III}(X_3) = \min_{\{X_2\}} [f_{II}(X_2) + V_{III}(X_2, X_3)]$$

Giving,

X_3	H	I	J	K
$f_{III}(X_3)$	13	12	11	13
Path $X_0X_1X_2$	ACE	ACE	ACF	ADG

f_{IV}: Minimum cost of stages I through IV as a function of X_4:

$$f_{IV}(X_4) = \min_{X_3} (f_{III}(X_3) + V_{IV}(X_3, X_4))$$

$$f_{IV}(L) = \min(13 + 9, 12 + 3, 11 + 7, 13 + \infty) = 15 \quad X_3 = I$$
$$X_3 = H \quad I \quad J \quad K$$

$$f_{IV}(M) = \min(13 + \infty, 12 + 6, 11 + 8, 13 + 5) = 18 \quad X_3 = I \text{ or } K$$
$$H \quad I \quad J \quad K$$

f_V: Minimum cost of I $\rightarrow$ V as a function of X_5:

$$f_V(N) = \min_{(X_4)} (f_{IV}(X_4) + V_V(X_4, X_5))$$

$$= \min (15 + 4, 18 + 3) = 19 \quad X_4 = L$$
$$X_4 = L \quad = M$$

Tracing back, the path of minimum cost is found as follows:

Stage 1	$\{X_i\}$	f_i
1	B, Ⓒ, D	5, ②, 3
2	Ⓔ, F, G	⑩, 6, 9
3	H, Ⓘ, J, K	13, ⑫, 11, 13
4	Ⓛ, M	⑮, 18
5	Ⓝ	⑲

It would be possible to carry out this procedure in the opposite direction just as easily.

An Allocation Problem

Table 5.7 lists the profits to be made in each of four ventures as a function of the investment in the particular venture. Given a limited amount of money to allocate, the problem is to find the optimal investment allocation. The only restriction is that investments must be made in integer amounts. For instance, if one had 10 units to invest and the policy were to put

3 in I

1 in II

5 in III

1 in IV

$$\text{Profit} = 0.65 + 0.25 + 0.65 + 0.20 = 1.75.$$

The problem is to find an allocation policy that yields the maximum profit. Let

X_1, X_2, X_3, X_4 be investments in I through IV

$V(X_1), V(X_2), V(X_3), V(X_4)$ be profits

$X_1 + X_2 + X_3 + X_4 = 10$ is the constraint. That is, 10 units must be invested.

TABLE 5.7 Profit Versus Investment

Investment amount	Profit from venture			
	I	II	III	IV
0	0	0	0	0
1	0.28	0.25	0.15	0.20
2	0.45	0.41	0.25	0.33
3	0.65	0.55	0.40	0.42
4	0.78	0.65	0.50	0.48
5	0.90	0.75	0.65	0.53
6	1.02	0.80	0.73	0.56
7	1.13	0.85	0.82	0.58
8	1.23	0.88	0.90	0.60
9	1.32	0.90	0.96	0.60
10	1.38	0.90	1.00	0.60

To transform this into a multistage problem, let the stages be

$$X_1, U_1, U_2, A$$

where $\quad U_1 = X_1 + X_2 \qquad U_1 \le A \qquad U_2 \le A$
$\qquad U_2 = U_1 + X_3 \qquad \{A\} = 0, 1, 2, 3, \dots, 10$
$\qquad A = U_2 + X_4$

The total profit is

$$f(X_1, X_2, X_3, X_4) = V_1(X_1) + V_2(X_2) + V_3(X_3) + V_4(X_4)$$

which can be written

$$f(X_1, U_1, U_2, A) = V_1(X_1) + V_2(U_1 - X_1) + V_3(U_2 - U_1) + V_4(A - U_2)$$

At the second stage, we can compute

$$f_2(U_1) = \max_{X_1 = 0, 1, \dots, U_1} [V_1(X_1) + V_2(U_1 - X_1)]$$

X_1, X_2, or U_1	$V_1(X_1)$	$V_2(X_2)$	$f_2(U_1)$	Optimal subpolicies I & II
0	1	0	0	(0, 0)
1	0.28	0.25	0.28	(1, 0)
2	0.45	0.41	0.53	(1, 1)
3	0.65	0.55	0.70	(2, 1)
4	0.78	0.65	0.90	(3, 1)
5	0.90	0.75	1.06	(3, 2)
6	1.02	0.80	1.20	(3, 3)
7	1.13	0.85	1.33	(4, 3)
8	1.23	0.88	1.45	(5, 3)
9	1.32	0.90	1.57	(6, 3)
10	1.38	0.90	1.68	(7, 3)

Next, at the third stage,

$$f_3(U_2) = \max_{U_1 = 0, 1, 2, \dots, U_2} [f_2(U_1) + V_3(U_2 - U_1)]$$

U_1, U_2 or X_3	$f_2(U_1)$	$V_3(X_3)$	$f_3(U_2)$	For I & II	For I, II & III
0	0	0	0	0, 0	0, 0, 0
1	0.28	0.15	0.28	1, 0	1, 0, 0
2	0.53	0.25	0.53	1, 1	1, 1, 0
3	0.70	0.40	0.70	2, 1	2, 1, 0
4	0.90	0.50	0.90	3, 1	3, 1, 0
5	1.06	0.62	1.06	3, 2	3, 2, 0
6	1.20	0.73	1.21	3, 3	3, 2, 1
7	1.33	0.82	1.35	4, 3	3, 3, 1
8	1.45	0.90	1.48	5, 3	4, 3, 1
9	1.57	0.96	1.60	6, 3	5, 3, 1 or 3, 3, 3
10	1.68	1.00	1.73	7, 3	4, 3, 3

Finally, the last stage is

$$f_4(A) = \max_{\{U_2\}} [f_3(U_2) + V_4(A - U_2)]$$

U_2, A or X_4	$f_3(U_2)$	$V_4(X_4)$	$f_4(A)$	Optimal subpolicy for I, II, III	Optimal policy
0	0	0	0	0, 0, 0	0, 0, 0, 0
1	0.28	0.20	0.28	1, 0, 0	1, 0, 0, 0
2	0.53	0.33	0.53	1, 1, 0	1, 1, 0, 0
3	0.70	0.42	0.73	2, 1, 0	1, 1, 0, 1
4	0.90	0.48	0.90	3, 1, 0	3, 1, 0, 0 or 2, 1, 0, 1
5	1.06	0.53	1.10	3, 2, 0	3, 1, 0, 1
6	1.21	0.56	1.26	3, 2, 1	3, 2, 0, 1
7	1.35	0.58	1.41	3, 3, 1	3, 2, 1, 1
8	1.48	0.60	1.55	4, 3, 1	3, 3, 1, 1
9	1.60	0.60	1.68	5, 3, 1 or 3, 3, 3	4, 3, 1, 1 or 3, 3, 1, 2
10	1.73	0.60	1.81	4, 3, 3	4, 3, 1, 2

Consider the procedure and solution:

1. It was not necessary to enumerate all possible solutions. Instead we used an orderly, stagewise search, the form of which was the same at each stage.
2. The solution was obtained not only for $A = 10$, but for the complete set of A's $\{A\} = 0, 1, 2, \ldots, 10$
3. The optimal policy contains only optimal subpolicies. For instance, $A = 10$, (4, 3, 1, 2) is the optimal policy. For stages I, II, III and $U_2 = 8$, (4, 3, 1) is the optimal subpolicy. For stages I and II and $U_1 = 7$, (4, 3) is the optimal subpolicy. For stage I only, $X_1 = 4$ fixes the policy.
4. Notice also, that by storing the intermediate results, we could work a number of different variations of the same problem with the data already computed.

Economic Dispatch of Thermal Systems Using Dynamic Programming

The problem of finding the economic dispatch of a system of thermal units can be expressed in such a fashion that it is amenable to solution using dynamic-programming methods. Assume:

1. All the units to be considered are to be on-line (i.e., the unit commitment problem is solved, or ignored.)
2. Losses are neglected.
3. It will be satisfactory to find the economic dispatch at discrete load steps rather than for continuous load levels.

In order to use the DP formulation, the units to be dispatched must be ordered. For this particular problem the loading order is arbitrary. Let i represent the order number of a unit and then define

$$P_i = \text{loading (MW) on unit } i$$

$$F_i(P_i) = \text{cost (R/h) for generating } P_i \text{ MW on unit } i$$

$$f_i(D) = \text{stage cost of supplying a demand of } D \text{ MW with } i \text{ units}$$

At each stage the recursive relationship that may be used to find the optimal economic operating schedule (i.e., operating policy) is

$$f_i(D) = \min_{\{P_i\}} \left[f_{i-1}(D - P_i) + F_i(P_i) \right]$$

It is easier to see by example. We will use a three-unit example with the following data.

Unit no.	Min MW	Max MW
1	100	500
2	100	500
3	200	1000

Unit costs in R per hour are given in Table 5.8. Infinite costs are assigned for outputs below the minimum and above the maximum ratings.

The method is quite appropriate to the computation of a whole range of schedules. Let us illustrate the process by finding the optimum schedule for a total demand, D, of 800 MW. The first stage with only one machine is obviously nothing more than a reproduction of the first column from Table 5.8. Therefore, let us proceed directly to stage 2.

$$f_2(D) = \min_{\{P_2\}} \left[f_1(D - P_2) + F_2(P_2) \right]$$

$$= \min_{\{P_2\}} \left[F_1(D - P_2) + F_2(P_2) \right]$$

TABLE 5.8   Unit Costs in R per hour

P_i (MW)	F_1	F_2	F_3
0	∞	∞	∞
100	500	400	∞
200	950	1000	1020
300	1400	1440	1450
400	1840	1800	1900
500	2320	2400	2350
600	∞	∞	2800
700			3240
800			3680
900			4130
1000			4570

TABLE 5.9 Stage 2 (Two Machines) Generation Dispatch

			$F_1(D - P_2) + F_2(P_2)$					Optimum Schedules	
		$P_2 = 0$	100	200	300	400	500		
D	$F_1(D)$	$F_2(P_2) = \infty$	400	1000	1440	1800	2400	f_2	P_2^*
0	∞	∞	→————————————————————————————→						
100	500		∞ →————————————————————————→						
200	950		900	∞ →——————————————————→				900	100
300	1400		1350	1500	∞ →——————————→			1350	100
400	1840		1800	1950	1940	∞ →——→		1800	100
500	2320		2240	2400	2390	2300	∞	2240	100
600	∞		2720	2840	2840	2750	2900	2720	100
700			∞	3320	3280	3200	3350	3200	400
800				∞	3760	3640	3800	3640	400
900					∞	4120	4240	4120	400
1000	↓	↓	↓	↓	↓	∞	4720	4720	500

P_2^* = value of P_2 that results in the minimum. The minimum costs and loading on unit 2 for serving the various load levels $\{D\}$ are contained in the two right-hand columns of Table 5.9. If we need a finer tuning of the schedule, we might be able to expand these tables in smaller steps in the range of the coarse optimum. This will work if the functions are smooth and convex in the neighborhood of the coarse optimum. If not, we need to cover the entire range with the finer grid.

For three machines the minimum cost is found from

$$f_3(D) = \min_{\{P_3\}} \left[F_3(P_3) + f_2(D - P_3) \right]$$

The table for three machines could easily be set up for a number of different demand levels. Instead, it is set up in Table 5.10 only for $D = 800$ MW. The data for $600 < P_3 < 200$ MW are omitted for obvious reasons. Therefore, $f_3(800) = 3690$ and $P_3^* = 300$ MW. The resulting demand on units 1 and 2 is 500 MW, and from Table 5.9, $P_2^* = 100$ MW so that $P_1^* = 400$ MW. Note that the schedules were finally found by looking up data in the three tables for the $f_i(D)$ in sequence. Therefore, in setting up a computerized routine only a moderate amount of data need be saved.

TABLE 5.10 Stage 3 (Three Machines) Generation Dispatch

		D = 800 MW		
P_3 (MW)	F_3 (R/h)	$800 - P_3$ (MW)	$f_2(800 - P_3)$ (R/h)	$F_3(P_3) + f_2(800 - P_3)$ (R/h)
200	1020	600	2720	3740
300	1450	500	2240	3690
400	1900	400	1800	3700
500	2350	300	1350	3700
600	2800	200	900	3700
700	⋮	⋮	⋮	∞
800				⋮
⋮				

The two significant items to note about this economic dispatching method are

1. **Nothing is demanded of the individual unit cost characteristics. They do not have to be convex, smooth, or even possess derivatives.**

2. **The order in which the units are considered is not important as long as all three units are assumed to be committed. (Note that unit 3 alone could serve the 800 MW demand at less cost.)**

The first point permits this method to be used for economic dispatching problems in which unit characteristics may be represented in detail to simulate increased friction losses when turbine values are first opened. The second point is worth making since it separates the economic dispatching problem from the unit commitment problem.

PROBLEMS

5.1 You have been assigned the job of building an oil pipeline from the West Coast of the United States to the East Coast. You are told that any one of the three West Coast sites is satisfactory and any of the East Coast sites is satisfactory. The numbers in Figure 5.18 represent relative cost in hundreds of millions R (R · 10^8). Find the cheapest West Coast to East Coast pipeline.

5.2 The Stagecoach Problem

A mythical salesman who had to travel west by stagecoach through unfriendly country wished to take the safest route. His starting point and destination were

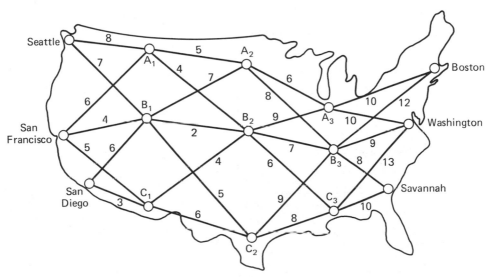

FIG. 5.18 Possible oil pipeline routes for Problem 5.1.

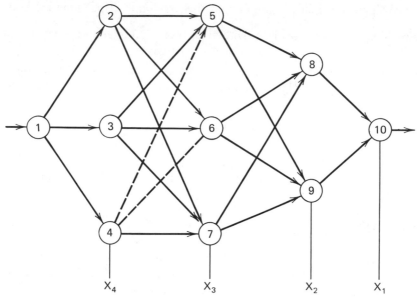

FIG. 5.19 Possible stagecoach routes for Problem 5.2.

fixed, but he had considerable choice as to which states he would travel through enroute. The possible stagecoach routes are shown in Figure 5.19. After some thought the salesman deduced a clever way of determining his safest route. Life insurance policies were offered to passengers, and since the cost of each policy was based on a careful evaluation of the safety of that run, the safest route should be the one with the cheapest policy. The cost of the standard policy on the stagecoach run from state i to state j, denoted as C_{ij}, is given in Figure 5.20. Find the safest path(s) for the salesman to take.

i \ $j=$	1	2	3	4	5	6	7	8	9	10
= 1		2	4	3						
2					7	4	6			
3					3	2	4			
4					4	1	5			
5								1	4	
6								6	3	
7								3	2	
8										3
9										4
10										

FIG. 5.20 Cost to go from state i to state j in Problem 5.2. Costs not shown are infinite.

5.3 Economic Dispatch Problem

Consider three generating units that do not have convex input-output functions. (This is the type of problem one encounters when considering valve points in the dispatch problem.)

Unit 1:

$$H_1(P_1) = \begin{cases} 80 + 8\,P_1 + 0.024\,P_1^2 & 20 \text{ MW} \le P_1 \le 60 \text{ MW} \\ 196.4 + 3\,P_1 + 0.075\,P_1^2 & 60 \text{ MW} \le P_1 \le 100 \text{ MW} \end{cases}$$

Generation limits are 20 MW $\le P_1 \le$ 100 MW.

Unit 2:

$$H_2(P_2) = \begin{cases} 120 + 6\,P_2 + 0.04\,P_2^2 & 20 \text{ MW} \le P_2 \le 40 \text{ MW} \\ 157.335 + 3.3333\,P_2 + 0.08333\,P_2^2 & 40 \text{ MW} \le P_2 \le 100 \text{ MW} \end{cases}$$

Generation limits are 20 MW $\le P_2 \le$ 100 MW

Unit 3:

$$H_3(P_3) = \begin{cases} 100 + 4.6666\,P_3 + 0.13333\,P_3^2 & 20 \text{ MW} \le P_3 \le 50 \text{ MW} \\ 316.66 + 2\,P_3 + 0.1\,P_3^2 & 50 \text{ MW} \le P_3 \le 100 \text{ MW} \end{cases}$$

Generation limits are 20 MW $\le P_3 \le$ 100 MW. Fuel costs = 1.5 R/MBtu for all units.

a. Plot the cost function for each unit (see Problem 3.1).

b. Plot the incremental cost function for each unit.

c. Find the most economical dispatch for the following total demands assuming all units are on line:

$$P_D = 100 \text{ MW}$$
$$P_D = 140 \text{ MW}$$
$$P_D = 180 \text{ MW}$$
$$P_D = 220 \text{ MW}$$
$$P_D = 260 \text{ MW}$$

Where $P_D = P_1 + P_2 + P_3$.

Solve using dynamic programming using discrete load steps of 20 MW starting at 20 MW through 100 MW for each unit.

d. Can you solve these dispatch problems without dynamic programming? If think you know how, try solving for $P_D = 100$ MW.

5.4 Unit Commitment

Given the unit data in Tables 5.11 and 5.12 use forward dynamic-programming to find the optimum unit commitment schedules covering the 8-h period.

TABLE 5.11 Unit Commitment Data for Problem 5.4

Unit	Max MW	Min MW	Incremental heat rate (Btu/kWh)	No-load energy input (MBtu/h)	Start-up energy (MBtu)
1	500	70	9,950	300	800
2	250	40	10,200	210	380
3	150	30	11,000	120	110
4	150	30	11,000	120	110

Load data (all time periods = 2 h):

Time period	Load (MW)
1	600
2	800
3	700
4	950

Start-up and shut-down rules

Unit	Minimum uptime (h)	Minimum downtime (h)
1	2	2
2	2	2
3	2	4
4	2	4

Fuel cost = 1.00 R/MBtu.

TABLE 5.12 Unit Combinations and Operating Cost for Problem 5.5

Combination	Unit 1	Unit 2	Unit 3	Unit 4	Operating cost (R/h) Load 600 MW	Load 700 MW	Load 800 MW	Load 950 MW
A	1	1	0	0	6505	7525	×	×
B	1	1	1	0	6649	7669	8705	×
C	1	1	1	1	6793	7813	8833	10475

1 = up; 0 = down

Table 5.12 gives all the combinations you need as well as the operating cost for each at the loads in the load data. An "X" indicates that a combination cannot supply the load. The starting conditions are: At the beginning of the first period units 1 and 2 are up, units 3 and 4 are down and have been down for 8 h.

5.5 Unit Commitment

Table 5.13 presents the unit characteristics and load pattern for a five-unit—four time period problem. Each time period is 2 h long. The input-output characteristics are approximated by a straight line from min to max generation so that the incremental heat rate is constant. Unit no-load and start-up costs are given in terms of heat energy requirements.

TABLE 5.13 The Unit Characteristic and Load Pattern for Problem 5.5

Unit	Max (MW)	Net full-load heat rate (Btu/kWh)	Incremental heat rate (Btu/kWh)	Min (MW)	No-load cost[a] (MBtu/h)	Start-up cost (MBtu)	Min up/down time (h)
1	200	11,000	9,900	50	220	400	8
2	60	11,433	10,100	15	80	150	8
3	50	12,000	10,800	15	60	105	4
4	40	12,900	11,900	5	40	0	4
5	25	13,500	12,140	5	34	0	4

Load pattern		Conditions
Hours	MW load	
1–2	250	1. Initially (prior to hour 1) only unit 1 is on and has been on for 4 hr.
3–4	320	2. Ignore losses, spinning reserve, etc. The only requirement is that the
5–6	110	generation be able to supply the load.
7–8	75	3. Fuel costs for all units may be taken as 1.40 ₽/MBtu

[a] See footnote on Table 5.4

a. Develop the priority list for these units and solve for the optimum unit commitment. Use a strict priority list with a search range of three ($X = 3$) and save no more than three strategies ($N = 3$). Ignore min up/min down times for units.

b. Solve the same commitment problem using the strict priority list with $X = 3$ and $N = 3$ as in (a), but obey the min up/min down time rules.

c. (*Optional*) Find the optimum unit commitment without use of a strict priority list (i.e., all 32 unit on/off combinations are valid). Restrict the search range to decrease your effort. Obey the min up/min down time rules.

When using a dynamic-programming method to solve a unit commitment problem with minimum up- and downtime rules, one must save an additional piece of information at each state each hour. This information simply tells whether any units are ineligible to be shut down or started up at that state. If such units exist at a particular state, the transition cost, S_{cost}, to a state that violates the start-up/shut-down rules should be given a value of infinity.

FURTHER READING

Some good introductory references to the unit commitment problem are found in references 1–3. A survey of the state-of-the-art (as of 1975) of unit commitment solutions is found in reference 4. References 5 and 6 provide a good look at two commercial unit commitment programs in present use.

References 7–11 deal with unit commitment as an integer-programming problem. Much of the pioneering work in this area was done by Garver (7), who also sounded a note of pessimism in a discussion of reference 8 written together with Happ in 1968. Further research 9–11 has refined the unit commitment solution by integer programming but has never really overcome the Garver-Happ limitations presented in the 1968 discussion, thus leaving dynamic programming as the only viable solution technique to large-scale unit commitment problems.

References 12–16 should be consulted for a deeper study of the mathematical programming technique known as dynamic programming, which was introduced in the Appendix of this chapter. The reader should see references 17 and 18 for a discussion of valve-point loading and for a thorough development of economic dispatch via dynamic programming.

1. Baldwin, C. J., Dale, K. M., Dittrich, R. F., "A Study of Economic Shutdown of Generating Units in Daily Dispatch," *AIEE Transactions on Power Apparatus and Systems* Vol. PAS-78, December 1959, pp. 1272–1284.

2. Burns, R. M., Gibson, C. A., "Optimization of Priority Lists for a Unit Commitment Program," IEEE Power Engineering Society Summer Meeting, Paper A75 453-1, 1975.

3. Davidson, P. M., Kohbrman, F. J., Master, G. L., Schafer, G. R., Evans, J. R., Lovewell, K. M., Payne, T. B., "Unit Commitment Start-Stop Scheduling in the Pennsylvania-New Jersey-Maryland Interconnection," 1967 PICA Conference Proceedings, 1967, pp. 127–132.

4. Gruhl, J., Schweppe, F., Ruane, M., "Unit Commitment Scheduling of Electric Power Systems," *Systems Engineering for Power: Status and Prospects*, Henniker, N. H., August 1975. U.S. Government Printing Office Washington D.C.

5. Pang, C. K., Chen, H. C., "Optimal Short-Term Thermal Unit Commitment," *IEEE Transactions on Power Apparatus and Systems*, Vol. PAS-95, July/August 1976, pp. 1336–1346.

6. Happ, H. H., Johnson, P. C., Wright, W. J., "Large Scale Hydro-Thermal Unit Commitment—Method and Results," *IEEE Transactions on Power Apparatus and Systems*, Vol. PAS-90, May/June 1971, pp. 1373–1384.

7. Garver, L. L., "Power Generation Scheduling by Integer Programming—Development of Theory," *AIEE Transactions on Power Apparatus and Systems*, February 1963, pp. 730–735.

8. Muckstadt, J. A., Wilson, R. C., "An Application of Mixed-Integer Programming Duality to Scheduling Thermal Generating Systems," *IEEE Transactions on Power Apparatus and Systems*. December 1968, pp. 1968–1978.

9. Ohuchi, A., Kaji, I., "A Branch-and-Bound Algorithm for Start-up and Shut-down Problems of Thermal Generating Units," *Electrical Engineering in Japan*, Vol. 95, No. 5, 1975, pp. 54–61.

10. Dillon, T. S., Egan, G. T., "Application of Combinational Methods to the Problems of Maintenance Scheduling and Unit Commitment in Large Power Systems," Proceedings of IFAC Symposium on Large Scale Systems Theory and Applications, Udine, Italy, 1976.

11. Dillon, T. S., Edwin, K. W., Kochs, H. D., Taud, R. J., "Integer Programming Approach to the Problem of Optimal Unit Commitment with Probabilistic Reserve Determination," *IEEE Transactions on Power Apparatus and Systems*, Vol. 97, November/December 1978, pp. 2154–2166.

12. Kaufmann, A., *Graphs, Dynamic Programming and Finite Games*, Academic Press, New York, 1967.

13. Kaufmann, A., Cruon, R., *Dynamic Programming: Sequential Scientific Management*, Academic Press, New York, 1967.

14. Howard, R. A., *Dynamic Programming and Markov Processes*, Wiley and Technology Press, New York, 1960.

15. Bellman, R. E., Dreyfus, S. E., *Applied Dynamics Programming*, Princeton University Press, Princeton, N.J., 1962.

16. Neuhauser, G. L., *Introduction to Dynamic Programming*, Wiley, New York, 1966.

17. Happ, H. H., Ille, W. B., Reisinger, R. M., "Economic System Operation Considering Valve Throttling Losses, I.—Method of Computing Valve-Loop Heat Rates on Multivalve Turbines," *IEEE Transactions on Power Apparatus and Systems*, Vol. PAS-82, February 1963, pp. 609–615.

18. Ringlee, R. J., Williams, D. D., "Economic Dispatch Operation Considering Valve Throttling Losses, II—Distribution of System Loads by the Method of Dynamic Programming," *IEEE Transactions on Power Apparatus and Systems*, Vol. PAS-82, February 1963, pp. 615–622.

chapter 6

Generation with Limited Energy Supply

6.1 INTRODUCTION

The economic operation of a power system requires that expenditures for fuel be minimized over a period of time. When there is no limitation on the fuel supply to any of the plants in the system, the economic dispatch can be carried out with only the present conditions as data in the economic dispatch algorithm. In such a case the fuel costs are simply the incoming price of fuel with, perhaps, adjustments for fuel handling and maintenance of the plant.

When the energy resource available to a particular plant (be it coal, oil, gas, water, or nuclear fuel) is a limiting factor in the operation of the plant, the entire economic dispatch calculation must be done differently. Each economic dispatch calculation must account for what happened before and what will happen in the future.

This chapter begins the development of solutions to the dispatching problem "over time." The techniques used are an extension of the familiar LaGrange formulation. Concepts involving slack variables and penalty functions are introduced to allow solution under certain conditions.

The example chosen to start is a fixed fuel supply that must be paid for whether or not it is consumed.

We might have started with a limited fuel supply of natural gas that must be used as boiler fuel because it has been declared as "surplus." The take-or-pay fuel supply contract is probably the simplest of these possibilities.

Alternatively, we might have started directly with the problem of economic scheduling of hydroelectric plants with their stored supply of water or with light-water-moderated nuclear reactors supplying steam to drive turbine generators. Hydroelectric plant scheduling involves the scheduling of water flows, impoundments (storage), and releases into what usually prove to be a rather complicated hydraulic network (namely, the water shed). The treatment of nuclear unit scheduling requires some understanding of the physics involved in the reactor core and is really beyond the scope of this current text (the methods useful for optimizing the unit outputs are, however, quite similar to those used in scheduling other limited energy systems).

6.2 TAKE-OR-PAY FUEL SUPPLY CONTRACT

Assume there are N normally fueled thermal plants plus one turbine generator fueled under a *"take-or-pay" agreement.* We will interpret this type of agreement as being one in which the utility agrees to use a minimum amount of fuel during a period (the "take") or, failing to use this amount, it agrees to pay the minimum charge. This last clause is the "pay" part of the "take-or-pay" contract.

While this unit's cumulative fuel consumption is below the minimum, the system excluding this unit should be scheduled to minimize the total fuel cost subject to the constraint that the total fuel consumption for the period for this particular unit is equal to the specified amount. Once the specified amount of fuel has been used, the unit should be scheduled normally. Let us consider a special case where the minimum amount of fuel consumption is also the maximum. The system is shown in Figure 6.1. We will consider the operation of the system over

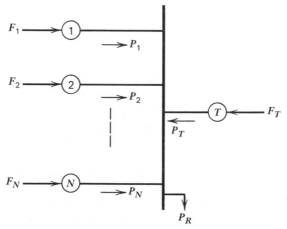

FIG. 6.1 $N + 1$ unit system with take-or-pay fuel supply at unit T.

j_{max} time intervals j where $j = 1, \ldots, j_{max}$, so that

$$P_{1j}, P_{2j}, \ldots, P_{Tj} \quad \text{(Power outputs)}$$

$$F_{1j}, F_{2j}, \ldots, F_{Nj} \quad \text{(Fuel cost rate)}$$

and

$$q_{T1}, q_{T2}, \ldots, q_{Tj} \quad \text{(Take-or-pay fuel input)}$$

are the power outputs, fuel costs, and take-or-pay fuel inputs.

where
$P_{ij} \triangleq$ power from i^{th} unit in the jth time interval

$F_{ij} \triangleq R/h$ cost for i^{th} unit during the jth time interval

$q_{Tj} \triangleq$ fuel input for unit T in j^{th} time interval

$F_{Tj} \triangleq R/h$ cost for unit T in j^{th} time interval

$P_{Rj} \triangleq$ total load in the jth time interval

$n_j \triangleq$ Number of hours in the jth time interval

Mathematically, the problem is as follow.

$$\min \sum_{j=1}^{j_{max}} n_j \sum_{i=1}^{N} F_{ij} + \sum_{j=1}^{j_{max}} n_j F_{Tj} \tag{6.1}$$

subject to

$$\phi = \sum_{j=1}^{j_{max}} n_j q_{Tj} - q_{TOT} = 0 \tag{6.2}$$

and

$$\psi_j = P_{Rj} - \sum_{i=1}^{N} P_{ij} - P_{Tj} = 0 \qquad \text{for } j = 1 \cdots j_{max} \tag{6.3}$$

or in words

We wish to determine the minimum production cost for units 1 to N subject to constraints that ensure that fuel consumption is correct and also subject to the set of constraints to ensure that power supplied is correct each interval.

Note that (for the present) we are ignoring high and low limits on the units themselves. It should also be noted that the term

$$\sum_{j=1}^{j_{max}} n_j F_{Tj}$$

is constant because the total fuel to be used in the "T" plant is fixed. Therefore, the total cost of that fuel will be constant and we can drop this term from the objective function.

The LaGrange function is

$$\mathscr{L} = \sum_{j=1}^{j_{max}} n_j \sum_{i=1}^{N} F_{ij} + \sum_{j=1}^{j_{max}} \lambda_j \left(P_{Rj} - \sum_{i=1}^{N} P_{ij} - P_{Tj} \right)$$

$$+ \gamma \left(\sum_{j=1}^{j_{max}} n_j q_{Tj} - q_{TOT} \right) \tag{6.4}$$

The independent variables are the powers P_{ij} and P_{Tj}, since $F_{ij} = F_i(P_{ij})$ and $q_{Tj} = q_T(P_{Tj})$.
For any given time period, $j = k$,

$$\frac{\partial \mathscr{L}}{\partial P_{ik}} = 0 = n_k \frac{dF_{ik}}{dP_{ik}} - \lambda_k \qquad \text{for } i = 1, \ldots, N \tag{6.5}$$

and

$$\frac{\partial \mathscr{L}}{\partial P_{Tk}} = -\lambda_k + \gamma n_k \frac{dq_{Tk}}{dP_{Tk}} = 0 \tag{6.6}$$

Note that if one analyzes the dimensions of γ, it would be R per unit of q (e.g., R/ft^3, R/bbl, R/ton). As such, γ has the units of a "fuel price" expressed in volume units rather than MBtu as we have used up to now. Because of this, γ is often referred to as a "pseudoprice" or "shadow price." In fact, once it is realized what is happening in this analysis it becomes obvious that we could solve fuel-limited dispatch problems by simply adjusting the price of the limited fuel(s), thus, the terms "pseudoprice" and "shadow price" are quite meaningful.

Since γ appears unsubscripted in Eq. 6.6, γ would be expected to be a constant value over all the time periods. This is true unless the fuel-limited machine is constrained by fuel-storage limitations. We will encounter such limitations in hydro-plant scheduling in Chapter 7. The appendix to Chapter 7 shows when to expect a constant γ and when to expect a discontinuity in γ.

Figure 6.2a shows how the load pattern may look. The solution to a fuel-limited dispatching problem will require dividing the load pattern into time intervals as in

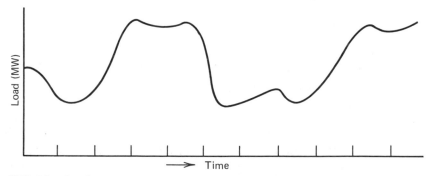

FIG. 6.2a   Load pattern.

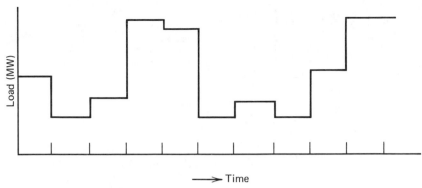

FIG. 6.2b Discrete load pattern.

Figure 6.2b and assuming load to be constant during each interval. Assuming all units are on-line for the period, the optimum dispatch could be done using a simple search procedure for γ as is shown in Figure 6.3. Note that the procedure shown in Figure 6.3 will only work if the fuel-limited unit does not hit either its high or its low limit in any time interval.

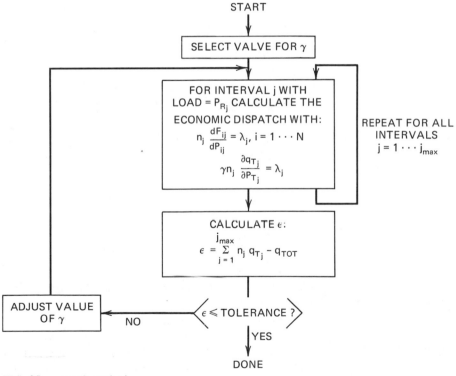

FIG. 6.3 γ search method.

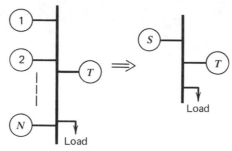

FIG. 6.4 Composite generator unit.

6.3 COMPOSITE GENERATION PRODUCTION COST FUNCTION

A useful technique to facilitate the take-or-pay fuel supply contract procedure is to develop a composite generation production cost curve for all the non-fuel-constrained units. For example, suppose there were N non-fuel-constrained units to be scheduled with the fuel-constrained unit as shown in Figure 6.4. Then a composite cost curve for units $1, 2, \ldots, N$ can be developed.

$$F_s(P_s) = F_1(P_1) + \cdots + F_N(P_N) \tag{6.7}$$

where

$$P_s = P_1 + \cdots + P_N$$

and

$$\frac{dF_1}{dP_1} = \frac{dF_2}{dP_2} = \cdots = \frac{dF_N}{dP_N} = \lambda$$

If one of the units hits a limit, its output is held constant, as in Chapter 3, Eq. 3.6.

A simple procedure to allow one to generate $F_s(P_s)$ consists of adjusting λ from $\lambda^{\min}$ to $\lambda^{\max}$ in specified increments, where

$$\lambda^{\min} = \min\left[\frac{dF_i}{dP_i}, i = 1 \cdots N\right]$$

$$\lambda^{\max} = \max\left[\frac{dF_i}{dP_i}, i = 1 \cdots N\right]$$

At each increment, calculate the total fuel consumption and the total power output for all the units. These points represent points on the $F_s(P_s)$ curve. The points may be used directly by assuming $F_s(P_s)$ consists of straight-line segments between the points or a smooth curve may be fit to the points using a least-squares fitting program. Be aware, however, that such smooth curves may have undesirable properties such as nonconvexity (e.g., the first derivative is not monotonically increasing). The procedure to generate the points on $F_s(P_s)$ is shown in Figure 6.5.

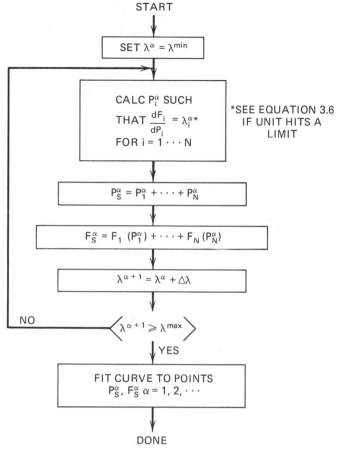

START

SET $\lambda^\alpha = \lambda^{min}$

CALC P_i^α SUCH
THAT $\dfrac{dF_i}{dP_i} = \lambda_i^{\alpha*}$
FOR $i = 1 \cdots N$

*SEE EQUATION 3.6
IF UNIT HITS A
LIMIT

$P_S^\alpha = P_1^\alpha + \cdots + P_N^\alpha$

$F_S^\alpha = F_1 (P_1^\alpha) + \cdots + F_N (P_N^\alpha)$

$\lambda^{\alpha + 1} = \lambda^\alpha + \Delta\lambda$

NO

$\lambda^{\alpha + 1} \geqslant \lambda^{max}$

YES

FIT CURVE TO POINTS
P_S^α, F_S^α $\alpha = 1, 2, \cdots$

DONE

FIG. 6.5   Procedure for obtaining composite cost curve.

EXAMPLE 6A

The three generating units from Example 3A are to be combined into a composite generating unit. The fuel costs assigned to these units will be

Fuel cost for unit 1 = 1.1 R/MBtu

Fuel cost for unit 2 = 1.4 R/MBtu

Fuel cost for unit 3 = 1.5 R/MBtu

Figure 6.6a shows the individual unit incremental costs, which range from 8.3886 to 14.847 R/MWh. A program was written based on Figure 6.5, and lambda was stepped from 8.3886 to 14.847.

At each increment the three units are dispatched to the same lambda and then outputs and generating costs are added as shown in Figure 6.5. The results are given

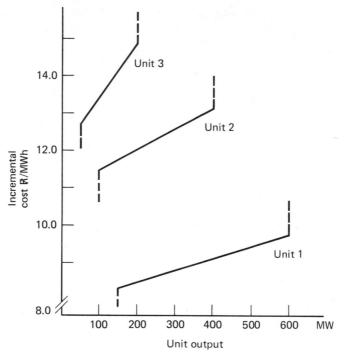

FIG. 6.6a Unit incremental costs.

TABLE 6.1 Lambda Steps Used in Constructing Composite Cost Curve for Example 6A

Step	lambda	P_s	F_s	F_s approx
1	8.3886	300.0	4077.12	4137.69
2	8.7115	403.4	4960.92	4924.39
3	9.0344	506.7	5878.10	5799.07
4	9.3574	610.1	6828.66	6761.72
5	9.6803	713.5	7812.59	7812.35
6	10.0032	750.0	8168.30	8204.68
7	11.6178	765.6	8348.58	8375.29
8	11.9407	825.0	9048.83	9044.86
9	12.2636	884.5	9768.28	9743.54
10	12.5866	943.9	10506.92	10471.31
12	12.9095	1019.4	11469.56	11436.96
13	13.2324	1088.4	12369.40	12360.58
14	13.5553	1110.67	12668.51	12668.05
15	13.8782	1133.00	12974.84	12979.63
16	14.2012	1155.34	13288.37	13295.30
17	14.5241	1177.67	13609.12	13615.09
18	14.8470	1200.00	13937.07	13938.98

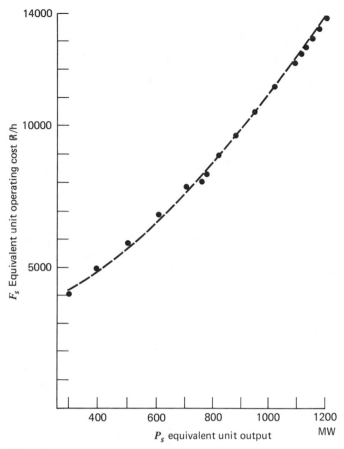

FIG. 6.6b Equivalent unit input/output curve.

in Table 6.1. The result, called F_s approx in Table 6.1 and shown in Figure 6.6b, was calculated by fitting a second-order polynomial to the P_s and F_s points using a least-squares fitting program. The equivalent unit function is

$$F_s \text{ approx}(P_s) = 2352.65 + 4.7151\,P_s + 0.0041168\,P_s^2$$
$$(\text{R/h}) \qquad\qquad 300 \text{ MW} \le P_s \le 1200 \text{ MW}$$

The reader should be aware that when fitting a polynomial to a set of points many choices can be made. The preceding function is a good fit to the total operating cost of the three units, but it is not that good at approximating the incremental cost. More advanced fitting methods should be used if one desires to match total operating cost as well as incremental cost. See Problem 6.2 for an alternative procedure.

EXAMPLE 6B

Find the optimal dispatch for a gas-fired steam plant given the following.

Gas-fired Plant:

$$H_T(P_T) = 300 + 6.0\,P_T + 0.0025\,P_T^2 \text{ MBtu/h}$$

Fuel cost for gas $= 2.0$ R/ccf (where 1 ccf $= 10^3$ ft^3)

The gas is rated at 1100 Btu/ft^3

$$50 \le P_T \le 400$$

Composite of Remaining Units:

$$H_s(P_s) = 200 + 8.5\,P_s + 0.002\,P_s^2 \text{ MBtu/h}$$

Equivalent fuel cost $= 0.6$ R/MBtu

$$50 \le P_s \le 500$$

The gas-fired plant must burn $40 \cdot 10^6$ ft^3 of gas. The load pattern is shown in Table 6.2. If the gas constraints are ignored, the optimum economic schedule for these two plants appears as is shown in Table 6.3. Operating cost of the composite unit over the entire 24-h period is 52,128.03 R. The total gas consumption is $21.8 \cdot 10^6$ ft^3. Since the gas-fired plant must burn $40 \cdot 10^6$ ft^3 of gas, the cost will be 2.0 R/1000 ft^3 $\times$ $40 \cdot 10^6$ ft^3, which is 80,000 R for the gas. Therefore, the total cost will be 132,128.03 R. The solution method shown in Figure 6.3 was used with γ values ranging from 0.500 to 0.875. The final value for γ is 0.8742 R/ccf with an optimal schedule as shown in Table 6.4. This schedule has a fuel cost for the composite unit of 34,937.47 R. Note that the gas unit is run much harder and that it

TABLE 6.2 Load Pattern

Time period	Load
1. 12 PM– 4 AM	400 MW
2. 4 AM– 8 AM	650 MW
3. 8 AM–12 AM	800 MW
4. 12 AM– 4 PM	500 MW
5. 4 PM– 8 PM	200 MW
6. 8 PM–12 PM	300 MW

Where: $n_j = 4, j = 1 \cdots 6$.

TABLE 6.3 Optimum Economic Schedule (Gas Constraints Ignored)

Time period	P_s	P_T
1	350	50
2	500	150
3	500	300
4	450	50
5	150	50
6	250	50

TABLE 6.4 Optimal Schedule (Gas Constraints Met)

Time period	P_s	P_T
1	197.3	202.6
2	353.2	296.8
3	446.7	353.3
4	259.7	240.3
5	72.6	127.4
6	135.0	165.0

does not hit either limit in the optimal schedule. Further note that the total cost is now

$$34{,}937.47\ R + 80{,}000\ R = 114{,}937.47\ R$$

so we have lowered the total fuel expense by properly scheduling the gas plant.

6.4 SOLUTION BY GRADIENT SEARCH TECHNIQUES

An alternative solution procedure to the one shown in Figure 6.3 makes use of Eqs. 6.5 and 6.6.

$$n_k \frac{dF_{ik}}{dP_{ik}} = \lambda_k$$

and

$$\lambda_k = \gamma n_k \frac{dq_{Tk}}{dP_{Tk}}$$

then

$$\gamma = \frac{\left(\dfrac{dF_{ik}}{dP_{ik}} \right)}{\dfrac{dq_{Tk}}{dP_{Tk}}} \tag{6.8}$$

For an optimum dispatch, γ will be constant for all hours j, $j = 1, \ldots, j_{\max}$.

We can make use of this fact to obtain an optimal schedule using the procedures shown in Figure 6.7a or Figure 6.7b. Both these procedures attempt to adjust fuel-limited generation so that γ will be constant over time. The algorithm shown in Figure 6.7a differs from the algorithm shown in Figure 6.7b in the way the problem is started and in the way various time intervals are selected for adjustment. The algorithm in Figure 6.7a requires an initial feasible but not optimal schedule and then finds an optimal schedule by "pairwise" trade-offs of fuel consumption while maintaining problem feasibility. The algorithm in Figure 6.7b does not require an

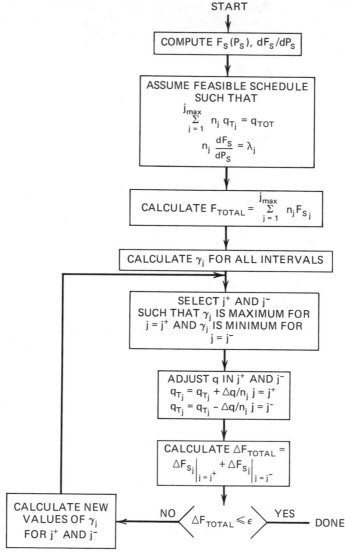

FIG. 6.7a Gradient method based on relaxation technique.

initial feasible fuel usage schedule but achieves this while optimizing. These two methods may be called gradient methods because q_{Tj} is treated as a vector and the γ_j's indicate the gradient of the objective function with respect to q_{Tj}.

EXAMPLE 6C

Use the method of Figure 6.7b to obtain an optimal schedule for the problem given in Example 6B. Assume that the starting schedule is the economic dispatch schedule shown in Example 6A.

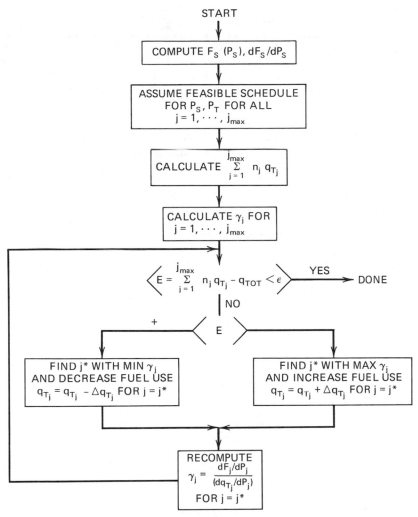

FIG. 6.7b Gradient method based on a simple search.

Initial Dispatch

Time period	1	2	3	4	5	6
P_S	350	500	500	450	150	250
P_T	50	150	300	50	50	50
γ	1.0454	1.0266	0.9240	1.0876	0.9610	1.0032

$\sum q_T = 21.84 \cdot 10^6 \text{ ft}^3$.

Since we wish to burn $40.0 \cdot 10^6 \text{ ft}^3$ of gas, the error is negative, therefore we must increase fuel usage in the time period having maximum γ, that is, period 4. As a start increase P_T to 150 MW and drop P_s to 350 MW in period 4.

Result of Step 1

Time period	1	2	3	4	5	6
P_s	350	500	500	350	150	250
P_T	50	150	300	150	50	50
γ	1.0454	1.0266	0.9240	0.9680	0.9610	1.0032

$\sum q_T = 24.2 \cdot 10^6 \text{ ft}^3$.

The error is still negative, so we must increase fuel usage in the period with maximum γ, which is now period 1. Increase P_T to 200 MW and drop P_s to 200 MW in period 1.

Result of Step 2

Time period	1	2	3	4	5	6
P_s	200	500	500	350	150	250
P_T	200	150	300	150	50	50
γ	0.8769	1.0266	0.9240	0.9680	0.9610	1.0032

$\sum q_T = 27.8 \cdot 10^6 \text{ ft}^3$.

and so on. After 11 steps, the schedule looks like:

Time period	1	2	3	4	5	6
P_s	200	350	450	250	75	140
P_T	200	300	350	250	125	160
γ	0.8769	0.8712	0.8772	0.8648	0.8767	0.8794

$\sum q_T = 40.002 \cdot 10^6 \text{ ft}^3$.

which is beginning to look similar to the optimal schedule generated in Example 6A.

6.5 HARD LIMITS AND SLACK VARIABLES

This section takes account of hard limits on the take-or-pay generating unit. The limits are

$$P_T \geq P_{T\min} \tag{6.9}$$

and

$$P_T \leq P_{T\max} \tag{6.10}$$

These may be added to the LaGrangian by the use of two constraint functions and two new variables called *slack variables* (see Appendix, Chapter 3). The constraint functions are

$$\psi_{1j} = P_{Tj} - P_{T\max} + S_{1j}^2 = 0 \tag{6.11}$$

and

$$\psi_{2j} = P_{Tmin} - P_{Tj} + S_{2j}^2 \quad = 0 \tag{6.12}$$

where S_{1j} and S_{2j} are slack variables that may take on any real value including zero. The new LaGrangian then becomes

$$\mathscr{L} = \sum_{j=1}^{jmax} n_j \sum_{i=1}^{N} F_{ij} + \sum_{j=1}^{jmax} \lambda_j \left(P_{Rj} - \sum_{i=1}^{N} P_{ij} - P_{Tj} \right)$$

$$+ \gamma \left(\sum_{j=1}^{jmax} n_j q_{Tj} - Q_{TOT} \right)$$

$$+ \sum_{j=1}^{jmax} \alpha_{1j}(P_{Tj} - P_{Tmax} + S_{1j}^2) + \sum_{j=1}^{jmax} \alpha_{2j}(P_{Tmin} - P_{Tj} + S_{2j}^2) \tag{6.13}$$

where α_{1j}, α_{2j} are LaGrange multipliers. Now, the first partial derivatives for the kth period are

$$\frac{\partial \mathscr{L}}{\partial P_{1k}} = 0 = n_j \frac{dF_i}{dP_{ik}} - \lambda_k$$

$$\frac{\partial \mathscr{L}}{\partial P_{Tk}} = 0 = -\lambda_k + \alpha_{1k} - \alpha_{2k} + \gamma n_j \frac{dq_{Tk}}{dP_{Tk}}$$

$$\frac{\partial \mathscr{L}}{\partial S_{1k}} = 0 = 2\alpha_{1k}S_{1k} \tag{6.14}$$

$$\frac{\partial \mathscr{L}}{\partial S_{2k}} = 0 = 2\alpha_{2k}S_{2k}$$

As was noted in the Appendix to Chapter 3, when the constrained variable (P_{Tk} in this case) is within bounds, the new LaGrange multipliers $\alpha_{1k} = \alpha_{2k} = 0$ and S_{1k} and S_{2k} are nonzero. When the variable is limited, one of the slack variables, S_{1k} or S_{2k}, becomes zero and the associated LaGrange multiplier will take on a non-zero value.

Suppose in some interval k, $P_{Tk} = P_{max}$, then $S_{1k} = 0$ and $\alpha_{1k} \neq 0$. Thus

$$-\lambda_k + \alpha_{1k} + \gamma n_j \frac{dq_{Tk}}{dP_{Tk}} = 0 \tag{6.15}$$

and if

$$\lambda_k > \gamma n_j \frac{dq_{Tk}}{dP_{Tk}}$$

the value of α_{2k} will take on the value just sufficient to make the equality true.

EXAMPLE 6D

Repeat Example 6B with the maximum generation on P_T reduced to 300 MW. Note that the optimum schedule in Example 6A gave a $P_T = 353.3$ MW in the third time

TABLE 6.5 Resulting Optimal Schedule with $P_{T\max} = 300$ MW

Time period j	P_{sj}	P_{Tj}	λ_j	$\gamma_{nj} \dfrac{\partial q_T}{\partial P_{Tj}}$	α_{2j}
1	183.4	216.6	5.54	5.54	0
2	350.0	300.0	5.94	5.86	0.08
3	500.0	300.0	25.00	5.86	19.14
4	345.4	254.6	5.69	5.69	0
5	59.5	140.5	5.24	5.24	0
6	121.4	178.6	5.39	5.39	0

period. When the limit is reduced to 300 MW, the gas-fired unit will have to burn more fuel in other time periods to meet the $40 \cdot 10^3$ ft^3 gas consumption constraint.

Table 6.5 shows the resulting optimal schedule where $\gamma = 0.2151$ and total cost = 122,984.83 ₽.

6.6 FUEL SCHEDULING BY LINEAR PROGRAMMING

Figure 6.8 shows the major elements in the chain making up the delivery system that starts with raw fuel suppliers and ends up in delivery of electric power to individual customers. The basic elements of the chain are

The Suppliers: Coal, oil, and gas companies with which the utility must negotiate contracts to acquire fuel. The contracts are usually written for a long term (10 to 20 yr) and may have stipulations such as the minimum and maximum limits on the quantity of fuel delivered over a specified time period. The time period may be as long as a year, a month, a week, a day, or even for a period of only a few minutes. Prices may change as subject to the renegotiation provisions of the contracts.

Transportation: Railroads, unit trains, river barges, gas pipeline companies, and such all present problems in scheduling of deliveries of fuel.

Inventory: Coal piles, oil storage tanks, underground gas storage facilities. Inventories must be kept at proper levels to forestall fuel shortages when load levels exceed forecast or suppliers or shippers are unable to deliver. Price fluctuations also complicate the decisions on when and how much to add or subtract from inventories.

The remainder of the system—generators, transmission, and loads—are covered in other chapters.

One of the most useful tools for solving large fuel-scheduling problems is linear programming (LP). If the reader is not familiar with LP, an easily understood algorithm is provided in the Appendix of this chapter.

Linear programming is an optimization procedure that minimizes a linear objective function with variables that are also subject to linear constraints. Because

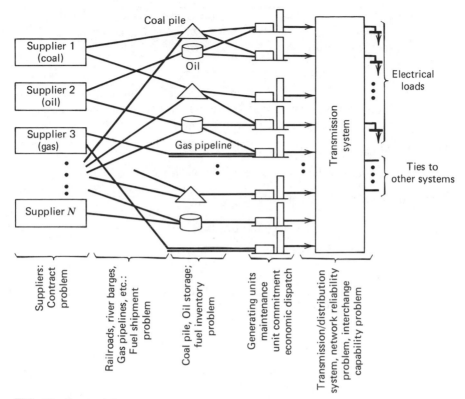

FIG. 6.8 Energy delivery system.

of this limitation, any nonlinear functions either in the objective or in the constraint equations will have to be approximated by linear or piecewise linear functions.

To solve a fuel-scheduling problem with linear programming, we must break the total time period involved into discrete time increments as was done in Example 6B. The LP solution will then consist of an objective function that is made up of a sum of linear or piecewise linear functions each of which is a function of one or more variables from only one time step. The constraints will be linear functions of variables from each time step. Some constraints will be made up of variables drawn from one time step whereas others will span two or more time steps. The best way to illustrate how to set up an LP to solve a fuel-scheduling problem will be to use an example.

EXAMPLE 6E

We are given two coal-burning generating units that must both remain on-line for a 3-wk period. The combined output from the two units is to supply the following loads (loads are assumed constant for 1 wk).

Week	Load (MW)
1	1200
2	1500
3	800

The two units are to be supplied by one coal supplier who is under contract to supply 40,000 tons of coal per week to the two plants. The plants have existing coal inventories at the start of the 3-wk period. We must solve for the following.

1. How should each plant be operated each week?
2. How should the coal deliveries be made up each week?

The data for the problem are as follows.

Coal: Heat value = 11,500 Btu/lb = 23 MBtu/ton

(1 ton = 2000 lb)

Coal can all be delivered to one plant or the other or it can be split, some going to one plant, some to the other, as long as the total delivery in each week is equal to 40,000 tons. The coal costs 30 R/ton or 1.3 R/MBtu

Inventories: Plant 1:   has an initial inventory of 70,000 tons, its final inventory is not restricted

Plant 2:   has an initial inventory of 70,000 tons, its final inventory is not restricted

Both plants have a maximum coal storage capacity of 200,000 tons of coal.

Generating Units:

Unit	Min (MW)	Max (MW)	Heat input at min (MBtu/h)	Heat input at max (MBtu/h)
1	150	600	1620	5340
2	400	1000	3850	8750

The input versus output function will be approximated by a linear function for each unit.

$$H_1(P_1) = 380.0 + 8.267 P_1$$
$$H_2(P_2) = 583.3 + 8.167 P_2$$

The unit cost curves are

$$F_1(P_1) = 1.3 \ \text{R/MBtu} \times H_1(P_1) = 495.65 + 10.78 P_1 \ (\text{R/h})$$
$$F_2(P_2) = 1.3 \ \text{R/MBtu} \times H_2(P_2) = 760.8 + 10.65 P_2 \ (\text{R/h})$$

The coal consumption q(tons/h) for each unit is

$$q_1(P_1) = \frac{1}{23} \left(\frac{\text{Tons}}{\text{MBtu}} \right) \times H_1(P_1) = 16.52 + 0.3594\, P_1 \text{ tons/h}$$

$$q_2(P_2) = \frac{1}{23} \left(\frac{\text{Tons}}{\text{MBtu}} \right) \times H_2(P_2) = 25.36 + 0.3551\, P_2 \text{ tons/h}$$

To solve this problem with linear programming, assume that the units are to be operated at a constant rate during each week and that the coal deliveries will each take place at the beginning of each week. Therefore, we will set up the problem with 1-wk time periods and the generating unit cost functions and coal consumption functions will be multipled by 168 h to put them on a "per week" basis, then

$$F_1(P_1) = 83{,}269.2 + 1811\, P_1 \text{ R/wk}$$
$$F_2(P_2) = 127{,}814.4 + 1789\, P_2 \text{ R/wk}$$
$$q_1(P_1) = 2775.4 + 60.4\, P_1 \text{ tons/wk}$$
$$q_2(P_2) = 4260.5 + 59.7\, P_2 \text{ tons/wk}$$

$$(6.16)$$

We are now ready to set up the objective function and the constraints for our linear programming solution.

Objective Function: To minimize the operating cost over the 3-wk period. The objective function is

$$\text{Minimize } Z = F_1[P_1(1)] + F_2[P_2(1)] + F_1[(P_1(2)] + F_2[P_2(2)]$$
$$+ F_1[P_1(3)] + F_2[P_2(3)] \tag{6.17}$$

where $P_i(j)$ is the power output of the ith unit during the jth week, $j = 1 \cdots 3$.

Constraints: During each time period, the total power delivered from the units must equal the scheduled load to be supplied, then

$$P_1(1) + P_2(1) = 1200$$
$$P_1(2) + P_2(2) = 1500 \tag{6.18}$$
$$P_1(3) + P_2(3) = 800$$

Similarly, the coal deliveries, D_1 and D_2, made to plant 1 and plant 2, respectively, during each week must sum to 40,000 tons. Then

$$D_1(1) + D_2(1) = 40{,}000$$
$$D_1(2) + D_2(2) = 40{,}000 \tag{6.19}$$
$$D_1(3) + D_2(3) = 40{,}000$$

The volume of coal at each plant at the beginning of each week plus the delivery of coal to that plant minus the coal burned at the plant will give the coal remaining at the beginning of the next week. Letting V_1 and V_2 be the volume of coal in each coal pile at the beginning of the week, respectively, we have the following set of equations governing the two coal piles.

$$V_1(1) + D_1(1) - q_1(1) = V_1(2)$$
$$V_2(1) + D_2(1) - q_2(1) = V_2(2)$$
$$V_1(2) + D_1(2) - q_1(2) = V_1(3)$$
$$V_2(2) + D_2(2) - q_2(2) = V_2(3) \tag{6.20}$$
$$V_1(3) + D_1(3) - q_1(3) = V_1(4)$$
$$V_2(3) + D_2(3) - q_2(3) = V_2(4)$$

where $V_i(j)$ is the volume of coal in the ith coal pile at the beginning of the jth week.

To set these equations up for the linear-programming solution, substitute the $q_1(P_1)$ and $q_2(P_2)$ equations from 6.16 into the equations of 6.20. In addition, all constant terms are placed on the right of the equal sign and all variable terms on the left, this leaves the constraints in the standard form for inclusion in the LP. The result is

$$D_1(1) - 60.4\,P_1(1) - V_1(2) = 2775.4 - V_1(1)$$
$$D_2(1) - 59.7\,P_2(1) - V_2(2) = 4260.5 - V_2(1)$$
$$V_1(2) + D_1(2) - 60.4\,P_1(2) - V_1(3) = 2775.4$$
$$V_2(2) + D_2(2) - 59.7\,P_2(2) - V_2(3) = 4260.5 \tag{6.21}$$
$$V_1(3) + D_1(3) - 60.4\,P_2(3) - V_1(4) = 2775.4$$
$$V_2(3) + D_2(3) - 59.7\,P_2(3) - V_2(4) = 4260.5$$

NOTE: $V_1(1)$ and $V_2(1)$ are constants that will be set when we start the problem.

The constraints from Eqs. 6.18, 6.19, and 6.21 are arranged in a matrix, as shown in Figure 6.9. Each variable is given an upper and lower bound in keeping with the "upper bound" solution shown in the Appendix of this chapter. The $P_1(t)$ and $P_2(t)$ variables are given the upper and lower bounds corresponding to the upper and lower limits on the generating units. $D_1(t)$ and $D_2(t)$ are given upper and lower bounds of 40,000 and zero. $V_1(t)$ and $V_2(t)$ are given upper and lower bounds of 200,000 and zero.

Solution: The solution to this problem was carried out with a computer program written to solve the upper bound LP problem using the algorithm shown in the Appendix. The first problem solved had coal storage at the beginning of the first

FIG. 6.9 Linear-programming constraint matrix for Example 6E.

Problem Variable →	D1(1)	P1(1)	D2(1)	P2(1)	V1(2)	D1(2)	P1(2)	V2(2)	D2(2)	P2(2)	V1(3)	D1(3)	P1(3)	V2(3)	D2(3)	P2(3)	V1(4)	V2(4)	Constraint Units
LP Variable →	X_1	X_2	X_3	X_4	X_5	X_6	X_7	X_8	X_9	X_{10}	X_{11}	X_{12}	X_{13}	X_{14}	X_{15}	X_{16}	X_{17}	X_{18}	
(Week 1 / Week 2 / Week 3 / Final Conditions)																			
Constraint 1		1		1															1200
2	1		1																40000
3	1	−60.4			−1														$2775.4 - V_1(1)$
4			1	−59.7				−1											$4260.5 - V_2(1)$
5							1			1									1600
6						1			1										40000
7					1	1	−60.4				−1								2775.4
8								1	1	−59.7				−1					4260.5
9													1			1			1600
10												1			1				40000
11											1	1	−60.4				−1		2775.4
12														1	1	−59.7		−1	4260.5
Variable min.	0	150	0	400	0	0	150	0	0	400	0	0	150	0	0	400	0	0	
Variable max.	40000	600	40000	1000	200000	40000	600	200000	40000	1000	200000	40000	600	200000	40000	1000	200000	200000	

Week 1: X_1–X_4 Week 2: X_5–X_{10} Week 3: X_{11}–X_{16} Final Conditions: X_{17}–X_{18}

week of

$$V_1(1) = 70,000 \text{ tons}$$

$$V_2(1) = 70,000 \text{ tons}$$

The solution is

t	P_1	P_2	D_1	D_2	V_1	V_2
1	200	1000	0	40000	70000	70000
2	500	1000	0	40000	55144.6	46039.5
3	150	650	0	40000	22169.2	22079.0
4					10333.8	19013.5

Total operating cost = 6,913,449.0 R

In this case, there are no constraints on the coal deliveries to either plant and the system can run in the most economic manner. Since unit 2 has a lower incremental cost, it is run at its maximum when possible. Furthermore, since no restrictions were placed on the coal pile levels at the end of the third week, the coal deliveries could have been shifted a little from unit 2 to unit 1 with no effect on the generation dispatch.

The next case solved was purposely structured to create a fuel shortage at unit 2. The beginning inventory at plant 2 was set to 50,000 tons, and a requirement was imposed that at the end of the third week the coal pile at unit 2 be no less than 8000 tons. The solution was made by changing the right-hand side of the fourth constraint from -65739.5 (i.e., $4260.5 - 70,000$) to -45739.5 (i.e., $4260.5 - 50,000$) and placing a lower bound on $V_2(4)$ (i.e., variable X_{18}) of 8000. The solution is

t	P_1	P_2	D_1	D_2	V_1	V_2
1	200	1000	0	40000	70000	50000
2	600	900	0	40000	55144.6	26039.5
3	200.5	599.5	0	40000	16129.2	8049.0
4					1241.93	8000.0

Total operating cost = 6,916,760.00 R

Note that this solution requires unit 2 to drop off its generation in order to meet the end-point constraint on its coal pile. In this case, all the coal must be delivered to plant 2 to minimize the overall cost.

The final case was constructed to show the interaction of the fuel deliveries and the economic dispatch of the generating units. In this case, the initial coal piles were set to 10,000 tons and 150,000 tons, respectively. Furthermore, a restriction of 30,000 tons minimum in the coal pile at unit 1 at the end of the third week was imposed.

To obtain the most economic operation of the two units over the 3-wk period, the coal deliveries will have to be adjusted to ensure both plants have sufficient coal. The solution was obtained by setting the right-hand side of the third and fourth constraint equations to -7224.6 and -145739.5, respectively, as well as imposing a lower bound of 30,000 on $V_1(4)$ (i.e., variable X_{17}).

The solution is

t	P_1	P_2	D_1	D_2	V_1	V_2
1	200	1000	4855.4	35144.6	10000	150000
2	500	1000	34810.8	5189.2	0	121184.1
3	150	650	40000	0	1835.4	62412.8
4					30000	19347.3

Total operating cost = 6,913,449.00 R

The LP was able to find a solution that allowed the most economic operation of the units while still directing enough coal to unit 1 to allow it to meet its end-point coal pile constraint. Note that in practice we would probably not wish to let the coal pile at unit 1 go to zero. This could be prevented by placing an appropriate lower bound on all the volume variables (i.e., X_5, X_8, X_{11}, X_{14}, X_{17}, and X_{18}).

This example has shown how a fuel-management problem can be solved with linear programming. The important factor in being able to solve very large fuel-scheduling problems is to have a linear-programming code capable of solving large problems having perhaps tens of thousands of constraints and as many or more problem variables. Using such codes, elaborate fuel-scheduling problems can be optimized out over several years and play a critical role in utility fuel-management decisions.

APPENDIX
Linear Programming

Linear programming is perhaps the most widely applied mathematical programming technique. Simply stated, linear programming seeks to find the optimum value of a linear objective function while meeting a set of linear constraints. That is, we wish to find the optimum set of x's that minimize the following objective function.

$$Z = c_1 x_1 + c_2 x_2 + \cdots + c_N x_N$$

Subject to a set of linear constraints:

$$a_{11} x_1 + a_{12} x_2 + \cdots + a_{1N} x_N \leq b_1$$
$$a_{21} x_1 + a_{22} x_2 + \cdots + a_{2N} x_N \leq b_2$$
$$\vdots$$

In addition, the variables themselves may have specified upper and lower limits.

$$x_i^{min} \leq x_i \leq x_i^{max} \qquad i = 1 \cdots N$$

There are a variety of solutions to the LP problem. Many of these solutions are tailored to a particular type of problem. This appendix will not try to develop the theory of alternate LP solution methods. Rather, it will present a simple LP algorithm that can be used (or programmed on a computer) to solve the applicable power-system sample problems given in this text.

The algorithm is presented in its simplest form. There are alternative formulations, and these will be indicated when appropriate. If the student has access to a standard LP program, such a standard program may be used to solve any of the problems in this book.

The LP technique presented here is properly called an *upper-bounding dual linear programming algorithm*. The "upper-bounding" part of its name refers to the fact that variable limits are handled implicitly in the algorithm. Dual refers to the theory behind the way in which the algorithm operates. For a complete explanation of the primal and dual algorithms, refer to the references cited at the end of this chapter.

In order to proceed in an orderly fashion to solve a dual upper-bound linear programming problem, we must first add what is called a *slack variable* to each constraint. The slack variable is so named because it equals the difference or slack between a constraint and its limit. By placing a slack variable into an inequality constraint, we can transform it into an equality constraint. For example, suppose we are given the following constraint.

$$2x_1 + 3x_2 \leq 15 \tag{6A.1}$$

We can transform this constraint to an equality constraint by adding a slack variable, x_3.

$$2x_1 + 3x_2 + x_3 = 15 \tag{6A.2}$$

If x_1 and x_2 were to be given values such that the sum of the first two terms in Eq. 6A.2 added to less than 15, we could still satisfy Eq. 6A.2 by setting x_3 to the difference. For example, if $x_1 = 1$, $x_2 = 3$, then $x_3 = 4$ would satisfy Eq. 6A.2. We can go even further, however, and restrict the values of x_3 so that Eq. 6A.2 still acts as an inequality constraint such as Eq. 6A.1. Note that when the first two terms of Eq. 6A.2 add to exactly 15, x_3 must be set to zero. By restricting x_3 to always be a positive number, we can force Eq. 6A.2 to yield the same effect as Eq. 6A.1. Thus:

$$\left. \begin{array}{c} 2x_1 + 3x_2 + x_3 = 15 \\ 0 \leq x_3 \leq \infty \end{array} \right\} \text{ is equivalent to: } 2x_1 + 3x_2 \leq 15$$

Note that we could change a "less than or equal to" constraint to a "greater than or equal to" constraint by changing the sign on the slack variable. Then

$$\left. \begin{array}{c} 2x_1 + 3x_2 - x_3 = 15 \\ 0 \leq x_3 \leq \infty \end{array} \right\} \text{ is equivalent to: } 2x_1 + 3x_2 \geq 15$$

Because of the way the dual upper-bounding algorithm is initialized, we will always require slack variables in every constraint. In the case of an equality constraint, we will add a slack variable and then require its upper and lower bounds to both equal zero.

To solve our linear programming algorithm, we must arrange the objective function and constraints in a tabular form as follows.

$$a_{11}x_1 + a_{12}x_2 + \cdots + x_{\text{slack}_1} \qquad\qquad = b_1$$

$$a_{21}x_1 + a_{22}x_2 + \cdots \qquad\quad + x_{\text{slack}_2} \quad = b_2 \qquad\qquad \text{(6A-3)}$$

$$c_1x_1 + c_2x_2 + \qquad\qquad\qquad\qquad -Z = 0$$

$$\underbrace{\qquad\qquad\qquad\qquad}_{\text{Basis variables}}$$

Because we have added slack variables to each constraint, we automatically have arranged the set of equations into what is called *canonical* form. In canonical form, there is at least one variable in each constraint whose coefficient is zero in all the other constraints. These variables are called the *basis variables*. The entire solution procedure for the linear programming algorithm centers on performing "pivot" operations that can exchange a nonbasis variable for a basis variable. A pivot operation may be shown by using our tableau in Eq. 6A.3. Suppose we wished to exchange variable x_1, a nonbasis variable for x_{slack_2}, a slack variable. This could be accomplished by "pivoting" on column 1 row 2. To carry out the pivoting operation we execute the following steps.

Pivoting on Column 1 Row 2

Step 1 Multiply row 2 by $1/a_{21}$
That is, each $a_{2j}, j = 1 \cdots N$ in row 2 becomes

$$a'_{2j} = \frac{a_{2j}}{a_{21}} \qquad j = 1 \cdots N$$

and

$$b_2 \text{ becomes } b'_2 = \frac{b_2}{a_{21}}.$$

Step 2 For each row i ($i \neq 2$), multiply row 2 by a_{i1} and subtract from row i. That is, each coefficient a_{ij} in row i ($i \neq 2$) becomes

$$a'_{ij} = a_{ij} - a_{i1}a'_{2j} \qquad j = 1 \cdots N$$

and

$$b_i \text{ becomes } b'_i = b_i - a_{i1}b'_2$$

Step 3 Last of all, we also perform the same operations in step 2 on the cost row. That is, each coefficient c_j becomes

$$c'_j = c_j - c_1a'_{2j} \qquad j = 1 \cdots N$$

The result of carrying out the pivot operation will look like

$$a'_{12}x_2 + \cdots x_{\text{slack}_1} + a'_{1s_2}x_{\text{slack}_2} = b'_1$$

$$x_1 + a'_{22}x_2 + \cdots + a'_{2s_2}x_{\text{slack}_2} = b'_2$$

$$c'_2x_2 + \cdots + c'_{s2}x_{\text{slack}_2} - Z = Z'$$

Notice that the new basis for our tableau is formed by variable x_1 and x_{slack_1}. x_{slack_2} no longer has nonzero coefficients in row 1 or the cost row.

The dual upper-bounding algorithm proceeds in simple steps wherein variables that are in the basis are exchanged for variables out of the basis. When an exchange is made, a pivot operation is carried out at the appropriate row and column. The nonbasis variables are held equal to either their upper or their lower value while the basis variables are allowed to take any value without respect to their upper or lower bounds. The solution terminates when all the basis variables are within their respective limits.

In order to use the dual upper-bound LP algorithm, follow these rules.

Start:

1. Each variable that has a nonzero coefficient in the cost row (i.e., the objective function) must be set according to the following rule.

$$\text{If } C_j > 0, \quad \text{set } x_j = x_j^{\text{min}}$$

$$\text{If } C_j < 0, \quad \text{set } x_j = x_j^{\text{max}}$$

2. If $C_j = 0$, x_j may be set to any value, but for convenience set it to its minimum also.

3. Add a slack variable to each constraint. Using the x_j values from steps 1 and 2, set the slack variables to make each constraint equal to its limit.

Variable exchange:

1. Find the basis variable with the greatest violation, this determines the row to be pivot on. Call this row R. If there are no limit violations among the basis variables, we are done. The most violated variable leaves the basis and is set equal to the limit that was violated.

2. Select the variable to enter the basis using one of the following column selection procedures.

Column Selection Procedure P1 (Most violated variable below its minimum)

Given constraint row R, whose basis variable is below its minimum and is the worst violation. Pick column S, so that, $c_S/(-a_{R,S})$ is minimum for all S that meet the following rules.

a. S is not in the current basis.

b. $a_{R,S}$ is not equal to zero.

c. If x_S is at its minimum, then $a_{R,S}$ must be negative and c_S must be positive or zero.

d. If x_S is at its maximum, then $a_{R,S}$ must be positive and c_S must be negative or zero.

Column Selection Procedure P2 (Most violated variable above its maximum)

Given constraint row R, whose basis variable is above its maximum and is the worst violation. Pick column S, so that, $c_S/a_{R,S}$ is the minimum for all S that meet the following rules:

a. S is not in the current basis.

b. $a_{R,S}$ is not already zero.

c. If x_S is at its minimum, then $a_{R,S}$ must be positive and c_S must be positive or zero.

d. If x_S is at its maximum, then $a_{R,S}$ must be negative and c_S must be negative or zero.

3. When a column has been selected, pivot at the selected row R (from step 1) and column S (from step 2). The pivot column's variable, S goes into the basis.

If no column fits the column selection criteria, we have an infeasible solution. That is, there are no values for $x_1 \cdots x_N$ that will satisfy all constraints simultaneously. In some problems, the cost coefficient c_S associated with column S will be zero for several different values of S. In such a case, $c_S/a_{R,S}$ will be zero for each such S and none of them will be the minimum. The fact that c_S is zero means that there will be no increase in cost if any of the S's are pivoted into the basis; therefore the algorithm is indifferent to which one is chosen.

Setting the Variables After Pivoting

1. All nonbasis variables, except x_S remain as they were before pivoting.

2. The most violated variable is set to the limit that was violated.

3. Since all nonbasis variables are determined, we can proceed to set each basis variable to whatever value is required to make the constraints balance. Note that this last step may move all the basis variables to new values, and some may now end up violating their respective limits (including the x_S variable).

Go back to step 1 of the variable exchange procedure.

These steps are shown in flowchart form in Figure 6.10.

To help you understand the procedures involved, a sample problem is solved using the dual upper-bounding algorithm. The sample problem, shown in Figure 6.11,

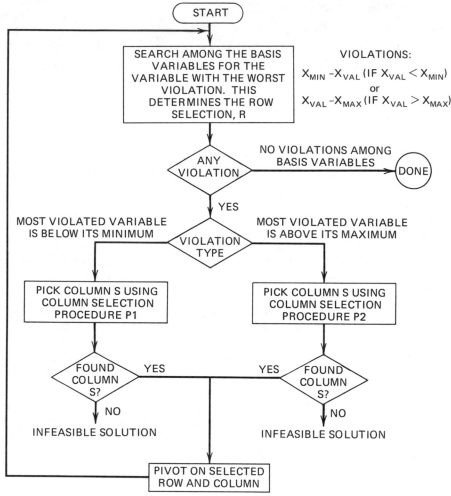

FIG. 6.10 Dual upper-bound linear programming algorithm.

consists of a two-variable objective with one equality constraint and one inequality constraint.

First, we must put the equations into canonical form by adding slack variables x_3 and x_4. These variables are given limits corresponding to the type of constraint into which they are placed. x_3 is the slack variable in the equality constraint, so its limits are both zero; x_4 is in an inequality constraint so it is restricted to be a positive number. To start the problem, the objective function must be set to the minimum value it can attain, and the algorithm will then seek the minimum constrained solution by increasing the objective just enough to reach the constrained solution. Thus we set x_1 and x_2 both at their minimum values since the cost coefficients are

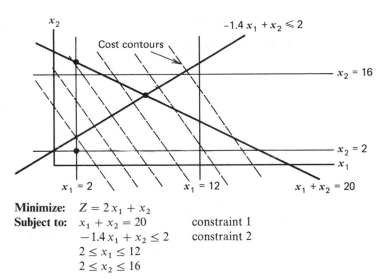

Minimize: $Z = 2x_1 + x_2$

Subject to: $x_1 + x_2 = 20$ constraint 1

$-1.4x_1 + x_2 \leq 2$ constraint 2

$2 \leq x_1 \leq 12$

$2 \leq x_2 \leq 16$

FIG. 6.11 Sample LP problem.

both positive. These conditions are shown here

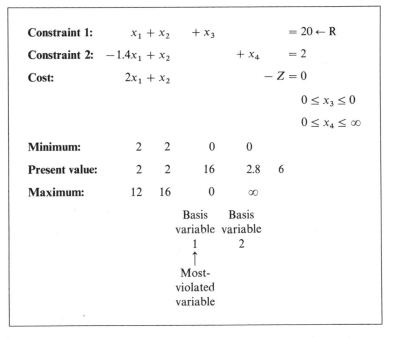

We can see from these conditions that variable x_3 is the worst-violated variable and that it presently exceeds its maximum limit of zero. Thus we must use column procedure P2 on constraint number 1. This is summarized as follows.

Using selection procedure P2 on constraint 1:

$$i = 1 \quad a_1 > 0 \quad x_1 = x_1^{min} \quad c_1 > 0 \quad \text{then } \frac{c_1}{a_1} = \frac{2}{1} = 2$$

$$i = 2 \quad a_2 > 0 \quad x_2 = x_2^{min} \quad c_1 > 0 \quad \text{then } \frac{c_2}{a_2} = \frac{1}{1} = 1$$

$$\min c_i/a_i \text{ is 1 at } i = 2$$

Pivot at column 2 row 1

To carry out the required pivot operations on column 2 row 1, we need merely subtract the first constraint from the second constraint and from the objective function. This results in:

Constraint 1:	x_1	$+ x_2 + x_3$			$= 20$
Constraint 2:	$-2.4 x_1$	$- x_3$	$+ x_4$		$= -18 \leftarrow R$
Cost:	x_1	$- x_3$		$- Z = -20$	
Minimum:	2	2 0	0		
Present value:	2	18 0	-13.2 22		
Maximum:	12	16 0	∞		

Basis   Basis
variable   variable
1   2
↑
Most-
violated
variable

We can see now that the variable with the worst violation is x_4 and that x_4 is below its minimum. Thus we must use selection procedure P1 as follows.

Using selection procedure P1 on constraint 2

$$i = 1 \quad a_1 < 0 \quad x_1 = x_1^{min} \quad c_1 > 0 \quad \text{then } \frac{c_1}{-a_1} = \frac{1}{-(-2.4)} = 0.4166$$

$$i = 3 \quad a_3 < 0 \quad x_3 = x_3^{min} = x_3^{max} \quad c_3 < 0 \quad \text{then } x_3 \text{ is not eligible}$$

Pivot at column 1 row 2

Which after pivoting results in

Constraint 1:		$x_2 + 0.5833\, x_3 + 0.4166\, x_4$	$= \quad 12.5$	
Constraint 2:	x_1	$+ 0.4166\, x_3 - 0.4166\, x_4$	$= \quad 7.5$	
Cost:		$- 1.4166\, x_3 + 0.4166\, x_4$	$- Z = -27.5$	
Minimum:	2	2	0	0
Present value:	7.5	12.5	0	$0 \ -27.5$
Maximum:	12	16	0	∞
	Basis	Basis		
	variable	variable		
	1	2		

At this point we have no violations among the basis variables, so the algorithm can stop at the optimum.

$$\left.\begin{array}{l} x_1 = 7.5 \\ x_2 = 12.5 \end{array}\right\} \ \text{cost} = 27.5$$

See Figure 6.11 to verify that this is the optimum. The dots in Figure 6.11 show the solution points beginning at the starting point $x_1 = 2$, $x_2 = 2$, cost $= 6.0$ then going to $x_1 = 2$, $x_2 = 18$, cost $= 22.0$ and finally to the optimum $x_1 = 7.5$, $x_2 = 12.5$, cost $= 27.5$

How does this algorithm work? At each step two decisions are made.

1. Select the most-violated variable.
2. Select a variable to enter the basis.

The first decision will allow the procedure to eliminate, one after the other, those constraint violations that exist at the start as well as those that happen during the variable-exchange steps. The second decision (using the column selection procedures) guarantees that the rate of increase in cost to move the violated variable to its limit is minimized. Thus the algorithm starts from a minimum cost, infeasible solution (constraints violated), toward a minimum cost, feasible solution, by minimizing the rate of cost increase at each step.

PROBLEMS

6.1 Three units are on-line all 720 h of a 30-day month. Their characteristics are as follows.

$$H_1 = 225 + 8.47\,P_1 + 0.0025\,P_1^2, \quad 50 \le P_1 \le 350$$
$$H_2 = 729 + 6.20\,P_2 + 0.0081\,P_2^2, \quad 50 \le P_2 \le 350$$
$$H_3 = 400 + 7.50\,P_3 + 0.0025\,P_3^2, \quad 50 \le P_3 \le 450$$

In these equations the H_i are in MBtu/h and the P_i are in MW.

Fuel costs for units 2 and 3 are 0.60ℝ/MBtu. Unit 1, however, is operated under a take-or-pay fuel contract where 60,000 tons of coal are to be burned and/or paid for each 30-day period. This coal costs 12ℝ/ton delivered and has an average heat content of 12,500 Btu/lb (1 ton = 2000 lb).

The system monthly load-duration curve may be approximated by three steps as follows.

Load (MW)	Hours duration	Energy (MWh)
800	50	40,000
500	550	275,000
300	120	36,000
Total	720 h	351,000 Mwh

a. Compute the economic schedule for the month assuming all three units are on-line all the time and that the coal must be consumed. Show the MW loading for each load period, the MWh of each unit, and the value of gamma (the psuedo-fuel cost).

b. What would be the schedule if unit 1 was burning the coal at 12 ℝ/ton with no constraint to use 60,000 tons? Assume the coal may be purchased on the spot market for that price and compute all the data asked for in (a). In addition, calculate the amount of coal required for the unit.

6.2 Refer to Example 6A, where three generating units are combined into a single composite generating unit. Repeat the example, except develop an equivalent incremental cost characteristic using only the incremental characteristics of the three units. Using this composite incremental characteristic plus the zero-load intercept costs of the three units, develop the total cost characteristic of the composite. (Suggestion: Fit the composite incremental cost data points using a linear approximation and a least-squares fitting algorithm.)

6.3 Refer to Problem 3.8, where three generator units have input/output curves specified as a series of straight-line segments. Can you develop a composite input/output curve for the three units? Assume all three units are on-line and that the composite input/output curve has as many linear segments as needed.

6.4 Refer to Example 6E. The first problem solved in Example 6E left the end-point restrictions at zero to 200,000 tons for both coal piles at the end of the 3-wk period. Resolve the first problem ($V_1(1) = 70,000$ and $V_2(1) = 70,000$) with the added restriction that the final volume of coal at plant 2 at the end of the third week be at least 20,000 tons.

6.5 Refer to Example 6E. In the second case solved with the LP algorithm (starting volumes equal to 70,000 and 50,000 for plant 1 and plant 2, respectively), we restricted the final volume of the coal pile at plant 2 to be 8000 tons. What is the optimum schedule if this final volume restriction is relaxed (i.e., the final coal pile at plant 2 could go to zero)?

6.6 An oil-fired power plant (Figure 6.12) has the following fuel consumption curve.

$$q(\text{bbl/h}) = \begin{cases} 50 + P + 0.005\,P^2 & \text{for } 100 \le P \le 500 \text{ MW} \\ 0 & \text{for } P = 0 \end{cases}$$

The plant is connected to an oil storage tank with a maximum capacity of 4000 bbl. The tank has an initial volume of oil of 3000 bbl. In addition, there is a pipeline supplying oil to the plant. The pipeline terminates in the same storage tank and must be operated by contract at 500 bbl/h. The oil-fired power plant supplies energy into a system along with other units. The other units have an equivalent cost curve of

$$F_{eq} = 300 + 6\,P_{eq} + 0.0025\,P_{eq}^2$$

$$50 \le P_{eq} \le 700 \text{ MW}$$

The load to be supplied is given as follows.

Period	Load (MW)
1	400
2	900
3	700

Each time period is 2 h in length. Find the oil-fired plant's schedule using dynamic programming such that the operating cost on the equivalent plant is minimized and the final volume in the storage tank is 2000 bbl at the end of the third period. When solving, you may use 2000, 3000, and 4000 bbl as the storage volume states for the tank. The q versus P function values you will need are included in the following table.

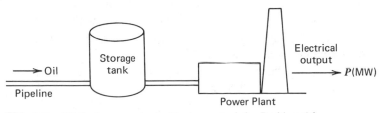

FIG. 6.12 Oil-fired power plant with storage tank for Problem 6.6.

q(bbl/h)	P(MW)
0	0
200	100.0
250	123.6
500	216.2
750	287.3
1000	347.2
1250	400.0
1500	447.7
1800	500.0

The plant may be shut down for any of the 2-h periods with no start-up or shut-down costs.

FURTHER READING

There has not been a great deal of research work on fuel scheduling as specifically applied to power systems. However, the fuel-scheduling problem for power systems is not really that much different from other "scheduling" problems, and for this type of problem, a great deal of literature exists.

References 1–4 are representative of current efforts in applying scheduling techniques to the power system fuel-scheduling problem. References 5–8 are textbooks on linear programming that the authors have used. There are many more texts that cover LP and its variations. The reader is encouraged to study LP independently of this text if a great deal of use is to be made of LP. Many computing equipment and independent software companies have excellent LP codes that can be used rather than writing one's own code. Reference 8 is the basis for the algorithm in the appendix to this chapter.

1. Trefny, F. J., Lee, K. Y., "Economic Fuel Dispatch," *IEEE Transactions on Power Apparatus and Systems* Vol. 100, July 1981, 3468–3477.

2. Seymore, G. F., "Fuel Scheduling for Electric Power Systems," In A. M. Erisman, K. W. Noves, M. H. Dwarakanath (eds), *Electric Power Problems: The Mathematical Challenge*, SIAM, Philadelphia, 1980, pp. 378–392.

3. Lamont, J. W. Lesso, W. G. "An Approach to Daily Fossil Fuel Management," In A. M. Erisman, K. W. Noves, M. H. Dwarakanath (eds.), *Electric Power Problems: The Mathematical Challenge*, SIAM, Philadelphia, 1980, pp. 414–425.

4. Lamont, J. W., Lesso, W. G., Rantz, M., "Daily Fossil Fuel Management," 1979 *PICA Conference Proceedings*, IEEE Publication, 79CH1381-3-PWR, pp. 228–235.

5. Lasdon, L. S., *Optimization Theory for Large Systems*, Macmillan, New York, 1970.

6. Hadley, G., *Linear Programming*, Addison-Wesley, Reading, Mass., 1962.

7. Wagner, H. M., *Principle of Operations Research with Application to Managerial Decisions*, Prentice-Hall, Englewood Cliffs, N.J., 1975.

8. Wagner, H. M., "The Dual Simplex Algorithm for Bounded Variables," *Naval Research Logistics Quarterly*, Vol. 5, 1958, pp. 257–261.

chapter 7

Hydrothermal
Coordination

7.1 INTRODUCTION

The systematic coordination of the operation of a system of hydroelectric generation plants is usually more complex than the scheduling of an all-thermal generation system. The reason is both simple and important. That is, the hydroelectric plants may very well be coupled both electrically (i.e., they all serve the same load) and hydraulically (i.e., the water outflow from one plant may be a very significant portion of the inflow to one or more other, downstream plants).

No two hydroelectric systems in the world are alike. They are all different. The reason for the differences are the natural differences in the watersheds, the differences in the manmade storage and release elements used to control the water flows, and the very many different types of natural and manmade constraints imposed on the operation of hydroelectric systems. River systems may be simple with relatively few tributaries (e.g., the Connecticut River) with dams in series (hydraulically) along the river. River systems may encompass thousands of acres, extend over vast multinational areas, and include many tributaries and complex arrangements of storage reservoirs (e.g., the Columbia River basin in the Pacific Northwest).

Reservoirs may be developed with very large storage capacity with a few high head-plants along the river. Alternatively, the river may have been developed with a larger number of dams and reservoirs, each with smaller storage capacity. Water may be intentionally diverted through long raceways that tunnel through an entire mountain range (e.g., the Snowy Mountain scheme in Australia). In European developments, auxiliary reservoirs, control dams, locks, and even separate systems for pumping water back upstream have been added to rivers.

However, the one single aspect of hydroelectric plants that differentiates the coordination of their operation more than any other is the many, and highly varied, constraints. In many hydro systems the generation of power is an adjunct to the control of flood waters or the regular, scheduled release of water for irrigation. Recreation centers may have developed along the shores of a large reservoir so that only small surface water elevation changes are possible. Water release in a river may well have to be controlled so that the river is navigable at all times. Sudden changes with high-volume releases of water may be prohibited because the release may result in a large wave traveling downstream with potentially damaging effects. Fish ladders may be needed. Water releases may be dictated by international treaty.

TO REPEAT: All hydro systems are different.

7.1.1 Long-range Hydro-scheduling

The coordination of the operation of hydroelectric plants involves, of course, the scheduling of water releases. The *long-range hydro-scheduling problem* involves the long-range forecasting of water availability and the scheduling of reservoir water releases (i.e., "drawdown") for an interval of time that depends on the reservoir capacities.

Typical long-range scheduling goes anywhere from 1 wk to 1 yr or several years. For hydro schemes with a capacity of impounding water over several seasons, the long-range problem involves meteorological and statistical analyses.

Nearer term water inflow forecasts might be based on snow melt expectations and near-term weather forecasts. For the long-term drawdown schedule, a basic policy selection must be made. Should the water be used under the assumption that it will be replaced at a rate based on the statistically expected (i.e., mean value) rate, or should the water be released using a "worst case" prediction. In the first instance, it may well be possible to save a great deal of electric energy production expense by displacing thermal generation with hydro generation. If, on the other hand, a worst case policy was selected, the hydro plants would be run so as to minimize the risk of violating any of the hydrological constraints (e.g., running reservoirs too low, not having enough water to navigate a river). Conceivably, such a schedule would hold back water until it became quite likely that even worst case rainfall (runoff, etc.) would still give ample water to meet the constraints.

Long-range scheduling involves optimizing a policy in the context of unknowns such as load, hydraulic inflows, and unit availabilities (steam and hydro). These unknowns are treated statistically, and long-range scheduling involves optimization of statistical variables. Useful techniques include

1. Dynamic programming, where the entire long-range operation time period is simulated (e.g., 1 yr) for a given set of conditions.
2. Composite hydraulic models, which can represent several reservoirs.
3. Statistical production cost models.

The problems and techniques of long-range hydro-scheduling are outside the scope of this text, so we will end the discussion at this point and continue with short-range hydro-scheduling.

7.1.2 Short-range Hydro-scheduling

Short-range hydro-scheduling (1 day to 1 wk) involves the hour-by-hour scheduling of all generation on a system to achieve minimum production cost for the given time period. In such a scheduling problem, the load, hydraulic inflows, and unit availabilities are assumed known. A set of starting conditions (e.g., reservoir levels) is given, and the optimal hourly schedule that minimizes a desired objective while meeting hydraulic, steam, and electric system constraints is sought. Part of the hydraulic constraints may involve meeting "end-point" conditions at the end of the scheduling interval in order to conform to a long-range, water release schedule previously established.

7.2 HYDROELECTRIC PLANT MODELS

To understand the requirements for the operation of hydroelectric plants, one must appreciate the limitations imposed on operation of hydro resources by flood control, navigation, fisheries, recreation, water supply, and other demands on the water bodies and streams as well as the characteristics of energy conversion from the potential energy of stored water to electric energy. The amount of energy available in a unit of stored water, say a cubic foot, is equal to the product of the weight of the water stored (in this case, 62.4 lb) times the height (in feet) that the water would fall. One thousand cubic feet of water falling a distance of 42.5 ft has the energy equivalent to 1 kWh. Correspondingly, 42.5 ft^3 of water falling 1000 ft also has the energy equivalent to 1 kWh.

Consider the sketch of a reservoir and hydroelectric plant shown in Figure 7.1. Let us consider some overall aspects of the falling water as it travels from the reservoir through the penstock to the inlet gates through the hydraulic turbine down the draft tube and out the tailrace at the plant exit. The power that the water can produce is equal to the rate of water flow in cubic feet per second times a conversion coefficient that takes into account the net head (the distance through which the water falls less the losses in head caused by the flow) times the conversion efficiency of the turbine generator. A flow of 1 ft^3/sec falling 100 ft has the power equivalent of

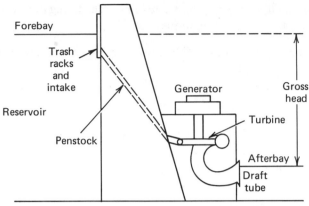

FIG. 7.1 Hydro plant components.

approximately 8.5 kW. If the flow-caused loss in head was 5%, or 5 ft, then the power equivalent for a flow of 1 ft^3 of water per second with the net drop of $100 - 5$ or 95 ft would have the power equivalent of slightly more than 8 kW ($8.5 \times 95\%$). Conversion efficiencies of turbine generators are typically in the range of 85 to 90% at the best efficiency operating point for the turbine generator so 1 ft^3/sec falling 100 ft would typically develop about 7 kW at most.

Let us return to our description of the hydroelectric plant as illustrated in Figure 7.1. The hydroelectric project consists of a body of water impounded by a dam, the hydro plant, and the exit channel or lower water body. The energy available for conversion to electrical energy of the water impounded by the dam is a function of the gross head; that is, the elevation of the surface of the reservoir less the elevation of the afterbay, or downstream water level below the hydroelectric plant. The head available to the turbine itself is slightly less than the gross head due to the friction losses in the intake, penstock, and draft tube. This is usually expressed as the *net head* and is equal to the gross head less the flow losses (measured in feet of head). The flow losses can be very significant for low head (10 to 60 ft) plants and for plants with long penstocks (several thousand feet). The water level at the afterbay is influenced by the flow out of the reservoir including plant release and any spilling of water over the top of the dam or through bypass raceways. During flooding conditions such as spring runoff, the rise in afterbay level can have a significant and adverse effect on the energy and capacity or power capability of the hydro plant.

The type of turbine used in a hydroelectric plant depends primarily on the design head for the plant. By far, the largest number of hydroelectric projects use reaction-type turbines. Only two types of reaction turbines are now in common use. For medium heads (that is, in the range from 60 to 1000 ft), the Francis turbine is used exclusively. For the low head plants (that is, for design heads in the range of 10 to 60 ft), the propeller turbine is used. The more modern propeller turbines have adjustable pitch blading (called *Kaplan turbines*) to improve the operating efficiency over a wide range in plant net head. Typical turbine performance results in an efficiency at full gate loading of between 85 to 90%. The Francis turbine and the

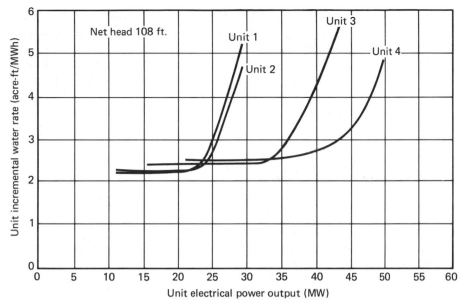

FIG. 7.2   Incremental water rate versus power output.

adjustable propeller turbine may operate at 65 to 125% of rated net head as compared to 90 to 110% for the fixed propeller.

Another factor affecting operating efficiency of hydro units is the MW loading. At light unit loadings, efficiency may drop below 70% (these ranges are often restricted by vibration and cavitation limits) and at full gate may rise to about 87%. If the best use of the hydro resource is to be obtained, operation of the hydro unit near its best efficiency gate position and near the designed head is necessary. This means that unit loading and control of reservoir forebay are necessary to make efficient use of hydro resources. Unit loading should be near best efficiency gate position, and water release schedules must be coordinated with reservoir inflows to maintain as high head on the turbines as the limitations on forebay operations will permit.

Typical plant performance for a medium head, four-unit plant in South America is illustrated in Figure 7.2. The incremental "water rate" is expressed in acre-feet per megawatt hour.* The rise in incremental water rate with increasing unit output results primarily from the increased hydraulic losses with the increased flow. A composite curve for multiple unit operation at the plant would reflect the mutual effects of hydraulic losses and rise in afterbay with plant discharge. Very careful attention must be given to the number of units run for a given required output. One unit run at best efficiency will usually use less water than two units run at half that load.

---

* An *acre-foot* is a common unit of water volume. It is the amount of water that will cover 1 acre to a depth of 1 ft (43,560 ft³). It also happens to be nearly equal to half a cubic foot per second flow for a day (43,200). An acre-foot is equal to 1.2335×10^3 m³.

High head plants (typically over 1000 ft) use impulse or Pelton turbines. In such turbines the water is directed into spoon-shaped buckets on the wheel by means of one or more water jets located around the outside of the wheel.

In the text that follows we will assume a characteristic giving the relationship between water flow through the turbine, q, and power output, P(MW), where q is expressed in ft^3/sec or acre-ft/h. Furthermore, we will not be concerned with what type of turbine is being used or the characteristics of the reservoir other than such limits as the reservoir head or volume and various flows.

7.3 SCHEDULING PROBLEMS

7.3.1 Types of Scheduling Problems

In the operation of a hydroelectric power system, three general categories of problems arise. These depend on the balance between the hydroelectric generation, the thermal generation, and the load.

Systems without any thermal generation are fairly rare. The economic scheduling of these systems is really a problem in scheduling water releases to satisfy all the hydraulic constraints and meet the demand for electrical energy. Techniques developed for scheduling hydrothermal systems may be used in some systems by assigning a pseudo-fuel cost to some hydroelectric plant. Then the schedule is developed by minimizing the production "cost" as in a conventional hydrothermal system. In all hydroelectric systems the scheduling could be done by simulating the water system and developing a schedule that leaves the reservoir levels with a maximum amount of stored energy. In geographically extensive hydroelectric systems, these simulations must recognize water travel times between plants.

Hydrothermal systems where the hydroelectric system is by far the largest component may be scheduled by economically scheduling the system to produce the minimum cost for the thermal system. These are basically problems in scheduling energy. A simple example is illustrated in the next section where the hydroelectric system cannot produce sufficient energy to meet the expected load.

The largest category of hydrothermal systems include those where there is a closer balance between the hydroelectric and thermal generation resources and those where the hydroelectric system is a small fraction of the total capacity. In these systems, the schedules are usually developed to minimize thermal-generation production costs recognizing all the diverse hydraulic constraints that may exist. The main portion of this chapter is concerned with systems of this type.

7.3.2 Scheduling Energy

Suppose, as in Figure 7.3, we have two sources of electrical energy to supply a load, one hydro and another steam. The hydro plant can supply the load by itself for a limited time. That is, for any time period j,

$$P_{Hj}^{max} \geq P_{Lj} \qquad j = 1 \cdots j_{max} \tag{7.1}$$

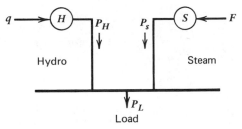

FIG. 7.3 Two-unit hydrothermal system.

However, the energy available from the hydro plant is insufficient to meet the load.

$$\sum_{j=1}^{j_{max}} P_{Hj}n_j \le \sum_{j=1}^{j_{max}} P_{Lj}n_j \qquad n_j = \text{number of hours in period } j$$

(7.2)

$$\sum_{j=1}^{j_{max}} n_j = T_{max} = \text{Total interval}$$

We would like to use up the entire amount of energy from the hydro plant in such a way that the cost of running the steam plant is minimized. The steam-plant energy required is

$$\sum_{j=1}^{j_{max}} P_{Lj}n_j - \sum_{j=1}^{j_{max}} P_{Hj}n_j = E \qquad (7.3)$$

| Load | Hydro | Steam |
| energy | energy | energy |

We will not require the steam unit to run for the entire interval of T_{max} hours. Therefore,

$$\sum_{j=1}^{N_s} P_{Sj}n_j = E \qquad N_s = \text{number of periods the steam plant is run} \qquad (7.4)$$

Then

$$\sum_{j=1}^{N_s} n_j \le T_{max}$$

The scheduling problem becomes

$$\text{Min } F_T = \sum_{j=1}^{N_s} F(P_{Sj})n_j \qquad (7.5)$$

Subject to

$$\sum_{j=1}^{N_s} P_{Sj}n_j - E = 0 \qquad (7.6)$$

and the LaGrange function is

$$\mathcal{L} = \sum_{j=1}^{N_s} F(P_{Sj})n_j + \alpha\left(E - \sum_{j=1}^{N_s} P_{sj}n_j\right) \qquad (7.7)$$

Then

$$\frac{\partial \mathcal{L}}{\partial P_{sj}} = \frac{dF(P_{sj})}{dP_{sj}} - \alpha = 0 \qquad \text{for } j = 1 \cdots N_s$$

or (7.8)

$$\frac{dF(P_{sj})}{dP_{sj}} = \alpha \qquad \text{for } j = 1 \cdots N_s$$

This means that the steam plant should be run at constant incremental cost for the entire period it is on. Let this optimum value of steam-generated power be P_s^*, which is the same for all time intervals the steam unit is on. This type of schedule is shown on Figure 7.4.

The total cost over the interval is

$$F_T = \sum_{j=1}^{N_s} F(P_s^*)n_j = F(P_s^*) \sum_{j=1}^{N_s} n_j = F(P_s^*)T_s \qquad (7.9)$$

where

$$T_s = \sum_{j=1}^{N_s} n_j = \text{the total run time for the steam plant}$$

Let the steam-plant cost be expressed as

$$F(P_s) = A + BP_s + CP_s^2 \qquad (7.10)$$

Then

$$F_T = (A + BP_s^* + CP_s^{*2})T_s \qquad (7.11)$$

Also note that

$$\sum_{j=1}^{N_s} P_{sj}n_j = \sum_{j=1}^{N_s} P_s^* n_j = P_s^* T_s = E \qquad (7.12)$$

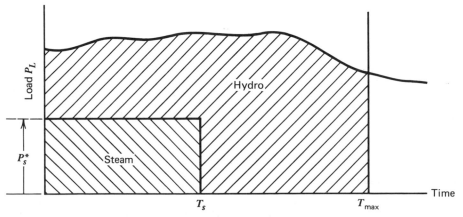

FIG. 7.4 Resulting optimal hydrothermal schedule.

Then

$$T_s = \frac{E}{P_s^*} \qquad (7.13)$$

and

$$F_T = (A + BP_s^* + CP_s^{*2})\left(\frac{E}{P_s^*}\right) \qquad (7.14)$$

Now we can establish the value of P_s^* by minimizing F_T.

$$\frac{\mathrm{d}F_T}{\mathrm{d}P_s^*} = \frac{-AE}{P_s^{*2}} + CE = 0 \qquad (7.15)$$

or

$$P_s^* = \sqrt{A/C} \qquad (7.16)$$

which means the unit should be operated at its maximum efficiency point long enough to supply the energy needed, E. Note if

$$F(P_s) = A + BP_s + CP_s^2 = f_c \cdot H(P_s) \qquad (7.17)$$

where f_c is the fuel cost. Then the heat rate is

$$\frac{H(P_s)}{P_s} = \frac{1}{f_c}\left(\frac{A}{P_s} + B + CP_s\right) \qquad (7.18)$$

and the heat rate has a minimum when

$$\frac{d}{dP_s}\left(\frac{H(P_s)}{P_s}\right) = 0 = \frac{-A}{P_s^2} + C \qquad (7.19)$$

giving best efficiency at

$$P_s = \sqrt{A/C} = P_s^* \qquad \longleftarrow \qquad (7.20)$$

EXAMPLE 7A

A hydro plant and a steam plant are to supply a constant load of 90 MW for 1 week (168 h). The unit characteristics are

Hydro Plant: $\qquad q = 300 + 15\,P_H$ acre-ft/h

$$0 \le P_H \le 100 \text{ MW}$$

Steam Plant: $\qquad H_s = 53.25 + 11.27\,P_s + 0.0213\,P_s^2$

$$12.5 \le P_s \le 50 \text{ MW}$$

→ Found by Dividing Production cost equation by P_s
and take the derivative with respect to P_s and
then set $= 0$.

Part 1

Let the hydro plant be limited to 10,000 MWh of energy. Solve for T_s^*, the run time of the steam unit. The load is $90 \times 168 = 15{,}120$ MWh, requiring 5120 MWh to be generated by the steam plant.

→ The steam plant's maximum efficiency is at $\sqrt{53.25/0.0213} = 50$ MW. Therefore, the steam plant will need to run for 5120/50 or 102.4 h. The resulting schedule will require the steam plant to run at 50 MW and the hydro plant at 40 MW for the first 102.4 h of the week and the hydro plant at 90 MW for the remainder.

Part 2

Instead of specifying the energy limit on the steam plant, let the limit be on the volume of water that can be drawn from the hydro plants' reservoir in 1 wk. Suppose the maximum drawdown is 250,000 acre-ft, how long should the steam unit run?

To solve this we must account for the plant's q versus P characteristic. A different flow will take place when the hydro plant is operated at 40 MW than when it is operated at 90 MW. In this case,

$$q_1 = [300 + 15(40)] \times T_s \text{ acre-ft}$$

$$q_2 = [300 + 15(90)] \times (168 - T_s) \text{ acre-ft}$$

and

$$q_1 + q_2 = 250{,}000 \text{ acre-ft}$$

Solving for T_s we get 36.27 h.

7.4 THE SHORT-TERM HYDROTHERMAL SCHEDULING PROBLEM

A more general and basic short-term hydrothermal scheduling problem requires that a given amount of water be used in such a way as to minimize the cost of running the thermal units. We will use Figure 7.5 in setting up this problem.

The problem we wish to set up is the general, short-term hydrothermal scheduling problem where the thermal system is represented by an equivalent unit, P_s, as was done in Chapter 6. In this case, there is a single hydroelectric plant, P_H. We assume that the hydro plant is not sufficient to supply all the load demands during the period and that there is a maximum total volume of water that may be discharged throughout the period of T_{max} hours.

In setting up this problem and the examples that follow, we assume all spillages, s_j, are zero. The only other hydraulic constraint we will impose initially is that the total volume of water discharged must be exactly as defined. Therefore, the

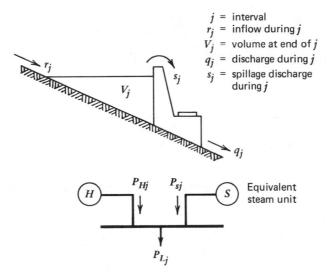

j = interval
r_j = inflow during j
V_j = volume at end of j
q_j = discharge during j
s_j = spillage discharge during j

Equivalent steam unit

FIG. 7.5 Hydrothermal system with hydraulic constraints.

mathematical scheduling problem may be set up as follows:

Problem:
$$\text{Min } F_T = \sum_{j=1}^{j_{max}} n_j F_j \qquad (7.21)$$

Subject to:
$$\sum_{j=1}^{j_{max}} n_j q_j = q_{TOT} \qquad \text{Total water discharge}$$

total steam power

$$P_{Lj} - P_{Hj} - P_{sj} = 0 \qquad \text{Load balance for } j = 1 \cdots j_{max}$$

where:
$$n_j = \text{length of } j^{th} \text{ interval}$$

$$\sum_{j=1}^{j_{max}} n_j = T_{max}$$

and the loads are constant in each interval. Other constraints could be imposed, such as:

$$V_j = V_{j-1} + (r_j - q_i - S_j) n_j$$

$V_j\|_{j=0} = V_s$	Starting volume
$V_j\|_{j=j_{max}} = V_E$	Ending volume
$q_{min} \le q_j \le q_{max}$	Flow limits for $j = 1 \cdots j_{max}$
$q_j = Q_j$	Fixed discharge for a particular hour

$$V_{min} \le V_j \le V_{max}$$

$$\sum_{j=1}^{j_{max}} n_j q_j = Q_{TOT}$$

Assume constant head operation and assume a q versus P characteristic is available as shown in Figure 7.6 so that

$$q = q(P_H) \qquad (7.22)$$

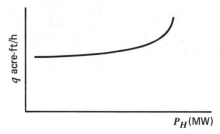

FIG. 7.6 Hydroelectric unit input-output characteristic for constant head.

We now have a similar problem to the take-or-pay fuel problem. The LaGrange function is

$$\mathcal{L} = \sum_{j=1}^{j_{max}} [n_j F(P_{sj}) + \lambda_j(P_{Lj} - P_{Hj} - P_{sj})]$$

$$+ \gamma \left[\sum_{j=1}^{j_{max}} n_j q_j(P_{Hj}) - q_{TOT} \right] \tag{7.23}$$

and for a specific interval $j = k$,

$$\frac{\partial \mathcal{L}}{\partial P_{sk}} = 0$$

gives

$$n_k \frac{dF_{sk}}{dP_{sk}} = \lambda_k \tag{7.24}$$

and

$$\frac{\partial \mathcal{L}}{\partial P_{Hk}} = 0$$

gives

$$\gamma n_k \frac{dq_k}{dP_{Hk}} = \lambda_k \tag{7.25}$$

This is solved using the same techniques shown in Chapter 6.

Suppose we add the network losses to the problem. Then at each hour,

$$P_{Lj} + P_{loss\,j} - P_{Hj} - P_{sj} = 0 \tag{7.26}$$

and the LaGrange function becomes

$$\mathcal{L} = \sum_{j=1}^{j_{max}} [n_j F(P_{sj}) + \lambda_j(P_{Lj} + P_{loss\,j} - P_{Hj} - P_{sj})]$$

$$+ \gamma \left[\sum_{j=1}^{j_{max}} n_j q_j(P_{Hj}) - q_{TOT} \right] \tag{7.27}$$

---

with resulting coordination equations (hour k):

$$n_k \frac{dF(P_{sk})}{dP_{sk}} + \lambda_k \frac{\partial P_{loss\,k}}{\partial P_{sk}} = \lambda_k \qquad (7.28)$$

$$\gamma n_k \frac{dq(P_{Hk})}{dP_{Hk}} + \lambda_k \frac{\partial P_{loss\,k}}{\partial P_{Hk}} = \lambda_k \qquad (7.29)$$

This gives rise to a more complex scheduling solution requiring three loops as shown in Figure 7.7. In this solution procedure ε_1 and ε_2 are the respective tolerances on the load balance and water balance relationships.

Note that this problem ignores volume and hourly discharge rate constraints. As a result, the value of γ will be constant over the entire scheduling period as long as the

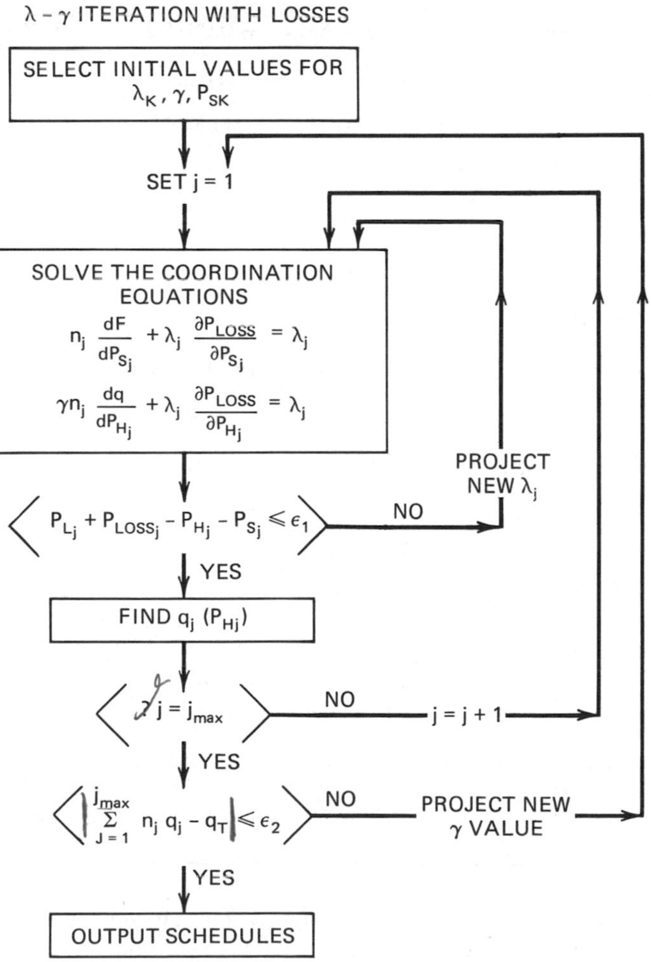

FIG. 7.7 A λ-γ iteration scheme for hydrothermal scheduling.

units remain within their respective scheduling ranges. The value of γ would change if a constraint (i.e., $V_j = V_{max}$, etc.) were encountered. This would require that the scheduling logic recognize such constraints and take appropriate steps to adjust γ so that the constrained variable does not go beyond its limit. The appendix to this chapter gives a proof that γ is constant when no storage constraints are encountered. As usual, in any gradient method, care must be exercised to allow constrained variables to move off their constraints if the solution so dictates.

EXAMPLE 7B

A load is to be supplied from a hydro plant and a steam system whose characteristics are given here.

Equivalent Steam System: $\quad H = 500 + 8.0\, P_s + 0.0016\, P_s^2 \quad$ (MBtu/h)

$$\text{Fuel cost} = 1.15 \ \text{R/MBtu}$$

$$150 \text{ MW} \le P_s \le 1500 \text{ MW}$$

Hydro Plant: $\quad q = 330 + 4.97\, P_H$ acre-ft/h

$$0 \le P_H \le 1000 \text{ MW}$$

$$q = 5300 + 12(P_H - 1000) + 0.05(P_H - 1000)^2 \text{ acre-ft/h}$$

$$1000 < P_H < 1100 \text{ MW}$$

The hydro plant is located a good distance from the load. The electrical losses are

$$P_{loss} = 0.00008\, P_h^2 \text{ MW}$$

The load to be supplied is connected at the steam plant and has the following schedule:

12 midnight–12 noon	1200 MW
12 noon –12 midnight	1500 MW

The hydro unit's reservoir is limited to a drawdown of 100,000 acre-ft over the entire 24-h period. Inflow to the reservoir is to be neglected. The optimal schedule for this problem was found using a program written using Figure 7.7. The results are

Time period	P steam	P hydro	Hydro discharge acre-ft/h
12 midnight–12 noon	567.4	668.3	3651.5
12 noon –12 midnight	685.7	875.6	4681.7

The optimal value for γ is 2.028378 R/MW. The storage in the hydro plant's reservoir goes down in time as shown in Figure 7.8. No natural inflows or spillage are assumed to occur.

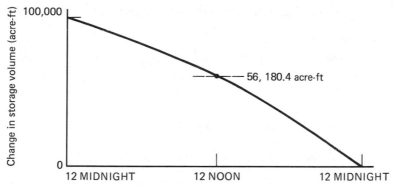

FIG. 7.8 Change in storage volume (= cumulative discharge) versus time for Example 7B.

7.5 SHORT-TERM HYDRO-SCHEDULING: A GRADIENT APPROACH

The following is an outline of a first-order gradient approach, as shown in Figure 6.7a, to the problem of finding the optimum schedule for a hydrothermal power system. We assume a single equivalent thermal unit with a convex input-output curve and a single hydro plant.

Let: j = the interval = 1, 2, 3, . . . , j_{max}

V_j = storage volume at the end of interval j

q_j = discharge rate during interval j

r_j = inflow rate into the storage reservoir during interval j

P_{sj} = steam generation during j^{th} interval.

s_j = spillage discharge rate during interval j

$P_{loss\,j}$ = losses, assumed here to be zero

P_{Lj} = received power during the jth interval (load)

P_{Hj} = hydro generation during the jth hour

Next, we let the discharge from the hydro plant be a function of the hydro power output only. That is, a constant head characteristic is assumed. Then,

$$q_j(P_{Hj}) = q_j$$

so that to a first order,*

$$\Delta q_j = \frac{dq_j}{dP_H} \Delta P_{Hj}$$

The total cost for fuel over the $j = 1, 2, 3, . . . , j_{max}$ intervals is

$$F_T = \sum_{j=1}^{j_{max}} n_j F_j(P_{sj})$$

* ΔP_s and ΔF designate changes in the quantities P_s and F.

This may be expanded in a Taylor series to give the change in fuel cost for a change in steam-plant schedule.

$$\Delta F_T = \sum_{j=1}^{j\text{max}} n_j \left(F'_j \Delta P_{sj} + \frac{1}{2} F''_j (\Delta P_{sj})^2 + \cdots \right)$$

To the first order this is

$$\Delta F_T = \sum_{j=1}^{j\text{max}} n_j F'_j \Delta P_{sj}$$

In any given interval, the electrical powers must balance.

$$P_{Lj} - P_{sj} - P_{Hj} = 0$$

so that,

$$\Delta P_{sj} = -\Delta P_{Hj}$$

or

$$\Delta P_{sj} = -\frac{\Delta q_j}{\dfrac{dq_j}{dP_{Hj}}}$$

Therefore,

$$\Delta F_T = -\sum_{j=1}^{j\text{max}} n_j \left(\frac{\dfrac{dF_j}{dP_{sj}}}{\dfrac{dq_j}{dP_{Hj}}} \right) \Delta q_j = -\sum_{j=1}^{j\text{max}} n_j \gamma_j \Delta q_j$$

where

$$\gamma_j = \frac{\dfrac{dF_j}{dP_{sj}}}{\dfrac{dq_j}{dP_{Hj}}}$$

The variables γ_j are the incremental water values in the various intervals and give an indication of how to make the "moves" in the application of the first-order technique. That is, the "steepest descent" to reach minimum fuel cost (or the best period to release a unit of water) is the period with the maximum value of gamma. The values of water release, Δq_j, must be chosen to stay within the hydraulic constraints. These may be determined by use of the *hydraulic continuity equation.*

$$V_j = V_{j-1} + (r_j - q_j - s_j)n_j$$

to compute the reservoir storage each interval. We must also observe the storage limits,

$$V_{\min} \le V_j \le V_{\max}$$

We will assume spillage is prohibited so that all $s_j = 0$, even though there may well be circumstances where allowing $s_j > 0$ for some j might reduce the thermal system cost.

The discharge flow may be constrained both in rate and in total. That is,

$$q_{min} \leq q_j \leq q_{max}$$

and

$$\sum_{j=1}^{j_{max}} n_j q_j = q_{TOT}$$

The flowchart in Figure 6.7a illustrates the application of this method. Figure 7.9 illustrates a typical trajectory of storage volume versus time and illustrates the special rules that must be followed when constraints are taken. Whenever a constraint is reached, that is, storage V_j is equal to V_{min} or V_{max}, one must choose intervals in a more restricted manner than as shown in Figure 6.7a. This is summarized here.

1. **No Constraints Reached**
   Select the pair of intervals j^- and j^+ anywhere from $j = 1 \cdots j_{max}$.

2. **A Constraint is Reached**
   Option A: Choose the j^- and j^+ within one of the subintervals. That is, choose both j^- and j^+ from periods 1 or 2 or 3 in Figure 7.9. This will guarantee that the constraint is not violated. For example, choosing a time j^+ within period 1 to increase release and choosing j^- also in period 1 to decrease release will mean no net release change at the end of subinterval 1, so the V_{min} constraint will not be violated.
   Option B: Choose j^- and j^+ from different subintervals so that the constraint is no longer reached. For example, choosing j^+ within period 2 and j^- within period 1 will mean the V_{min} and V_{max} limits are no longer reached at all.

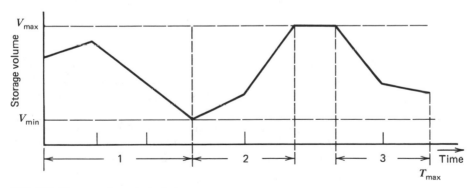

FIG. 7.9 Storage volume trajectory.

Other than these special rules, one can apply the flowchart of Figure 6.7a exactly as shown (while understanding that q is water rather than fuel as in Figure 6.7a).

EXAMPLE 7C

Find an optimal hydro schedule using the gradient technique of Section 7.5. The hydro plant and equivalent steam plant are the same as Example 7B, with the following additions.

Load Pattern:

Load Pattern:	First day	12 midnight–12 noon	1200 MW	
		12 noon –12 midnight	1500 MW	
	Second day	12 midnight–12 noon	1100 MW	
		12 noon –12 midnight	1800 MW	
	Third day	12 midnight–12 noon	950 MW	
		12 noon –12 midnight	1300 MW	

Hydro Reservoir: 1. 100,000 acre-ft at the start.

   2. Must have 60,000 acre-ft at the end of schedule.

   3. Reservoir volume is limited as follows:

$$60,000 \text{ acre-ft} \leq V \leq 120,000 \text{ acre-ft}.$$

   4. There is a constant inflow into the reservoir of 2000 acre-ft/h over the entire 3-day period

The initial schedule has constant discharge, thereafter each update or "step" in the gradient calculations was carried out by entering the j^+, j^- and Δq into a computer terminal that then recalculated all period γ's, flows, and so forth. The results of running this program are shown in Figure 7.10. Note that the column labeled VOLUME gives the reservoir volume at the *end* of each 12-h period. Note that after the fifth step the volume schedule reaches its bottom limit at the end of period 4. The subsequent steps require a choice of j^+ and j^- from either $\{1, 2, 3$ and $4\}$ or from $\{5, 6\}$. (P_s, P_H are MW, gamma is $\mathbb{R}$/acre-ft, volume is in acre-ft, discharge is in acre-ft/h.)

Note that the "optimum" schedule is undoubtedly located between the last two iterations. If we were to release less water in any of the first four intervals and more during 5 or 6, the thermal system cost would increase. We can theoretically reduce our operating costs a few fractions of an $\mathbb{R}$ by leveling the γ values in each of the two subintervals, $\{1, 2, 3, 4\}$ and $\{5, 6\}$, but the effort is probably not worthwhile.

INITIAL SCHEDULE (CONSTANT DISCHARGE)

J	Ps	PH	GAMMA	VOLUME	DISCHARGE
1	752.20	447.80	2.40807	93333.3	2555.555
2	1052.20	447.80	2.63020	86666.7	2555.555
3	652.20	447.80	2.33402	80000.0	2555.555
4	1352.20	447.80	2.85233	73333.4	2555.555
5	502.20	447.80	2.22296	66666.7	2555.555
6	852.20	447.80	2.48211	60000.1	2555.555

TOTAL OPERATING COST FOR ABOVE SCHEDULE =   719725.50 R

ENTER JMAX,JMIN,DELQ
4,5,1000

J	Ps	PH	GAMMA	VOLUME	DISCHARGE
1	752.20	447.80	2.40807	93333.3	2555.555
2	1052.20	447.80	2.63020	86666.7	2555.555
3	652.20	447.80	2.33402	80000.0	2555.555
4	1150.99	649.01	2.70335	61333.4	3555.555
5	703.41	246.59	2.37194	66666.7	1555.555
6	852.20	447.80	2.48211	60000.1	2555.555

TOTAL OPERATING COST FOR ABOVE SCHEDULE =   713960.75 R

ENTER JMAX,JMIN,DELQ
4,3,400

J	Ps	PH	GAMMA	VOLUME	DISCHARGE
1	752.20 /	447.80	2.40807	93333.3	2555.555
2	1052.20	447.80	2.63020	86666.7	2555.555
3	732.69	367.31	2.39362	84800.0	2155.555
4	1070.51	729.49	2.64376	61333.4	3955.555
5	703.41	246.59	2.37194	66666.7	1555.555
6	852.20	447.80	2.48211	60000.1	2555.555

TOTAL OPERATING COST FOR ABOVE SCHEDULE =   712474.00 R

ENTER JMAX,JMIN,DELQ
4,5,100

J	Ps	PH	GAMMA	VOLUME	DISCHARGE
1	752.20	447.80	2.40807	93333.3	2555.555
2	1052.20	447.80	2.63020	86666.7	2555.555
3	732.69	367.31	2.39362	84800.0	2155.555
4	1050.39	749.61	2.62886	60133.4	4055.555
5	723.53	226.47	2.38684	66666.7	1455.555
6	852.20	447.80	2.48211	60000.1	2555.555

TOTAL OPERATING COST FOR ABOVE SCHEDULE =   712165.75 R

ENTER JMAX,JMIN,DELQ
2,5,10

J	Ps	PH	GAMMA	VOLUME	DISCHARGE
1	752.20	447.80	2.40807	93333.3	2555.555
2	1050.19	449.81	2.62871	86546.7	2565.555
3	732.69	367.31	2.39362	84680.0	2155.555
4	1050.39	749.61	2.62886	60013.4	4055.555
5	725.54	224.46	2.38833	66666.7	1445.555
6	852.20	447.80	2.48211	60000.1	2555.555

TOTAL OPERATING COST FOR ABOVE SCHEDULE =   712136.75 R

ENTER JMAX,JMIN,DELQ
4,5,1.111

J	Ps	PH	GAMMA	VOLUME	DISCHARGE
1	752.20	447.80	2.40807	93333.3	2555.555
2	1050.19	449.81	2.62871	86546.7	2565.555
3	732.69	367.31	2.39362	84680.0	2155.555
4	1050.17	749.83	2.62870	60000.0	4056.666
5	725.77	224.23	2.38849	66666.7	1444.444
6	852.20	447.80	2.48211	60000.0	2555.555

TOTAL OPERATING COST FOR ABOVE SCHEDULE =   712133.50 R

```
ENTER JMAX,JMIN,DELQ
2,3,800
```

J	Ps	PH	GAMMA	VOLUME	DISCHARGE
1	752.20	447.80	2.40807	93333.3	2555.555
2	889.22	610.78	2.50953	76946.7	3365.555
3	893.65	206.35	2.51280	84680.0	1355.555
4	1050.17	749.83	2.62870	60000.0	4056.666
5	725.77	224.23	2.38849	66665.7	1444.444
6	852.20	447.80	2.48211	50000.0	2555.555

```
TOTAL OPERATING COST FOR ABOVE SCHEDULE =        711020.75 R
ENTER JMAX,JMIN,DELQ
4,1,750
```

J	Ps	PH	GAMMA	VOLUME	DISCHARGE
1	903.11	296.89	2.51981	102333.3	1805.555
2	889.22	610.78	2.50953	85946.7	3365.555
3	893.65	206.35	2.51280	93680.0	1355.555
4	899.26	900.74	2.51696	60000.0	4806.665
5	725.77	224.23	2.38849	66665.7	1444.444
6	852.20	447.80	2.48211	60000.1	2555.555

```
TOTAL OPERATING COST FOR ABOVE SCHEDULE =        710040.75 R
ENTER JMAX,JMIN,DELQ
6,5,400
```

J	Ps	PH	GAMMA	VOLUME	DISCHARGE
1	903.11	296.89	2.51981	102333.3	1805.555
2	889.22	610.78	2.50953	85946.7	3365.555
3	893.65	206.35	2.51280	93680.0	1355.555
4	899.26	900.74	2.51696	60000.0	4806.665
5	806.25	143.75	2.44809	71466.7	1044.444
6	771.72	528.28	2.42252	60000.1	2955.555

```
TOTAL OPERATING COST FOR ABOVE SCHEDULE =        709877.38 R
```

FIG. 7.10 Computer printout for Example 7C.

7.6 HYDRO UNITS IN SERIES (HYDRAULICALLY COUPLED)

Consider now a hydraulically coupled system consisting of three reservoirs in series (see Figure 7.11). The discharge from any upstream reservoir is assumed to flow directly into the succeeding downstream plant with no time lag. The hydraulic continuity equations are

$$V_{1j} = V_{1j-1} + (r_{1j} - s_{1j} - q_{1j})n_j$$

$$V_{2j} = V_{2j-1} + (q_{1j} + s_{1j} - s_{2j} - q_{2j})n_j$$

$$V_{3j} = V_{3j-1} + (q_{2j} + s_{2j} - s_{3j} - q_{3j})n_j$$

where r_j = inflow

V_j = reservoir volume

s_j = spill rate over the dam's spillway

q_j = hydro plant discharge

n_j = number of hours in each scheduling period

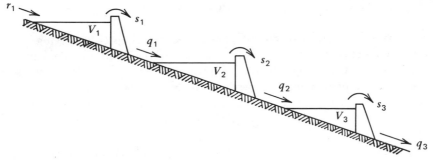

FIG. 7.11 Hydraulically coupled hydroelectric plants.

The object is to minimize

$$\sum_{j=1}^{j\text{max}} n_j F(P_{sj}) = \text{total cost} \tag{7.30}$$

Subject to the following constraints.

$$P_{\text{load } j} - P_{sj} - P_{H1j} - P_{H2j} - P_{H3j} = 0$$

and

$$V_{1j} - V_{1j-1} - (r_{1j} - s_{1j} - q_{1j})n_j = 0 \tag{7.31}$$

$$V_{2j} - V_{2j-1} - (q_{1j} + s_{1j} - s_{2j} - q_{2j})n_j = 0$$

$$V_{3j} - V_{3j-1} - (q_{2j} + s_{2j} - s_{3j} - q_{3j})n_j = 0$$

All equations in set 7.31 must apply for $j = 1 \cdots j_{\text{max}}$.

The LaGrange function would then appear as

$$\begin{aligned}
\mathscr{L} = \sum_{j=1}^{j\text{max}} &\left\{ [n_j F(P_{sj}) - \lambda_j(P_{\text{load } j} - P_{sj} - P_{H1j} - P_{H2j} - P_{H3j})] \right. \\
&+ \gamma_{1j}[V_{1j} - V_{1j-1} - (r_{1j} - s_{1j} - q_{1j})n_j] \\
&+ \gamma_{2j}[V_{2j} - V_{2j-1} - (q_{1j} + s_{1j} - s_{2j} - q_{2j})n_j] \\
&\left. + \gamma_{3j}[V_{3j} - V_{3j-1} - (q_{2j} + s_{2j} - s_{3j} - q_{3j})n_j] \right\}
\end{aligned}$$

Note that we could have included more constraints to take care of reservoir volume limits, end-point volume limits, and so forth, which would have necessitated using the Kuhn-Tucker conditions when limits were reached.

Hydro-scheduling with multiple-coupled plants is a formidable task. Lambda-gamma iteration techniques or gradient techniques can be used: in either case, convergence to the optimal solution can be slow. For these reasons, hydro scheduling for such systems is often done with dynamic programming, which will be covered in Section 7.8.

7.7 PUMPED-STORAGE HYDRO PLANTS

Pumped-storage hydro plants are designed to save fuel costs by serving the peak load (a high fuel-cost load) with hydro energy and then pumping the water back up into the reservoir at light load periods (a lower cost load). These plants may involve separate pumps and turbines or, more recently, reversible pump turbines. Their operation is illustrated by the two graphs in Figure 7.12. The first is the composite thermal system input/output characteristic and the second is the load cycle.

The pumped-storage plant is operated until the added pumping cost exceeds the savings in thermal costs due to the *peak shaving operations*. Figure 7.12 illustrates the operation on a daily cycle. If

$$\left. \begin{array}{l} e_g = \text{generation, MWh} \\ e_p = \text{pumping load, MWh} \end{array} \right\} \text{ for the same volume of water}$$

then the cycle efficiency is

$$\eta = \frac{e_g}{e_p} \qquad (\eta \text{ is typically about 0.67})$$

Storage reservoirs have limited storage capability and typically provide 4 to 8 or 10 h of continuous operation as a generator. Pumped-storage plants may be operated on a daily or weekly cycle. When operated on a weekly cycle, pumped-storage plants will start the week (say a Monday morning in the United States) with a full reservoir. The plant will then be scheduled over a weekly period to act as a generator during high load hours and to refill the reservoir partially, or completely, during off-peak periods.

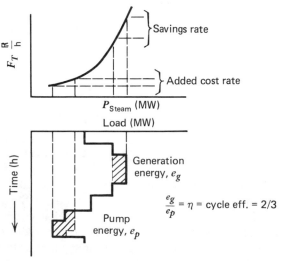

FIG. 7.12   Thermal input-output characteristic and typical daily load cycle.

Frequently, special interconnection arrangements may facilitate pumping operations if arrangements are made to purchase low cost, off-peak energy. In some systems, the system operator will require a complete daily refill of the reservoir when there is any concern over the availability of capacity reserves. In those instances, economy is secondary to reliability.

7.7.1 Pumped-Storage Hydro-scheduling with a λ-γ Iteration

Assume

1. Constant head hydro operation.
2. An equivalent steam unit with convex input/output curve.
3. A 24-h operating schedule, each time interval equals 1 h.
4. In any one interval, the plant is either pumping or generating or idle (idle will be considered as just a limiting case of pumping or generating).
5. Beginning and ending storage reservoir volumes are specified.
6. Pumping can be done continuously over the range of pump capability.
7. Pump and generating ratings are the same.
8. There is a constant cycle efficiency, η.

The problem is set up ignoring reservoir volume constraints to show that the same type of equations can result as those that arose in the conventional hydro case. Figure 7.13 shows the water flows and equivalent electrical system. In some interval, j,

$$r_j = \text{inflow (acre-ft/}h)$$

$$V_j = \text{volume at end of interval (acre-ft)}$$

$$q_j = \text{discharge if generating (acre-ft/}h)$$

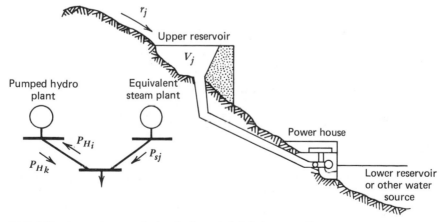

FIG. 7.13 Pumped-storage hydraulic flows and electric system flows.

or

$$w_j = \text{pumping rate if pumping (acre-ft/h)}$$

Intervals during the day are classified into two sets.

$$\{k\} = \text{intervals of generation}$$

$$\{i\} = \text{intervals of pumping}$$

The reservoir constraints are to be monitored in the computational procedure. The initial and final volumes are

$$V_0 = V_s$$

$$V_{24} = V_e$$

The problem is to minimize the sum of the hourly costs for steam generation over the day while observing the constraints. This total fuel cost for a day is (note that we have dropped n_j here since $n_j = 1$ h):

$$F_T = \sum_{j=1}^{24} F_j(P_{sj})$$

We consider the two sets of time intervals.

1. $\{k\}$: **Generation Intervals:**   The electrical and hydraulic constraints are

$$P_{Lk} + P_{\text{loss } k} - P_{sk} - P_{Hk} = 0$$

$$V_k - V_{k-1} - r_k + q_k = 0$$

These give rise to a LaGrange function during a generation hour (interval k) of

$$E_k = F_k + \lambda_k(P_{Lk} + P_{\text{loss } k} - P_{sk} - P_{Hk}) + \gamma_k(V_k - V_{k-1} - r_k + q_k) \quad (7.32)$$

2. $\{i\}$: **Pump Intervals:**   Similarly for a typical pumping interval, i,

$$P_{Li} + P_{\text{loss } i} - P_{si} + P_{Hi} = 0$$

$$V_i - V_{i-1} - r_i - w_i = 0 \quad (7.33)$$

$$E_i = F_i + \lambda_i(P_{Li} + P_{\text{loss } i} - P_{si} + P_{Hi}) + \gamma_i(V_i - V_{i-1} - r_i - w_i)$$

Therefore, the total LaGrange function is

$$E = \sum_{\{k\}} E_k + \sum_{\{i\}} E_i + \varepsilon_s(V_0 - V_s) + \varepsilon_e(V_{24} - V_e) \quad (7.34)$$

where the end-point constraints on the storage have been added.

In this formulation, the hours in which no pumped hydro takes place may be considered as pump (or generate) intervals with

$$P_{Hi} = P_{Hk} = 0$$

To find the minimum of $F_T = \sum F_j$, we set the first partial derivatives of E to zero.

1. $\{k\}$: Generation Intervals:

$$\frac{\partial E}{\partial P_{sk}} = 0 = -\lambda_k \left(1 - \frac{\partial P_{loss}}{\partial P_{sk}} \right) + \frac{dF_k}{dP_{sk}}$$

$$\frac{\partial E}{\partial P_{Hk}} = 0 = -\lambda_k \left(1 - \frac{\partial P_{loss}}{\partial P_{Hk}} \right) + \gamma_K \frac{dq_k}{dP_{Hk}}$$

(7.35)

2. $\{i\}$: Pump intervals:

$$\frac{\partial E}{\partial P_{si}} = 0 = -\lambda_i \left(1 - \frac{\partial P_{loss}}{\partial P_{si}} \right) + \frac{dF_i}{dP_{si}}$$

$$\frac{\partial E}{\partial P_{Hi}} = 0 = +\lambda_i \left(1 + \frac{\partial P_{loss}}{\partial P_{Hk}} \right) - \gamma_i \frac{dw_i}{dP_{Hi}}$$

(7.36)

For the $\partial E/\partial V$, we can consider any interval of the entire day, for instance, the ℓth interval, which is not the first or 24th hour.

$$\frac{\partial E}{\partial V_\ell} = 0 = \gamma_\ell - \gamma_{\ell+1}$$

and for $\ell = 0$ and $= 24$

$$\frac{\partial E}{\partial V_0} = 0 = -\gamma_1 + \varepsilon_s \quad \text{and} \quad \frac{\partial E}{\partial V_{24}} = 0 = \gamma_{24} + \varepsilon_e \tag{7.37}$$

From Eq. 7.37, it may be seen that γ is a constant. Therefore, it is possible to solve the pumped-storage scheduling problem by means of a λ-γ iteration over the time interval chosen. It is necessary to monitor the calculations to prevent a violation of the reservoir constraints or else to incorporate them in the formulation.

It is also possible to set up the problem of scheduling the pumped-storage hydro plant in a form that is very similar to the gradient technique used for scheduling conventional hydro plants.

7.7.2 Pumped-Storage Scheduling by a Gradient Method

The interval designations and equivalent electrical system are the same as those shown previously. This time, losses will be neglected. Take a 24-h period and start the schedule with no pumped-storage hydro activity initially. Assume that the steam system is operated each hour such that

$$\frac{dF_j}{dP_{sj}} = \lambda_j \quad j = 1, 2, 3, \ldots, 24$$

That is, the single, equivalent steam-plant source is realized by generating an economic schedule for the load range covered by the daily load cycle.

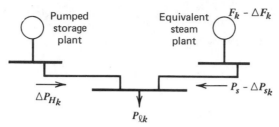

FIG. 7.14 Incremental increase in hydro generation in hour k.

Next, assume the pumped-storage plant generates a small amount of power, ΔP_{Hk}, at the peak period k. These changes are shown in Figure 7.14. The change in steam-plant cost is

$$\Delta F_k = \frac{\partial F_k}{\partial P_{sk}} \cdot \Delta P_{sk} = -\frac{dF_k}{dP_{sk}} \Delta P_{Hk} \quad \text{or} \quad \Delta F_k = -\lambda_k \Delta P_{Hk} \quad (7.38)$$

which is the savings due to generating ΔP_{Hk}.

Next, we assume that the plant will start the day with a given reservoir volume and we wish to end with the same volume. The volume may be measured in terms of the MWh of generation of the plant. The overall operating cycle has an efficiency, η. For instance, if $\eta = 2/3$; 3 MWh of pumping are required to replace 2 MWh of generation water use. Therefore, to replace the water used in generating the ΔP_{Hk} power, we need to pump an amount $(\Delta P_{Hk}/\eta)$.

To do this, search for the lowest cost ($=$ lowest load) interval, i, of the day to do the pumping. This changes the steam system cost by an amount

$$\Delta F_i = \frac{\partial F_i}{\partial P_{si}} \Delta P_{si} = \frac{dF_i}{dP_{si}} \left(\frac{\Delta P_{Hk}}{\eta} \right) = \frac{\lambda_i}{\eta} \Delta P_{Hk} \quad (7.39)$$

The total cost change over the day is then

$$\Delta F_T = \Delta F_k + \Delta F_i$$

$$= \Delta P_{Hk} \left(\frac{\lambda_i}{\eta} - \lambda_k \right) \quad (7.40)$$

Therefore, the decision to generate in k and replace the water in i is economic if ΔF_T is negative (a decrease in cost), this is true if

$$\lambda_k > \frac{\lambda_i}{\eta} \quad (7.41)$$

There are practical considerations to be observed, such as making certain that the generation and pump powers required are less than or equal to the pump or generation capacity in any interval. The whole cycle may be repeated until

1. It is no longer possible to find periods k and i such that $\lambda_k > \lambda_i/\eta$.
2. The maximum or minimum storage constraints have been reached.

When implementing this method, it may be necessary also to do pumping in more than one interval to avoid power requirements greater than the unit rating. This can be done; then the criteria would be

$$\lambda_k > (\lambda_i + \lambda_{i'})/\eta$$

Figure 7.15 shows the way in which a single pump-generate step could be made. In this figure, the maximum capacity is taken as 1500 MW, where the pumped-storage unit is generating or pumping.

These procedures assume that commitment of units does not change as a result of the operation of the pumped-storage hydro plant. It does not presume that the equivalent steam-plant characteristics are identical in the 2 h because the same techniques can be used when different thermal characteristics are present in different hours.

Longer cycles may also be considered. For instance, you could start a schedule for a week and perhaps find that you were using the water on the weekday peaks and filling the tank on weekends. In the case where a reservoir constraint was reached, you would split the week into two parts and see if you could increase the overall

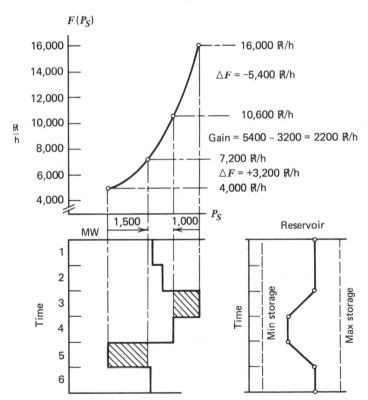

FIG. 7.15 Single step in gradient iteration for a pumped-storage plant. Cycle efficiency is two-thirds. Storage is expressed in equivalent MWh of generation.

savings by increasing the plant use. Another possibility may be to schedule each day of a week on a daily cycle. Multiple, uncoupled pumped-storage plants could also be scheduled in this fashion. The most reasonable looking schedules would be developed by running the plants through the scheduling routines in parallel. (Schedule a little on plant 1, then shift to plant 2, etc.) In this way, the plants will all share in the peak shaving. Hydraulically coupled pumped-storage plants and/or pump-back plants combined with conventional hydro plants may be handled similarly.

EXAMPLE 7D

A pumped-storage plant is to operate so as to minimize the operating cost of the steam units to which it is connected. The pumped-storage plant has the following characteristics.

Generating: q positive when generating, P_H is positive and $0 \le P_H \le +300$ MW

$$q(P_H) = 200 + 2 P_H \text{ acre-ft } (P_H \text{ in MW})$$

Pumping: q negative when pumping, P_P is negative and $-300 \text{ MW} \le P_P \le 0$

$$q(P_p) = -600 \text{ acre-ft/hr with } P_p = -300 \text{ MW}$$

Operating Restriction: The pumped hydro plant will be allowed to operate only at -300 MW when pumping. Cycle efficiency $\eta = 0.6667$ [the efficiency has already been built into the $q(P_H)$ equations].

The equivalent steam system has the cost curve

$$F(P_s) = 3877.5 + 3.9795 P_s + 0.00204 P_s^2 \text{ R/h } (200 \text{ MW} \le P_s \le 2500 \text{ MW})$$

Find the optimum pump-generate schedule using the gradient method for the following load schedule and reservoir constraint.

Load Schedule (each period is 4 h long)

Period	MW Load
1	1600
2	1800
3	1600
4	500
5	500
6	500

The reservoir starts at 8000 acre-ft and must be at 8000 acre-ft at the end of the sixth period.

Initial Schedule

Period	MW load	P_s	λ	Hydro pump/gen. (+ = gen, − = pump)	Reservoir volume at end of period
1	1600	1600	10.5	0	8000
2	1800	1800	11.3	0	8000
3	1600	1600	10.5	0	8000
4	500	500	6.02	0	8000
5	500	500	6.02	0	8000
6	500	500	6.02	0	8000

Starting with $k = 2$ and $i = 4$:

$$\lambda_2 = 11.3 \qquad \lambda_4 = 6.02 \qquad \lambda_4/\eta = 9.03$$

Therefore, it will pay to generate at as much as possible during the second period as long as the pump can restore the equivalent acre-ft of water during the fourth period. Therefore, the first schedule adjustment will look like the following.

Period	MW load	P_s	λ	Hydro pump/gen.	Reservoir volume at end of period
1	1600	1600	10.5	0	8000
2	1800	1600	10.5	+200	5600
3	1600	1600	10.5	0	5600
4	500	800	7.24	−300	8000
5	500	500	6.02	0	8000
6	500	500	6.02	0	8000

Next, we can choose to generate another 200 MW from the hydro plant during the first period and restore the reservoir during the fifth period.

Period	MW load	P_s	λ	Hydro pump/gen.	Reservoir volume at end of period
1	1600	1400	9.69	+200	5600
2	1800	1600	10.5	+200	3200
3	1600	1600	10.5	0	3200
4	500	800	7.24	−300	5600
5	500	800	7.24	−300	8000
6	500	500	6.02	0	8000

Finally, we can generate in the third period and replace the water in the sixth period.

Period	MW load	P_s	λ	Hydro pump/gen.	Reservoir volume at end of period
1	1600	1400	9.69	+200	5600
2	1800	1600	10.50	+200	3200
3	1600	1400	9.69	+200	800
4	500	800	7.24	−300	3200
5	500	800	7.24	−300	5600
6	500	800	7.24	−300	8000

A further savings can be realized by "flattening" the steam generation for the first three periods. Note that the costs for the first three periods as shown in the preceding table would be:

Period	P_s	Cost R	λ	Hydro pump/gen.
1	1400	53,788.80	9.69	+200
2	1600	61,868.40	10.50	+200
3	1400	53,788.80	9.69	+200
4, 5, 6	800	100,400.40	7.24	−300
		269,846.40		

If we run the hydro plant at full output during the peak (period 2) and then reduce the amount generated during periods 1 and 3, we will achieve a savings.

Period	P_s	Cost R	λ	Hydro pump/gen.
1	1450	55,747.50	9.90	+150
2	1500	57,747.00	10.10	+300
3	1450	55,747.50	9.90	+150
4, 5, 6		100,400.40	7.24	−300
		269,642.40		

The final reservoir schedule would be:

Period	Reservoir volume
1	6000
2	2800
3	800
4	3200
5	5600
6	8000

7.8 DYNAMIC PROGRAMMING SOLUTION TO THE HYDROTHERMAL SCHEDULING PROBLEM

Dynamic programming may be applied to the solution of the hydrothermal scheduling problem. The multiplant, hydraulically coupled systems offer computational difficulties that make it difficult to use that type of system to illustrate the benefits of applying DP to this problem. Instead we will illustrate the application with the single hydro plant operated in conjunction with a thermal system. Figure 7.16 shows a single, equivalent steam plant, P_s, and a hydro plant with storage, P_H, serving a single series of loads, P_L. Time intervals are denoted by j, where j runs between 1 and j_{max}.

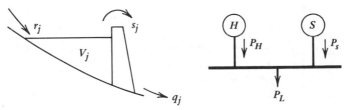

FIG. 7.16 Hydrothermal system model used in dynamic programming illustration.

Let: r_j = net inflow rate during period j

V_j = storage volume at the end of period j

q_j = flow rate through the turbine during period j

P_{Hj} = power output during period j

s_j = spillage rate during period j

P_{sj} = steam-plant output

P_{Lj} = load level

F_j = fuel cost rate for period j

Both starting and ending storage volumes, V_0 and V_{jmax}, are given as are the period loads. The steam plant is assumed to be on for the entire period. Its input-output characteristic is

$$F_j = a + bP_{sj} + cP_{sj}^2 \text{ R/h} \tag{7.42}$$

The water use rate characteristic of the hydroelectric plant is

$$q_j = d + gP_{Hj} + hP_{Hj}^2, \text{ acre-ft/h for } P_{Hj} > 0 \tag{7.43}$$

and

$$= 0 \quad \text{for } P_{Hj} = 0$$

The coefficients a through h are constants. We will take the units of water flow rate as acre-ft/h. If each interval, j, is n_j hours long, the volume in storage changes as

$$V_j = V_{j-1} + n_j(r_j - q_j - s_j) \tag{7.44}$$

Spilling water will not be permitted (i.e., all $s_j = 0$).
If V_i and V_k denote two different volume states, and

$$V_{j-1} = V_i$$

$$V_j = V_k$$

then the rate of flow through the hydro unit during interval j is

$$q_j = \frac{(V_i - V_k)}{n_j} + r_j$$

where q_j must be nonnegative and is limited to some maximum flow rate, q_{max}, which corresponds to the maximum power output of the hydro unit. The scheduling

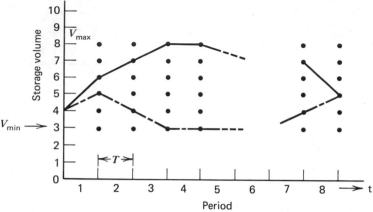

FIG. 7.17   Trajectories for hydro-plant operation.

problem involves finding the minimum cost trajectory (i.e., the volume at each stage). As indicated in Figure 7.17, numerous feasible trajectories may exist.

The DP algorithm is quite simple.

Let:

$\{i\}$ = the volume states at the start of the period j

$\{k\}$ = the states at the end of j

$TC_k(j)$ = the total cost from the start of the scheduling period to the end of period j for the reservoir storage state V_k

$PC(i, j - 1, k, j)$ = production cost of the thermal system in period j to go from an initial volume of V_i to an end of period volume V_k.

The forward DP algorithm is then,

$$TC_k(0) = 0$$

and

$$TC_k(j) = \min_{\{i\}} \left[TC_i(j - 1) + PC(i, j - 1; k, j) \right] \qquad (7.44)$$

We must be given the loads and natural inflows. The discharge rate through the hydro unit is, of course, fixed by the initial and ending storage levels and this in turn establishes the values of P_H and P_s. The computation of the thermal production cost follows directly.

There may well be volume states in the set V_k that are unreachable from some of the initial volume states V_i because of the operating limits on the hydro plants. There are many variations on the hydraulic constraints that may be incorporated in the DP computation. For example, the discharge rates may be fixed during certain intervals to allow fish ladders to operate or to provide water for irrigation.

Using the volume levels as state variables restricts the number of hydro power output levels that are considered at each stage since the discharge rate fixes the value of power. If a variable head plant is considered, it complicates the calculation of the power level as an average head must be used to establish the value of P_H. This is relatively easy to handle.

EXAMPLE 7E

It is, perhaps, better to use a simple numerical example than to attempt to discuss the DP application generally. Let us consider the two-plant case just described with the steam-plant characteristics as shown in Figure 7.18 with $F = 700 + 4.8 P_s + P_s^2/2000$, R/h, and $dF/dP_s = 4.8 + P_s/1000$, R/MWh, for P_s in MW and $200 \le P_s \le 1200$, MW. The hydro unit is a constant head plant shown in Figure 7.19 with

$$q = 260 + 10 P_H \qquad \text{for } P_H > 0, \qquad q = 0 \text{ for } P_H = 0$$

where

$$P_H \text{ is in MW}$$

and

$$0 \le P_H \le 200 \text{ MW}$$

The discharge rate is in acre-ft/h. There is no spillage, and both initial and final volumes are 10,000 acre-ft. The storage volume limits are 6000 and 18,000 acre-ft. The natural inflow is 1000 acre-ft/h.

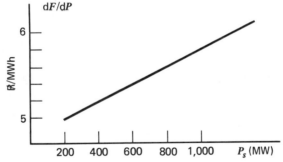

FIG. 7.18   Steam plant incremental cost function.

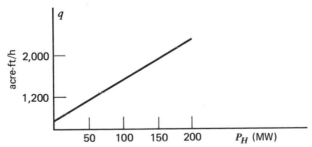

FIG. 7.19   Hydro plant q versus P_H function.

The scheduling problem to be examined is for a 24-h day with individual periods taken as 4 h each. (n_j = 4.0 h) The loads and natural inflows into the storage pond are

Period j	Load $P(j)$ (MW)	Inflow rate $r(j)$ (acre-ft/h)
1	600	1000
2	1000	1000
3	900	1000
4	500	1000
5	400	1000
6	300	1000

Procedure

If this were an actual scheduling problem, we might start the search using a coarse grid on both the time interval and the volume states. This would permit the future refinement of the search for the optimal trajectory after a crude search had established the general neighborhood. Finer grid steps bracketing the range of the coarse steps around the initial optimal trajectory could then be used to establish a better path. The method will work well for problems with convex (concave) functions. For this example we will limit our efforts to 4-h time steps and storage volume steps that are 2000 acre-ft apart.

During any period, the discharge rate through the hydro unit is

$$q_j = \frac{(V_{j-1} - V_j)}{4} + 1000 \tag{7.45}$$

The discharge rate must be nonnegative and not greater than 2260 acre-ft/h. For this problem we may use the equation that relates P_H, the plant output, to the discharge rate, q. In a more general case we may have to deal with tables that relate P_H, q, and the net hydraulic head.

The DP procedure may be illustrated for the first two intervals as follows. We take the storage volume steps at 6000, 8000, 10,000, ..., 18,000 acre-ft. The initial set of volume states is limited to 10,000 acre-ft. (In this example, volumes will be expressed in 1000 acre-ft to save space.) The table here summarizes the calculations for $j = 1$; the graph in Figure 7.20 shows the trajectories. We need not compute the data for greater volume states since it is possible to do no more than shut the unit down and allow the natural inflow to increase the amount of water stored.

	$j = 1$	$P_L(1) = 600$ MW	$\{i\} = 10$	
V_k	q	P_H	P_s	$TC_k(j)$
14	0	0	600	R15,040
12	500	24	576	14,523
10	1000	74	526	13,453
8	1500	124	476	12,392
6	2000	174	426	11,342

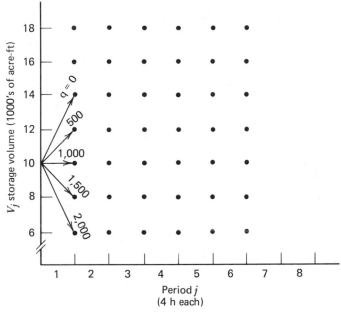

FIG. 7.20 Initial trajectories for DP example.

The tabulation for the second and succeeding intervals is more complex since there are a number of initial volume states to consider. A few are shown in the following table and illustrated in Figure 7.21.

$j = 2$	$P_L = 1000$ MW			$\{i\} = [6, 8, 10, 12, 14]$	
V_k	V_i	q	P_H	P_s	$TC_k(j)$
18	14	0	0	1000	R39,040[a]
16	14	500	24	976	38,484[a]
16	12	0	0	1000	38,523
14	14	1000	74	926	37,334[a]
14	12	500	24	976	37,967
14	10	0	0	1000	37,453
12	14	1500	124	876	36,194[a]
12	12	1000	74	926	36,818
12	10	500	24	976	36,897
12	8	0	0	1000	36,392
⋮	⋮		⋱		⋮
6	10	2000	174	826	33,477[a]
6	8	1500	124	876	33,546
6	6	1000	74	926	33,636

[a] Denotes the minimum cost path.

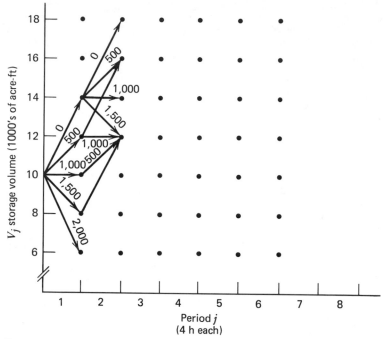

FIG. 7.21   Second-stage trajectories for DP example.

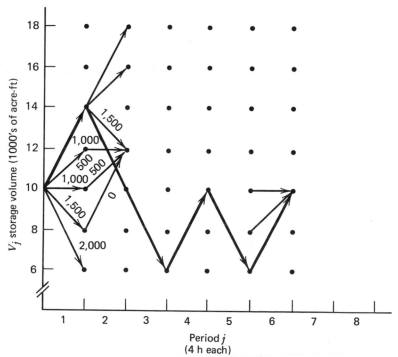

FIG. 7.22   Final trajectory for hydrothermal-scheduling example.

Finally, in the last period,

V_k	V_i	q	P_H	P_s	$TC_k(j)$
$j = 6$	$P_L = 300$ MW	$\{i\} = [6, 8, 10, 12, 14]$			
10	10	1000	74	226	82,240.61
10	8	500	24	276	82,260.21
10	6	0	0	300	81,738.46

are the only feasible combinations since the end volume is set at 10 and the minimum loading for the thermal plant is 200 MW.

The final, minimum cost trajectory for the storage volume is plotted in Figure 7.22. This path is determined to a rather coarse grid of 2000 acre-ft by 4-h steps in time and could be easily recomputed with finer increments.

7.8.1 Extension to Other Cases

The DP method is amenable to application in more complex situations. Longer time steps make it useful to compute seasonal *rule curves*, the long-term storage plan for a system of reservoirs. Variable-head cases may be treated. A sketch of the type of characteristics encountered in variable-head plants is shown on Figure 7.23. In this case, the variation in maximum plant output may be as important as the variation in water use rate as the net head varies.

7.8.2 Dynamic Programming Solution to Multiple Hydro-plant Problem

Suppose we are given the hydrothermal system shown in Figure 7.24. We have the following hydraulic equations when spilling is constrained to zero.

$$V_{1j} = V_{1j-1} + r_{1j} - q_{1j}$$

$$V_{2j} = V_{2j-1} + q_{1j} - q_{2j}$$

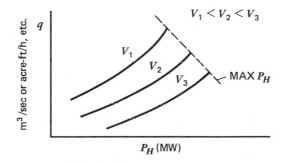

Variable head plant
$q = q(P_H, \bar{V})$
$\bar{V}$ = average volume used to represent
 the effect of the hydraulic head

FIG. 7.23 Input-output characteristic for variable-head hydroelectric plant.

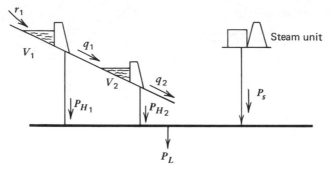

FIG. 7.24 Hydrothermal system with hydraulically coupled hydroelectric plants.

and the electrical equation

$$P_{H1}(q_{1j}) + P_{H2}(q_{2j}) + P_{sj} - P_{Lj} = 0$$

There are a variety of ways to set up the DP solution to this program. Perhaps the most obvious would be to again let the reservoir volumes, V_1 and V_2, be the state variables and then run over all feasible combinations. That is, let V_1 and V_2 both be divided into N volume steps $S_1 \cdots S_2$. Then the DP must consider N^2 steps at each time interval as shown on Figure 7.25.

This procedure might be a reasonable way to solve the multiple hydro-plant scheduling problem if the number volume steps were kept quite small. However, this is not practical when a realistic schedule is desired. Considered for example a reservoir volume that is divided into 10 steps ($N = 10$). If there were only one hydro plant, there would be 10 states at each time period resulting in a possible 100 paths to be investigated at each stage. If there were two reservoirs with 10 volume steps, there

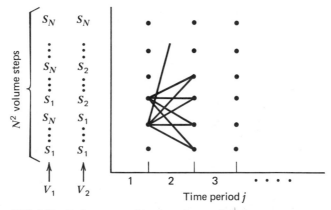

FIG. 7.25 Trajectory combinations for coupled plants.

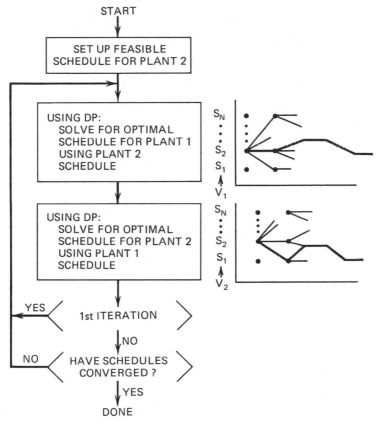

FIG. 7.26 Successive approximation solution.

would be 100 states at each time interval with a possibility of 10,000 paths to investigate at each stage.

This dimensionality problem can be overcome through the use of a procedure known as *successive approximation*. In this procedure, one reservoir is scheduled while keeping the other's schedule fixed, alternating from one reservoir to the other until the schedules converge. The steps taken in a successive approximation method appear in Figure 7.26.

EXAMPLE 7F

Two hydroelectric plants are located on the same river as shown in Figure 7.24. For this example, each hydro plant is identical to the one described in Example 7E. The equivalent fossil unit is also identical to that given in Example 7E. The inflow to plant number 1 is 1000 acre-ft/h and all the outflow from plant 1 flows into the reservoir of plant 2. We will schedule these two hydro plants using successive approximations. The reservoir conditions placed on the plants are

Plant 1: Initial reservoir volume = 6,000 acre-ft

Final reservoir volume = 16,000 acre-ft

Plant 2: Initial reservoir volume = 18,000 acre-ft

Final reservoir volume = 8,000 acre-ft

The initial schedule on plant 2 is set to

$$q_j = 0 \qquad \text{For } j = 1 \cdots 6$$

The load for the system is as follows.

Time period	Load (MW)
1	1100
2	1200
3	900
4	1100
5	1000
6	800

Time period = 4 hours

Iteration 1 (Plant 1)

The first half of iteration 1 schedules the optimal operation of plant 1 with the plant 2 schedule held as just given. Note that the volume upper limit constraint on plant 2 is ignored in this iteration.

Time period	Load (MW)	Steam P_s (MW)	P_{H1} (MW)	q_1 (acre-ft/h)	$V_1{}^a$ (acre-ft)	P_{H2} (MW)	q_2 (acre-ft/h)	$V_2{}^a$ (acre-ft)
1	1100	1100	0	0	10000	0	0	18000
2	1200	1026	174	2000	6000	0	0	26000
3	900	900	0	0	10000	0	0	26000
4	1100	976	124	1500	8000	0	0	32000
5	1000	1000	0	0	16000	0	0	32000
6	800	800	0	0	16000	0	0	32000

a Volumes are for reservoir volume at the end of the time period.

The total cost of supplying the power out of the steam plant is 139,528.8 R.

Iteration 1 (Plant 2)

The second half of iteration 1 schedules the optimal operation of plant 2 with the plant 1 schedule as calculated in the first half of iteration 1.

Time period	Load (MW)	Steam P_s (MW)	P_{H1} (MW)	q_1 (acre-ft/h)	V_1^a (acre-ft)	P_{H2} (MW)	q_2 (acre-ft/h)	V_2^a (acre-ft)
1	1100	901	0	0	10000	199	2250	9000
2	1200	852	174	2000	6000	174	2000	9000
3	900	900	0	0	10000	0	0	9000
4	1100	976	124	1500	8000	0	0	15000
5	1000	851	0	0	12000	149	1750	8000
6	800	800	0	0	16000	0	0	8000

[a] Volumes are for reservoir volume at the end of the time period.

The total cost of supplying the power out of the steam plant is now 127,504.9 R.

Iteration 2 (Plant 1)

We now reschedule plant 1 with the new plant 2 schedule as from the second half of iteration 1.

Time period	Load (MW)	Steam P_s (MW)	P_{H1} (MW)	q_1 (acre-ft/h)	V_1 (acre-ft)	P_{H2} (MW)	q_2 (acre-ft/h)	V_2 (acre-ft)
1	1100	901	0	0	10000	199	2250	9000
2	1200	914.5	111.5	1375	8500	174	2000	6500
3	900	900	0	0	12500	0	0	6500
4	1100	913.5	186.5	2125	8000	0	0	15000
5	1000	851	0	0	12000	149	1750	8000
6	800	800	0	0	16000	0	0	8000

Total cost = 127,487.0 R.

Iteration 2 (Plant 2)

Time period	Load (MW)	Steam P_s (MW)	P_{H1} (MW)	q_1 (acre-ft/h)	V_1 (acre-ft)	P_{H2} (MW)	q_2 (acre-ft/h)	V_2 (acre-ft)
1	1100	901	0	0	1000	199	2250	9000
2	1200	902	111.5	1375	8500	186.5	2125	6000
3	900	900	0	0	12500	0	0	6000
4	1100	913.4	186.5	2125	8000	0	0	14500
5	1000	863.5	0	0	12000	136.5	1625	8000
6	800	800	0	0	16000	0	0	8000

Total cost = 127,487.0 R.

As can be seen, each iteration of the successive approximations method yields a lower production cost. Further iterations yield no changes in the schedules.

Another useful technique in applying dynamic programming involves a change in the volume step size or quantization used in the DP process. The DP is started with large steps in volume and then, as the schedule converges, smaller steps are added

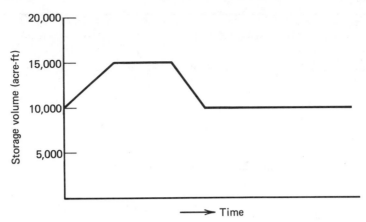

FIG. 7.27 Optimal schedule with 5000-acre-ft steps.

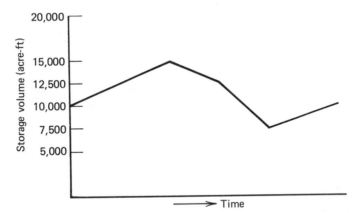

FIG. 7.28 Optimal schedule with steps added at 7500 and 12,500 acre-ft.

around the optimum schedule. For example, the optimal schedule shown in Figure 7.27 might have volume steps added as shown in Figure 7.28.

The advantage of using a variable step-size approach comes from not having to add the small increment steps over the entire range, thus keeping the DP search range reasonably small at each interval.

APPENDIX
Hydro-Scheduling With Storage Limitations

This appendix expands on the Lagrange equation formulation of the fuel-limited dispatch problem in Chapter 6 and the reservoir-limited hydro dispatch problem of Chapter 7. The expansion includes generator and reservoir storage limits and provides a proof that the "fuel cost" or "water cost" Lagrange multiplier γ will be constant unless reservoir storage limitations are encountered.

To begin, we will assume that we have a hydro unit and an equivalent steam unit supplying load as in Figure 7.5. Assume that the scheduling period is broken down into three equal time intervals with load, generation, reservoir inflow, and such constant within each period. In Chapter 6 (Section 6.2, Eqs. 6.1–6.6) and Chapter 7 (Section 7.4, Eqs. 7.22–7.29) we assumed that the total q was to be fixed at q_{TOT}, that is (see Section 7.4 for definition of variables),

$$q_{TOT} = \sum_{j=1}^{j_{max}} n_j q(P_{Hj}) \tag{7A.1}$$

In the case of a storage reservoir with an initial volume V_0, this constraint is equivalent to fixing the final volume in the reservoir. That is,

$$V_0 + n_1(r_1 - q(P_{H1})) = V_1 \tag{7A.2}$$

$$V_1 + n_2(r_2 - q(P_{H2})) = V_2 \tag{7A.3}$$

$$V_2 + n_3(r_3 - q(P_{H3})) = V_3 \tag{7A.4}$$

Substituting Eq. 7A.2 into Eq. 7A.3 and then substituting the result into Eq. 7A.4, we get

$$V_1 + \sum_{j=1}^{3} n_j r_j - \sum_{j=1}^{3} n_j q(P_{Hj}) = V_3 \tag{7A.5}$$

or

$$V_1 + \sum_{j=1}^{3} n_j r_j - q_{TOT} = V_3 \tag{7A.6}$$

Therefore, fixing q_{TOT} is equivalent to fixing V_3, the final reservoir storage. The optimization problem will be expressed as:

Minimize total steam plant cost: $\displaystyle\sum_{j=1}^{3} n_j F_s(P_{sj})$

Subject to equality constraints: $P_{Lj} - P_{sj} - P_{Hj} = 0 \qquad$ for $j = 1, 2, 3$

$$V_0 + n_1 r_1 - n_1 q(P_{H1}) = V_1$$

$$V_1 + n_2 r_2 - n_2 q(P_{H2}) = V_2$$

$$V_2 + n_3 r_3 - n_3 q(P_{H3}) = V_3$$

And subject to inequality constraints: $\quad V_j > V^{min} \qquad V_j < V^{max}$

$$P_{sj} > P_s^{min} \qquad P_{sj} < P_s^{max} \quad \text{for } j = 1, 2, 3$$

$$P_{Hj} > P_H^{min} \qquad P_{Hj} < P_H^{max}$$

We can now write a Lagrange equation to solve this problem.

$$\mathcal{L} = \sum_{j=1}^{3} n_j F_s(P_{sj}) + \sum_{j=1}^{3} \lambda_j(P_{Lj} - P_{sj} - P_{Hj})$$

$$+ \gamma_1(-V_0 - n_1 r_1 + n_1 q(P_{H1}) + V_1)$$

$$+ \gamma_2(-V_1 - n_2 r_2 + n_2 q(P_{H2}) + V_2)$$

$$+ \gamma_3(-V_2 - n_3 r_3 + n_3 q(P_{H3}) + V_3)$$

$$+ \sum_{j=1}^{3} \alpha_j^- (V^{\min} - V_j) + \sum_{j=1}^{3} \alpha_j^+ (V_j - V^{\max})$$

$$+ \sum_{j=1}^{3} \mu_{sj}^- (P_s^{\min} - P_{sj}) + \sum_{j=1}^{3} \mu_{sj}^+ (P_{sj} - P_s^{\max})$$

$$+ \sum_{j=1}^{3} \mu_{Hj}^- (P_H^{\min} - P_{Hj}) + \sum_{j=1}^{3} \mu_{Hj}^+ (P_{Hj} - P_H^{\max}) \tag{7A.7}$$

where:

n_j, P_{sj}, P_{Hj}, and $q(P_{Hj})$ are as defined in Section 7.4.

λ_j, γ_j, α_j^-, α_j^+, μ_{sj}^-, μ_{sj}^+, μ_{Hj}^-, μ_{Hj}^+ are Lagrange multipliers.

$V^{\min}$ and $V^{\max}$ are limits on reservoir storage.

$P_s^{\min}$, $P_s^{\max}$, $P_H^{\min}$, and $P_H^{\max}$ are limits on the generator output at the equivalent system and hydro plants, respectively.

We can set up the conditions for an optimum using the Kuhn-Tucker equations as shown in the Appendix to Chapter 3. The first set of conditions are

$$\frac{\partial \mathcal{L}}{\partial P_{sj}} = \frac{dF_s}{dP_{sj}} - \lambda_j - \mu_{sj}^- + \mu_{sj}^+ = 0 \tag{7A.8}$$

$$\frac{\partial \mathcal{L}}{\partial P_{Hj}} = -\lambda_j + \gamma_j n_j \frac{dq(P_{Hj})}{dP_{Hj}} - \mu_{Hj}^- + \mu_{Hj}^+ = 0 \tag{7A.9}$$

$$\frac{\partial \mathcal{L}}{\partial V_j} = \gamma_j - \gamma_{j+1} - \alpha_j^- + \alpha_j^+ = 0 \tag{7A.10}$$

The second and third set of conditions are just the original equality and inequality constraints. The fourth set of conditions are

$$\alpha_j^- (V^{\min} - V_j) = 0 \qquad \alpha_j^- \geq 0 \tag{7A.11}$$

$$\alpha_j^+ (V_j - V^{\max}) = 0 \qquad \alpha_j^+ \geq 0 \tag{7A.12}$$

$$\mu_{sj}^- (P_s^{\min} - P_{sj}) = 0 \qquad \mu_{Sj}^- \geq 0 \tag{7A.13}$$

$$\mu_{sj}^+ (P_{sj} - P_s^{\max}) = 0 \qquad \mu_{Sj}^+ \geq 0 \tag{7A.14}$$

$$\mu_{Hj}^- (P_H^{\min} - P_{Hj}) = 0 \qquad \mu_{Hj}^- \geq 0 \tag{7A.15}$$

$$\mu_{Hj}^+ (P_{Hj} - P_H^{\max}) = 0 \qquad \mu_{Hj}^+ \geq 0 \tag{7A.16}$$

If we assume that no generation limits are being hit, then μ_{sj}^-, μ_{sj}^+, μ_{Hj}^-, μ_{Hj}^+, for $j = 1, 2, 3$ are each equal to zero. The solution in Eqs. 7A.8, 7A.9, and 7A.10 is

$$\frac{dF_s}{dP_{sj}} = \lambda_j \qquad (7A.17)$$

$$\gamma_j n_j \frac{dq(P_{Hj})}{dP_{Hj}} = \lambda_j \qquad (7A.18)$$

$$\gamma_j - \gamma_{j+1} = \alpha_j^- - \alpha_j^+ \qquad (7A.19)$$

Now suppose the following volume-limiting solution exists.

$$V_1 > V^{\min} \quad \text{and} \quad V_1 < V^{\max}$$

then by Eq. 7A.11 and Eq. 7A.12

$$\alpha_1^- = 0 \quad \text{and} \quad \alpha_1^+ = 0$$

and

$$V_2 = V^{\min} \quad \text{and} \quad V_2 < V^{\max}$$

then

$$\alpha_2^- > 0 \qquad \alpha_2^+ = 0$$

Then clearly from Eq. 7A.19,

$$\gamma_1 - \gamma_2 = \alpha_1^- - \alpha_2^+ = 0$$

so

$$\gamma_1 = \gamma_2$$

and

$$\gamma_2 - \gamma_3 = \alpha_2^- - \alpha_2^+ > 0$$

so

$$\gamma_2 > \gamma_3$$

Thus we see that γ_j will be constant over time unless a storage volume limit is hit. Further note that this is true regardless of whether or not generator limits are hit.

PROBLEMS

7.1 Given the following steam and hydro plant characteristics:

Steam Plant

Incremental cost $= 2.0 + 0.002 P_s$ R/MWh and $100 \leq P_s \leq 500$ MW

Hydro Plant

Incremental water rate $= 50 + 0.02\, P_H$ ft^3/sec/MW $\qquad 0 \le P_H \le 500$ MW

Load

Time Period	Load
12 midnight to 9 AM	350 MW
9 AM–6 PM	700 MW
6 PM–12 midnight	350 MW

Assume

- The water input for $P_H = 0$ may also be assumed to be zero, that is

$$q(P_H) = 0 \text{ for } P_H = 0$$

- Neglect losses
- The thermal plant remains on line for the 24-h period.

Find

The optimum schedule of P_s and P_H over the 24-h period that meets the restriction that the total water used in 1250 million ft^3 of water, that is,

$$q_{TOT} = 1.25 * 10^9 \text{ ft}^3$$

7.2 Assume that the incremental water rate in problem 7.1 is constant at 60 ft^3/sec/MW and that the steam unit is not necessarily on all the time. Further assume that the thermal cost is

$$F(P_s) = 250 + 2\,P_s + P_s^2/1000$$

Repeat problem 7.1 with the same water constraint.

7.3 Gradient Method for Hydro-Thermal Scheduling

A thermal generation system has a composite fuel cost characteristic that may be approximated by

$$F = 700 + 4.8\,P_s + P_s^2/2000, \text{ R/h}$$

for

$$200 \le P_s \le 1200 \text{ MW}$$

The system load may also be supplied by a hydro unit with the following characteristics.

$q(P_H) = 0$ when $P_H = 0$

$q(P_H) = 260 + 10\ P_H$, acre-ft/h
     for $0 < P_H \leq 200$ MW

$q(P_H) = 2260 + 10(P_H - 200) + 0.028(P_H - 200)^2$, acre-ft/h
     for $200 < P_H \leq 250$ MW

The system load levels in chronological order are as follows.

Period	P_L, MW
1	600
2	1000
3	900
4	500
5	400
6	500

Each period is 4 h long.

7.3.1 Assume the thermal unit is on-line all the time and find the optimum schedule (the values of P_s and P_H for each period) such that the hydro plant uses 23,500 acre-ft of water. There are no other hydraulic constraints or storage limits, and you may turn the hydro unit off when it will help.

7.3.2 Now still assuming the thermal unit is on-line each period, use a gradient method to find the optimum schedule given the following conditions on the hydroelectric plant.

a. There is a constant inflow into the storage reservoir of 1000 acre-ft/h.

b. The storage reservoir limits are

$$V_{max} = 18{,}000 \text{ acre-ft}$$

and

$$V_{min} = 6{,}000 \text{ acre-ft}$$

c. The reservoir starts the day with a level of 10,000 acre-ft, and we wish to end the day with 10,500 acre-ft in storage.

7.4 Hydro-thermal Scheduling Using Dynamic Programming

Repeat Example 7E in the chapter except the hydroelectric unit's water rate characteristic is now one that reflects a variable head. This characteristic also exhibits a maximum capability that is related to the net head. That is,

$$q = 0 \qquad \text{for } P_H = 0$$

$$q = 260 + 10\ P_H\left(1.1 - \frac{\bar{V}}{100{,}000}\right), \quad \text{acre-ft/h}$$

for

$$0 < P_H \leq 200 \left(0.9 + \frac{\bar{V}}{100,000} \right), \text{ MW}$$

where

$$\bar{V} = \text{average reservoir volume}$$

For this problem assume constant rates during a period so that

$$\bar{V} = \frac{1}{2}(V_k + V_i)$$

where V_k = end of period volume

        V_i = start of period volume

    The required data are

Fossil Unit: On-line entire time

$$F = 770 + 5.28 \, P_s + 0.55 \times 10^{-3} \, P_s^2, \, \text{R/h}$$

for

$$200 \leq P_s \leq 1200 \text{ MW}$$

Hydro Storage and Inflow

$$r = 1000 \text{ acre-ft/h inflow}$$

$$6,000 \leq V \leq 18,000 \text{ acre-ft storage limits}$$

$$V = 10,000 \text{ acre-ft initially}$$

and

$$V = 10,000 \text{ acre-ft at end of period}$$

    Load for 4-h Periods

J: Period	Load (MW)
1	600
2	1000
3	900
4	500
5	400
6	300

Find the optimal schedule with storage volumes calculated at least to the nearest 500 acre-ft.

7.5 Pumped-Storage Plant Scheduling Problem

A thermal generation system has a composite fuel-cost characteristic as follows.

$$F = 250 + 1.5 \, P_s + P_s^2/200 \, \text{R/h}$$

for

$$200 \le P_s \le 1200 \text{ MW}$$

In addition, it has a pumped-storage plant with the following characteristics:

1. Maximum output as a generator = 180 MW
   (the unit may generate between 0 and 180 MW).
2. Pumping load = 200 MW
   (the unit may only pump at loads of 100 or 200 MW).
3. The cycle efficiency is 70%
   (that is, for every 70 MWh generated, 100 MWh of pumping energy are required).
4. The reservoir storage capacity is equivalent to 1600 MWh of generation.

The system load level in chronological order is the same as that in Problem 7.3.

a. Assume the reservoir is full at the start of the day and must be full at the end of the day. Schedule the pumped-storage plant to minimize the thermal system costs.
b. Repeat the solution to (a) assuming that the storage capacity of the reservoir is unknown and that it should be at the same level at the end of the day. How large should it be for minimum thermal production cost?

Note: In solving these problems you may assume that the pumped-storage plant may operate for partial time periods. That is, it does not have to stay at a constant output or pumping load for the entire 4-h load period.

FURTHER READING

The literature relating to hydrothermal scheduling is extensive. For the reader desiring a more complete guide to these references, we suggest starting with reference 1, which is a bibliography covering 1959 through 1972, prepared by a working group of the Power Engineering Society of IEEE.

References 2 and 3 contain examples of simulation methods applied to the scheduling of large hydroelectric systems. The five-part series of papers by Bernholz and Graham (4) presents a fairly comprehensive package of techniques for optimization of short-range hydrothermal schedules applied to the Ontario hydro system. Reference 5 is an example of optimal scheduling of the system on the Susquehanna River.

A theoretical development of the hydrothermal scheduling equations is contained in reference 6. This 1963 reference should be reviewed by any reader contemplating undertaking a research project in hydrothermal scheduling methods. It points out clearly the impact of the constraints and their effects on the pseudo, marginal value of the hydroelectric energy.

Reference 7 illustrates an application of gradient-search methods to the coupled plants in the Ontario system. Reference 8 illustrates the application of dynamic-programming techniques to this type of hydrothermal system in a tutorial fashion. Finally, references 9 and 10 contain examples of methods for scheduling pumped-storage hydroelectric plants in a predominantly thermal system.

This short reference list is only a sample. The reader should be aware that a literature search in hydrothermal-scheduling methods is a major undertaking. We suggest the serious student of this topic start with reference 1 and its predecessors and successors.

1. "Description and Bibliography of Major Economy-Security Functions, Parts I, II, and III," IEEE Working Group Report, *IEEE Transactions on Power Apparatus and Systems*, Vol. PAS-100, January 1981, pp. 211–235.

2. Bruderell, R. N., Gilbreath, J. H., "Economic Complementary Operation of Hydro Storage and Steam Power in the Integrated TVA System," *AIEE Transactions*, Vol. 78, June 1959, pp. 136–150.

3. Hildebrand, C. E., "The Analysis of Hydroelectric Power-Peaking and Poundage by Computer," *AIEE Transactions*, Vol. 79, Part III, December 1960, pp. 1023–1029.

4. Bernholz, B., Graham, L. J., "Hydrothermal Economic Scheduling," a five-part series:

   a. "Part I. Solution by Incremental Dynamic Programming," *AIEE Transactions*, Vol. 79, Part III, December 1960, pp. 921–932.

   b. "Part II. Extension of Basic Theory," *AIEE Transactions*, Vol. 81, Part III, January 1962, pp. 1089–1096.

   c. "Part III. Scheduling the Thermal System Using Constrained Steepest Descent," *AIEE Transactions*, Vol. 81, Part III, February 1962, pp. 1096–1105.

   d. "Part IV. A Continuous Procedure for Maximizing the Weighted Output of a Hydroelectric Generating Station," *AIEE Transactions*, Vol. 81, Part III, February 1962, pp. 1105–1107.

   e. "Part V. Scheduling a Hydrothermal System with Interconnections," *AIEE Transactions*, Vol. 82, Part III, June 1963, pp. 249–255.

5. Anstine, L. T., Ringlee, R. J., "Susquenhanna River Short-Range Hydrothermal Co-ordination," *AIEE Transactions*, Vol. 82 Part III, April 1963, pp. 185–191.

6. Kirchmayer, L. K., Ringlee, R. J., "Optimal Control of Thermal Hydro-System Operation," IFAC Proceedings, 1964, pp. 430/1–430/6.

7. Bainbridge, E. S., McNamee, J. M., Robinson, D. J., Nevison, R. D., "Hydrothermal Dispatch with Pumped Storage," *IEEE Transactions on Power Apparatus and Systems*, Vol. PAS-85, May 1966, pp. 472–485.

8. Engles, L., Larson, R. E., Peschon, J., Stanton, K. N., "Dynamic Programming Applied to Hydro and Thermal Generation Scheduling," A paper contained in the IEEE Tutorial Course Text, 76CH1107-2-PWR, 1976, IEEE, New York, N.Y.

9. Bernard, P. J., Dopazo, J. F., Stagg, G. W., "A Method for Economic Scheduling of a Combined Pumped Hydro and Steam-Generating System," *IEEE Transactions on Power Apparatus and Systems*, Vol. PAS-83, January 1964, pp. 23–30.

10. Kennedy, T., Mabuce, E. M., "Dispatch of Pumped Storage on an Interconnected Hydrothermal System," *IEEE Transactions on Power Apparatus and Systems*, Vol. PAS 84, June 1965, pp. 446–457.

chapter 8

Energy Production Cost Models for Fuel Budgeting and Planning

8.1 INTRODUCTION

Energy production cost models are computational models, usually implemented on a digital computer, designed to calculate future generation system production costs, requirements for energy imports, availability of energy for sales to other systems, and fuel consumption. They are widely used throughout the electric utility industry as an aid in long-range system planning, in fuel budgeting, and in system operation. The primary function of computing future system energy costs is accomplished by using computer models of expected load patterns and simulating the operation of the generation system to meet these loads.

The digital simulation of the generation system involves representation of:

1. Generating unit efficiency characteristics (input-output curves, etc.).
2. Fuel costs per unit of energy supplied.

3. System operating policies with regard to scheduling of unit operation and the economic dispatching of groups of units that are on-line.

4. Contracts for the purchases and sales of both energy and power capability.

When hydroelectric plants are a part of the power system, the production cost simulation will involve models of the policies used to operate these plants. The first production cost models were deterministic, in that the status of all units and energy resources was assumed known and the load is a single estimate. Stochastic models are currently used where the risk of sudden, random generating unit failures are considered as a part of the process and the future load to be served is considered to be a probability distribution.

This chapter will discuss the deterministic models briefly and go into some detail concerning the central techniques involved in the probabilistic model. It is not possible to delve into all the details involved in a typical modern computer program since these programs may be quite large with thousands of instructions and thousands of items of data. Any such discussion would be almost instantly out of date since new problems keep arising. For example, the original purpose of these production cost programs was primarily computation of future system operating costs. In recent years these models have been used to study such diverse areas as the possible effects of load management, the impact of fuel shortages, and the reliability of future systems.

The "universal" block diagram in Figure 8.1 shows the organization of a "typical" energy production cost program. The computation simulates the system operation on a chronological basis with system data input being altered at the start of each interval. These programs must be able to recognize and take into account in some fashion the need for scheduled maintenance outages. Logic may be incorporated in this type of program to simulate the maintenance outage allocation procedure actually used as well as to process maintenance schedules that are input to the program.

Expansion planning and fuel budgeting production cost programs require load models that cover weeks, months, and/or years. The expected load patterns may be modeled by the use of typical, normalized hourly load curves for the various types of days expected in each subinterval (i.e., month or week) or else by the use of load duration or load distribution curves.

A *load duration curve* expresses the period of time (say number of hours) in a fixed interval (day, week, month, or year) that the load is expected to equal or exceed a given megawatt value. It is usually plotted with the load on the vertical axis and the time period on the horizontal axis.

The scheduling of unit maintenance outages may involve time intervals as short as a day or as long as a month. The requirements for economic data such as unit, plant, and system consumption and fuel costs are usually on a monthly basis. When these time interval requirements conflict, as they often do, the load model must be created in the model for the smallest subinterval involved in the simulation.

Production cost programs may be found in many modern control centers as part of the overall "application program" structure where these production cost

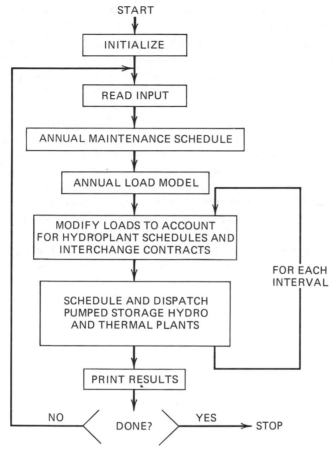

FIG. 8.1 Block diagram for a typical energy production cost program used for planning.

models are usually intended to produce shorter term computations of production costs (i.e., a few hours to the entire week) in order to facilitate negotiations for energy (or power) interchange between systems or else to compute cost savings in order to allocate economic benefits among pooled companies. In either application the production cost simulation is used to evaluate costs under two or more assumptions. For example, in interchange negotiations the system operators can evaluate the cost of producing the energy on the system versus the costs if it were to be purchased.

In U.S. power pools where units owned by several different utilities are dispatched by the control center, it is usually necessary to compute the production cost "savings" due to pooled operation. That is, each seller of energy is paid for the cost of producing the energy sold and may be given one-half the production cost "savings" of the system receiving the energy. One way of determining these savings is to simulate the production costs of each system supplying just its own load. In fact, in at least one U.S. pool this is called "own-load dispatch." These computed production costs can be compared with actual costs to arrive at the charges for transferring energy.

Production cost computations are also needed in fuel budgeting. This involves making computations to forecast the needs for future fuel supplies at specific plant sites. Arrangements for fuel supplies vary greatly among utilities. In some instances the utility may control the mining of coal or the production and transportation of natural gas; in others it may contract for fuel to be delivered to the plant. In many cases the utility will have made a long-term arrangement with a fuel supplier for the fuel needed for a specific plant. (Examples are mine-mouth coal plants or nuclear units.) In still other cases the utility may have to obtain fuel supplies on the open (i.e., "spot") market at whatever prices are prevailing at that time. In any case it is necessary to make a computation of the expected fuel supply requirements so that proper arrangements can be made sufficiently in advance of the requirements. This requires a forecast of specific quantities (and large quantities) of fuel at given future dates.

The operating center production cost needs may have a 7 day time horizon. The fuel budgeting time span may encompass 1 to 5 yr and might, in the case of the mine-mouth plant studies, extend out to the expected life of the plant. System expansion studies usually encompass a minimum of 10 yr and in many cases extend to 30 yr into the future. It is this difference in time horizon that makes different models and approaches suitable for different problems.

8.2 TYPES OF PRODUCTION COST PROGRAMS

Table 8.1 lists the major features that may vary from program to program and indicates, along the horizontal axis, the four major program uses of

1. Long-range planning.
2. Fuel budgeting.
3. Operations planning.
4. Weekly schedules.

Also indicated are the types of programs that have been found useful, so far, in each application. The type of load model used will determine, in part, the suitability of each program type for a given application.

Production cost models and digital computer program developments have historically followed the sequence in Table 8.1. This chapter will jump from the original model format where loads are represented using load-duration curves and generating units are block-loaded to the more recent, probabilistic production cost models. The types of production cost programs shown in Table 8.1, which utilize chronological load patterns, (i.e., load cycles) and deterministic scheduling methods, are computer implementations of the economic dispatching techniques and unit commitment methods explored in the previous seven chapters. That is, production costs and fuel consumption are computed repetitively assuming that the load cycles

TABLE 8.1 Energy Production Cost Programs

Load model	Interval considered	Economic dispatch procedure for thermal units	Long-range planning	Fuel budgeting	Operations planning	Weekly schedules
Total energy or load duration	Season or year	Block loading[a]	x			
Load duration or load cycles	Month or week	Incremental loading	x	x	x	
Load duration or load cycles	Month, week or days	Incremental loading with forced outages considered	x	x	x	
Load cycle	Weeks or days	Incremental loading (losses)	x	x	x	x

[a] The term "block loading" refers to the scheduling of complete units in economic order without regard to incremental cost. The procedure is illustrated in this section.

are known for an extended period into the future and that the availability of every unit can be predicted with 100% certainty for each subinterval of that future period. Any extended discussion of these procedures would be primarily a discussion of computer programming techniques and algorithms.

We will leave the pursuit of this to concentrate instead on the discussion of the probabilistic representation of the future loads and generating unit availabilities. In these computations, the expected values of production costs and fuel consumption are computed without the assumption of a perfectly known future.

The terminology used tends toward jargon and requires some explanation. In representing future loads, sometimes it is satisfactory to specify only the total energy generation for a period. This might be perfectly satisfactory if only total fuel consumption and production costs are of interest and neither capacity limitations nor chronological effects are important.

Where capacity limitations are of more concern, a *load-duration curve* might be used. Figure 8.2 shows an expected load pattern in (a), a histogram of load for a given time period in (b), and the load-duration curve constructed from it in (c). In practical developments, the density and distribution functions may be developed as histograms where each load level, L, denotes a range of loads. These last two curves are expressed in both hours and per unit probability versus the megawatts of load. Figure 8.3 shows the more conventional representation of a load-duration curve where the probability is now expressed in terms of the number of hours (out of the total hours in the period) that the load equals, or exceeds, a given level, L, MW. It is conventional in deterministic production cost analyses to show this curve with the

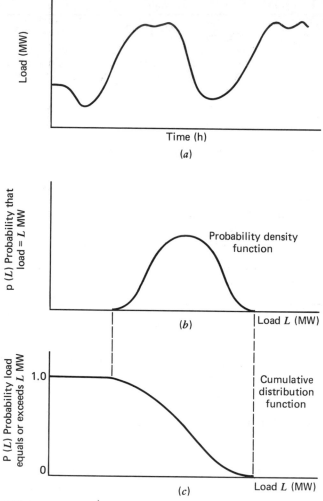

FIG. 8.2 Load probability functions.

load on the vertical axis. In the probabilistic calculations the form shown on Figure 8.2c is used.

In the simulation of the economic dispatch procedures with this type of load model, thermal units may be *block-loaded*. This means the units (or major segments of a unit) on the system are ordered in some fashion (usually cost) and are assumed to be fully loaded, or loaded up to the limitations of the load-duration curve. Figure 8.4 shows this procedure for a system where the internal peak load is 1700 MW. The units are considered to be loaded in a sequence determined by their average cost at full load in ₨/MWh. The amount of energy generated by each unit is equal to the area under the load-duration curve between the load levels in megawatts supplied by each unit.

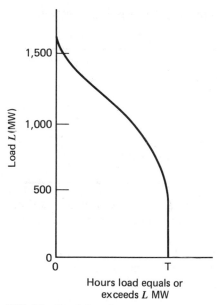

FIG. 8.3 Load-duration curve.

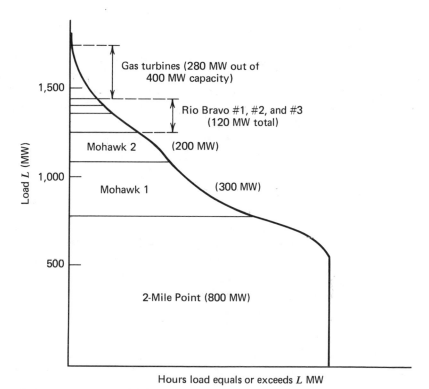

FIG. 8.4 Block-loaded units.

This system consists of three plants plus an array of gas-turbine generating units. These are

Unit	Maximum capability (MW)
2-mile point	800
Mohawk 1	300
Mohawk 2	200
Rio Bravo 1	75
Rio Bravo 2	25
Rio Bravo 3	20
8 gas turbines (each 50 MW)	400
Total	1820 MW

Note that in this system, the gas turbines are not used appreciably since the peak load is only 1700 MW and each unit is assumed to be available all the time during the interval.

Besides representing the thermal-generating plants, the various production cost programs must also simulate the effects of hydroelectric plants with and without water storage, contracts for energy and capacity purchases and sales, and pumped-storage hydroelectric plants. The action of all these results in a modified load to be served finally by the array of thermal units. The action of the thermal plants should be simulated to consider the security practices and policies of the power system as well as to simulate to some appropriate degree the economic dispatch procedures used on the system to control the unit output levels.

More complex production cost programs used to cover shorter time periods may duplicate the logic and procedures used in the control of the units. The most complex involve the procedures discussed in the previous three chapters on unit commitment and hydrothermal scheduling. These programs will usually use hourly forecasts of energy (i.e., the "hourly, integrated load" forecast) and thermal generating unit models that include incremental cost functions, start-up costs, and various other operating constraints.

EXAMPLE 8A

Let us consider the load-duration curve technique for a system of two units. Initially the random forced outages of the generating units will be neglected. Then we will incorporate consideration of these outages in order to show their effects on production costs and the ability of the small sample system to serve the load pattern expected. The load consists of the following.

x-load (MW)	Duration (h)	Energy (MWh)
100	20	2000
80	60	4800
40	20	800
Total = 100 h		7600 MWh

From these data we may construct a load-duration curve in tabular and graphic form. The load-duration curve shows the number of hours that the load equals or exceeds a given value.

x-load (MW)	Exact duration, $T\,p(x)$	$T\,P_n(k)$, hours that load equals or exceeds x
0	0	100
20	0	100
40	20	100
60	0	80
80	60	80
100	20	20
100+		0

The table has been created for uniform load-level steps of 20 MW each. The table also introduces the notation that is useful in regarding the load-duration curve as a form of probability distribution. The load density and distribution functions, $p(x)$ and $P_n(x)$, respectively, are probabilities. Thus $p(20) = 0$, $p(40) = 20/100 = 0.2$, $p(60) = 0$, and so forth, and $P_n(20) = P_n(40) = 1.0$, $P_n(60) = 0.8$, and so forth. The distribution function, $P_n(x)$, and the density, $p(x)$, are related as follows.

$$P_n(x) = 1 - \int_x^\infty p(x)\,dx \tag{8.1}$$

For discrete-density functions (or histograms) in tabular form, it is easiest to construct the distribution by cumulating the probability densities from the highest to the lowest values of the argument (the load levels).

The load-duration curve is shown in Figure 8.5 in a way that is convenient to use for the development of the probabilistic scheduling methods.

The two units of the generating system have the following characteristics.

Unit	Power output (MW)	Fuel input (10^6 Btu/h)	Fuel cost (R/10^6 Btu)	Fuel cost rate (R/h)	Incremental fuel cost (R/MWh)	Unit forced outage rate (per unit)
1	0	160	1	160	—	
	80	800	1	800	8	0.05
2	0	80	2	160	—	
	40	400	2	800	16	0.10

In addition to the usual input-output characteristics, the forced outage rates are given. This rate represents the fraction of time that the unit is not available due to a failure of some sort out of the total time that the unit should be available for service. In computing forced outage rates any periods where a unit is on scheduled outage for maintenance are excluded. The unit forced outage rates are initially neglected, and the two units are assumed to be available 100% of the time.

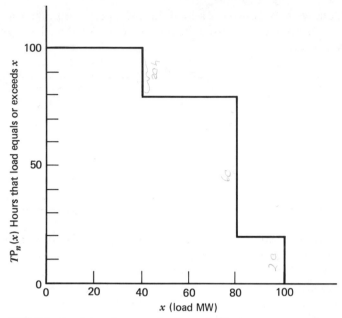

FIG. 8.5 Load-duration curve for Example 8A.

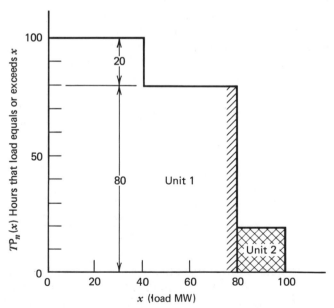

FIG. 8.6 Load-duration curve with block-loaded units.

Units are "block-loaded" with unit 1 being used first because of its lower average cost per MWh. The load-duration curve itself may be used to visualize the unit loadings. Figure 8.6 shows the two units block loaded.

Unit 1 is on-line for 100 h and is generating at an output level of 80 MW for 80 h and 40 MW for 20 h. Therefore, the production costs for unit 1 for this period are

$$= \text{hours on line} \times \text{no load fuel cost rate}$$
$$+ \text{energy generated} \times \text{incremental fuel cost rate}$$
$$= 100 \text{ h} \times 160 \text{ R/h} + (6400 + 800) \text{ MWh} \times 8 \text{ R/MWh}$$
$$= 16{,}000 \text{ R} + 57{,}600 \text{ R} = 73{,}600 \text{ R}$$

Similarly, unit 2 is required only 20 h in the interval and generates 400 MWh at a constant output level of 20 MW. Therefore its production costs for this period are

$$= 20 \text{ h} \times 160 \text{ R/h} + 400 \text{ MWh} \times 16 \text{ R/MWh} = 9600 \text{ R}$$

These data are summarized as follows.

Unit	Load (MW)	Hours duration	Energy (MWh)	Fuel used (10^6 Btu)	Fuel cost (R)
1	40	20	800	9,600	9,600
	80	80	6,400	64,000	64,000
			7,200	73,600	73,600
2	20	20	400	4,800	9,600
			7,600	78,400	83,200

Note that these two units can easily supply the expected loads. If a third unit were available it would not be used, except as standby reserve.

EXAMPLE 8B

Next let us consider the effects of the random forced outages of these units and compute the expected production costs. This example situation contains relatively few possible events so that the expected operation of each unit may be determined by enumeration of all the possible outcomes. For this procedure it is easiest at this point to utilize the load density rather than the load-distribution function.

Load level by load level the operation and generation of the two units are as follows.

1. Load = 40 MW; duration 20 h

Unit 1	On-line	20 h
	Operates	$0.95 \times 20 = 19$ h
	Output	40 MW
	Energy	$19 \times 40 = 760$ MWh

Unit 2   On-line   1 h
             Operates $0.9 \times 1 = 0.9$ h
             Output   40 MW
             Energy $0.9 \times 40 = 36$ MWh

Load energy   = 800 MWh
Generation   = 796 MWh
Unserved energy = 4 MWh
Shortages   40 MW for 0.1 h

2. Load = 80 MW; duration 60 h

Unit 1   On-line   60 h
             Operates $0.95 \times 60 = 57$ h
             Output   80 MW
             Energy $57 \times 80 = 4560$ MWh

Unit 2   On-line   60 h total
             Operates $0.9 \times 3 = 2.7$ h
             Output   40 MW
             Energy $2.7 \times 40 = 108$ MWh

Load energy   = 4800 MWh
Generation   = 4668 MWh
Unserved energy = 132 MWh
Shortages   80 MW for 0.3 h =   24 MWh
                     40 MW for 2.7 h = 108 MWh
                                   $\overline{132 \text{ MWh}}$

3. Load = 100 MW; duration 20 h

Unit 1   On-line   20 h
             Operates $0.95 \times 20 = 19$ h
             Output   80 MW
             Energy $19 \times 80 = 1520$ MWh

Unit 2   On-line   20 h
             Operates   as follows:

      a. Unit 1 on line and operating 19 h
          Unit 2 operates $0.9 \times 19 = 17.1$ h

          Output   20 MW
          Energy $17.1 \times 20 = 342$ MWh
          Shortage   20 MW for 1.9 h

      b. Unit 1 supposedly on-line, but not operating 1 h
          Unit 2 operates $0.9 \times 1 = 0.9$ h

          Output   40 MW
          Energy $0.9 \times 40 = 36$ MWh
          Shortages   100 MW for 0.1 h
                            60 MW for 0.9 h

Load energy   = 2000 MWHR
Generation   = 1898 MWHR
Unserved energy =   102 MWHR
Shortages   100 MW for 0.1 h =   10 MWh
                       60 MW for 0.9 h =   54 MWh
                       20 MW for 1.9 h =   38 MWh
                                          ‾‾‾‾‾‾‾‾
                                          102 MWh

Because this example is so small it has been necessary to make an arbitrary assumption concerning the commitment of the second unit. The assumption made is that the second unit will be on-line for any load level that equals or exceeds the capacity of the first unit. Thus the second unit is on-line for the 60-h duration of the 80 MW load. This assumption agrees with the algorithm developed later in the chapter.

The enumeration of the possible states is not quite complete. We have accounted for the periods when the load is satisfied and the times when there will be a real shortage of capacity. In addition, we need to separate the periods when the load is satisfied into periods where there is excess capability (more generation than load) and periods when the available capacity exactly matches the load (generation equals load). The latter periods are called *zero MW shortage* because there is no reserve capacity in that period. This information is needed in case an additional unit becomes available or emergency capacity needs to be purchased. This additional capacity would need to be operated during the entire period of a zero MW shortage because the occurrence of a real shortage is a random event depending on the failure of an operating generator.

For this example there are two such periods, one during the 40 MW load period and the other during the 80 MW load period. That is, the additional "zero MW shortage" conditions occur during those periods when the load is supplied precisely with no additional available capacity. Therefore, to the shortage events presented previously we add the following.

| | | | | Zero reserve |
Load	Duration	Unit 1	Unit 2	expected duration
1. 40 MW	20 h	Out	In	$0.05 \times 0.9 \times 20 = 0.9$
2. 80 MW	60 h	In	Out	$0.95 \times 0.1 \times 60 = 5.7$
				6.6 h

These "0 MW shortage" events are of significance in dispatching any additional supply since their total expected duration determines the number of hours they will be required.

All these events may be presented in an orderly fashion. Since each unit may be either on or off and there are three loads, the total number of possible events is $3 \times 2 \times 2 = 12$. These are summarized along with the consequence of each event in Table 8.2.

Now, having enumerated all the possible operating events, it is possible to compute the expected production costs and shortages. Recall from Example 8A

TABLE 8.2 Summary of All Possible States

Load (MW)	Duration (h)	Event no.	Unit 1		Unit 2		Combined event	
			Status[a]	Power (MW)	Status[a]	Power (MW)	Duration (h)	Consequence
40	20	1	1	40	1	0	17.1	Load satisfied; unit 2 not required
		2	1	40	0	0	1.9	Same as event no. 1
		3	0	0	1	40	0.9	Load satisfied; 0 MW shortage 0.9 h
		4	0	0	0	0	0.1	40 MW shortage 0.1 h
80	60	5	1	80	1	0	51.3	Load satisfied; unit 2 not required
		6	1	80	0	0	5.7	Load satisfied; 0 MW shortage 5.7 h
		7	0	0	1	40	2.7	40 MW shortage 2.7 h
		8	0	0	0	0	0.3	80 MW shortage 0.3 h
100	20	9	1	80	1	20	17.1	Load satisfied
		10	1	80	0	0	1.9	20 MW shortage 1.9 h
		11	0	0	1	40	.9	60 MW shortage 0.9 h
		12	0	0	0	0	.1	100 MW shortage 0.1 h

[a] Under "Status" a 1 denotes available and a 0 denotes unavailable.

that the operating cost characteristics of the two units are

$$F_1 = 160 + 8\,P_1,\ R/h$$

and

$$F_2 = 160 + 16\,P_2,\ R/h$$

and the fuel costs are 1 and 2 $R/10^6$ Btu, respectively. The calculated operating costs considering forced outages may be summarized as follows.

Unit	Hours on-line	Total expected operating hours	Expected energy generation (MWh)	Expected fuel use (10^6 Btu)	Expected production cost (R)
1	100	95.0	6840	69,920	69,920
2	81	72.9	522	10,008	20,016
	Totals		7362	79,928	89,936

The expected production costs for unit 1 are

$$= 95\ h \times 160\ R/h + 6840\ MWh \times 8\ R/MWh$$

and for unit 2

$$= 72.9\ h \times 160\ R/h + 552\ MWh \times 16\ R/MWh$$

Note that compared to the results of Example 8A, the fuel consumption has increased 1.95% over that found neglecting random forced outages and the total cost has increased 8.1%. This would be increased even more if the unserved energy, 238 MWh, were to be supplied by some high cost emergency source.

The expected unserved demands and energy may be summarized from the preceding data as shown in Table 8.3.

TABLE 8.3 Unserved Load

Unserved demand (MW)	Duration of shortage (h)	Unserved energy (MWh)	Duration of given shortages or more (h)
0	6.6	0	12.6
20	1.9	38	6.0
40	2.8	112	4.1
60	0.9	54	1.3
80	0.3	24	0.4
100	0.1	10	0.1
Totals	12.6	238	

The last column is the distribution of the need for additional capacity; $TP_n(x)$, referred to previously, computed after the two units have been scheduled. Data such as these are computed in probabilistic production cost programs to provide probabilistic measures of the generation system adequacy (i.e., reliability). If costs are assigned to the unsupplied demand and energy either to represent replacement costs for emergency purchases of capacity and energy or to represent the economic loss to society as a whole, when the system fails to meet the expected load, then these data will also provide an economic measure of the generation system.

This relatively simple example leads to a lengthy series of computations. The results point out the importance of considering random forced outages of generating units when production costs are being computed for prolonged future periods. The small size of this example tends to magnify the expected unserved demand distribution. In order reliably to supply a peak demand of 100 MW with a small number of units, the total capacity would be somewhere in the neighborhood of 200 MW. On the other hand, the relatively low forced outage rates of the units used in Example 8B tend to minimize the effects of outages on fuel consumption. Large steam turbine generators of 600 MW capacity, or more, frequently exhibit forced outage rates in excess of 10%.

It should also be fairly obvious at this point that the process of enumerating each possible state in order to compute expected operation, energy generation, and unserved demands cannot be carried much further without an organized and efficient scheduling method. For N_L load levels and N units, each of which may be on or off, there are $N_L \times 2^N$ possible events to enumerate. The next section will develop the type of procedure that is found in many probabilistic production cost programs.

8.3 PROBABILISTIC PRODUCTION COST PROGRAMS

Until the 1970s, production cost estimates were usually computed on the basis that the total generating capacity is always available except for scheduled maintenance outages. Operating experience indicates that the forced outage rate of thermal-generating units increases with the unit size. Power system energy production costs are adversely affected by this phenomena because the frequent long-duration outages of the more efficient base-load units require running the less efficient, more expensive plants at higher than expected capacity factors* and the importation of emergency energy. Some utility systems have reported the operation of peaking units for more than 150 h each month when these same units were originally justified under the assumption that they would be run only a few hours per month, if at all.

The period of time when the available generation is less than the expected load and the calculated quantities of power and energy required to be imported because of random failures (i.e., forced outages) are used as measures of system reliability or, more properly, generation system adequacy. The maximum emergency import

---

* *Capacity factor* is defined as follows.

$$\frac{\text{MWh generated by the unit}}{\text{(Number of hours in the period of interest)(Unit full-load MW capacity)}}$$

Thus a higher value (close to unity) indicates that a unit was run most of the time at full load. A lower value indicates the unit was loaded below full capacity most of the time or was shut down part of the time.

power and total energy imported are different dimensions of the same measure. These quantities and the expected shortage duration are useful as sensitive indicators of the need for additional capacity or interconnection capability.

8.3.1 Probability Methods and Uses in Generation Planning

Prior to developing the probabilistic production cost model procedure, we will digress to review basic probability methods and their application in evaluating generation-system reliability. The major application of probability methods in power systems has been primarily in the area of planning generating capacity requirements. This application, no matter what particular technique is used, assigns a probability to the generating capacity available, describes the load demands in some manner, and provides a numerical measure of the probability of failing to supply the expected power or energy demands. By defining a standard *risk level* (i.e., a standard or maximum probability of failure) and allowing system load demands to grow as a function of time, these probability methods may be utilized to calculate the time when new generating capacity will be required.

Three general categories of probability methods and measures have been developed and applied to the generation planning problem. These are

1. The loss-of-load method.
2. The loss-of-energy method.
3. The frequency and duration method.

The first measures reliability as the probability of meeting peak loads (or its converse, the failure probability). The second uses as a reliability measure the expected loss of energy. The frequency and duration method is based on a somewhat different approach. It calculates the expected frequencies of outages of various amounts of capacity and their corresponding expected durations. These calculated values are then used with appropriate, forecasted loads and reliability standards to establish capacity reserve margins.

The mathematical techniques used are straightforward applications of probability methods. First, to review combined probabilities, let

$$P(A) = \text{probability that event A ccurs}$$

$$P(B) = \text{probability that event B occurs}$$

$$P(A \cap B) = \text{joint probability that A and B occur together}$$

$$P(A \cup B) = \text{probability that either A occurs by itself, or B occurs by itself, or A and B occur together.}$$

Conditional probabilities will be omitted from this discussion. [A *conditional probability* is the probability that A will occur if B already has occurred and may be expressed $P(A/B)$].

A few needed rules from combinatorial probabilities are

1. If A and B are independent events (i.e., whether A occurs or not has no bearing on B), then the joint probability that A and B occur together is, $P(A \cap B) = P(A) \, P(B)$.

2. If the favorable result of an event is for A or B or both to occur, then the probability of this favorable result is $P(A \cup B) = P(A) + P(B) - P(A \cap B)$.

3. If in rule 2, A and B are "mutually exclusive" events (i.e., if one occurs, the other cannot), then $P(A \cap B) = 0$ and $P(A \cup B) = P(A) + P(B)$.

4. The number of combinations of n things taken r at a time is given by the formula

$$_nC_r = \frac{n!}{r!(n-r)!} \qquad (8.2)$$

5. In general, the probability of exactly r occurrences in n trials of an event that has a constant probability of occurrence p is

$$P_n(r) = {}_nC_r p^r q^{n-r} = \frac{n!}{r!(n-r)!} p^r q^{n-r} \qquad (8.3)$$

where $q = 1 - p$

Rule 5 is a generalized form of the binomial expansion, applying to all terms of the binomial $(p + q)^n$. This distribution has had widespread use in generating-system probability studies. For example, assume that a generation system is composed of four identical units and that each of these units has a probability of being in service at any randomly chosen time of p. The probability of its being out of service is $q = 1 - p$. Assume that each machine's behavior is independent of the others. Then, a table may be constructed showing the probability of having 4, 3, 2, 1, and none in service.

Number in service	Probability of occurrence
4	$P(4) = {}_4C_4 p^4 q^{4-4} = \dfrac{4!}{4!(4-4)!} p^4 = p^4$
3	$P(3) = {}_4C_3 p^3 q^{4-3} = \dfrac{4!}{3!(4-3)!} p^3 q = 4P^3 q$
2	$P(2) = {}_4C_2 p^2 q^{4-2} = \dfrac{4!}{2!(4-2)!} p^2 q^2 = 6p^2 q^2$
1	$P(1) = {}_4C_1 p^1 q^{4-1} = \dfrac{4!}{1!(4-1)!} pq^3 = 4pq^3$
0	$P(0) = {}_4C_0 p^0 q^{4-0} = \dfrac{4!}{0!(4-0)!} q^4 = q^4$

In this table, each of the probabilities is a term of the binomial expansion of the form:

$$_4C_m p^m q^{4-m}$$

where m is the number of units in service.

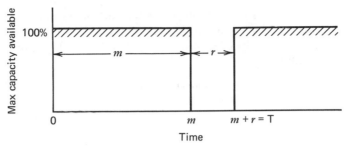

FIG. 8.7   Average availability cycle for a unit with two states.

These relationships assume a long-term average availability cycle as shown in Figure 8.7 for a given unit. In this long-term average cycle,

$$m = \text{average time available before failures}$$

$$r = \text{average repair time}$$

$$T = m + r = \text{mean time between failures}$$

Using these definitions for the generator taken as a binary state device,

$$p = \frac{m}{T} = \text{"innage rate" (per unit)}$$

$$q = 1 - p = \frac{r}{T} = \text{"outage rate" (per unit)}$$

Generating units may also be considered to be multistate devices when each state is characterized by the maximum available capacity and the probability of existence of that particular state. For instance, a large unit may have a forced reduction in output of, say, 20% of its rating when one boiler feed pump is out of service. This may happen 25% of the total time the unit is supposed to be available. In this case each unit state, (j) can be characterized by

$$C(j) = \text{maximum capacity available in state } (j)$$

$$p(j) = \text{probability that the unit is in state } (j)$$

where $\displaystyle\sum_{j=1}^{n} p(j) = 1.0$

$C(1) = 0$ (unit down)

$C(n) = 100\%$ capability (unit at full capacity)

In the probabilistic production cost calculations that follow we will attach other parameters to a state such as incremental cost for loading the unit between $C(j-1)$ and $C(j)$ MW.

The use of reliability techniques based on probability mathematics for generation planning frequently involves the construction of tables that show capacity on outage and the corresponding probability of that much, or more, capacity being on outage. The binomial probability distribution is cumbersome to use in practical computations. We will illustrate the simple numerical convolution using recursive techniques that are useful and efficient in handling units of various capacities and outage rates. The model of the generating capacity to be developed in this case is a table such as the following.

k	O_k generating capacity outage	Probability of occurrence of O_k or greater $= P_0(O_k)$
1	0 MW	1.000000
2	15 MW	0.950000
3	25 MW	0.813000
4	35 MW	0.095261
$\vdots$	$\vdots$	$\vdots$

On this table

$$k = \text{index showing the entry number}$$

$$O_k = \text{generating capacity outage, MW}$$

$$P_0(O_k) = \text{cumulative probability} = \text{probability of the occurrence}$$
$$\text{of an outage of } O_k, \text{ or larger}$$

This probability is a distribution rather than the density described with the binomial probability. It is a cumulative value rather than an exact probability (i.e., "exact" means probability density function).

Let each machine of the previously discussed hypothetical four-machine system be rated 10 MW, and let p(k) be the exact probability of occurrence of a particular event characterized by a given outage value. The table started previously may be expanded into Table 8.4. The function $P(O_k)$ is monotonic, and it should be obvious that the probability of having a zero or larger capacity outage is 1.0.

Since all generators do not have the same capacity or outage rate, the simple relationship for the binomial distribution in Table 8.4 does not hold in the general case. Beside the unit capability, the only other parameter associated with a generator in this technique is the average outage existence rate, q.

A simple recursive algorithm exists to add a unit to an existing outage probability table. Suppose an outage probability table exists that gives

$$P_0(x) \text{ versus } x$$

TABLE 8.4 Outage Probabilities

k	No. of machines in service	MW outage O_k	$p(k) = $ exact probability of outage O_k	$P(O_k) = $ probability of outage O_k, or larger
1	4	0	p^4	$p^4 + 4p^3q + 6p^2q^2 + 4pq^3 + q^4 \equiv 1.0$
2	3	10	$4p^3q$	$4p^3q + 6p^2q^2 + 4pq^3 + q^4$
3	2	20	$6p^2q^2$	$6p^2q^2 + 4pq^3 + q^4$
4	1	30	$4pq^3$	$4pq^3 + q^4$
5	0	40	q^4	q^4

Installed capacity = 40 MW.

where $P_0(x)$ = probability of x MW or more on outage

x = MW outage state

Now suppose you wish to add an "n-state" unit to the table that is described by

$$p(j) = \text{probability unit is in state } j$$

$$C(j) = \text{maximum capacity of state } j$$

$$C(n) = \text{capacity of unit}$$

$$O_j = C(n) - C(j) = \text{MW outage for state } j$$

Then the new table of outage probabilities may be found by a numerical convolution:

$$P_0'(x) = \sum_{j=1}^{n} p(j)P_0(x - O_j) \tag{8.4}$$

where $P_0(\leq 0) = 1.0$

This algorithm is an application of the combinational rules for independent, mutually exclusive "events." Each term of the algorithm is made up of (1) the event that the new unit is in state j with an outage O_j MW, and (2) the event that the "old" system has an outage of $(x - O_j)$ MW. The combined event therefore has an outage of x, or more, MW.

EXAMPLE 8C

Assume we have a generating system consisting of the following machines with their associated outage rate.

MW	Outage rate
10	0.02
10	0.02
10	0.02
10	0.02
5	0.02

The exact probability outage table for the first four units could be calculated using the binomial distribution directly and would result in the following table.

MW outage x	Exact probability $p(x)$	Cumulative probability $P_0(x)$
0	0.922368	1.000000
10	0.075295	0.077632
20	0.002305	0.002337
30	0.000032	0.000032
40	0	0

Now, the fifth machine can exist in either of two states: (1) it is in service with a probability of $p = 1 - q = 0.98$ and no additional system capacity is out, or (2) it is out of service with a probability of being in that state of $q = 0.02$ and 5 MW additional capacity is out of service.

The resulting outage-probability table will have additional outages because of the new combinations that have been added. This can be easily overcome by expanding the table developed for four machines to include these new outages. This is shown in Table 8.5 along with an example where the fifth, 5 MW, unit is added to the table.

TABLE 8.5   Adding Fifth Unit

x MW	$P_0(x)$	$0.98 \, P_0(x)$	$0.02 \, P_0(x - 5)$	$P'_0(x)$
0	1.000000	0.980000	0.020000	1.000000
5	0.077632	0.076079	0.020000	0.096079
10	0.077632	0.076079	0.001553	0.077632
15	0.002337	0.002290	0.001553	0.003843
20	0.002337	0.002290	0.000047	0.002337
25	0.000032	0.000031	0.000047	0.000078
30	0.000032	0.000031	0	0.000031
35	0	0	0	0
40	0	0	0	0
45	0	0	0	0

The correctness of this approach and resulting table may be seen by calculating the exact state probabilities for all possible combinations. That is,

MW out x	Exact probability $p(x)$
New machine in service	
$0 + 0$	$0.922368 \times 0.98 = 0.903921$
$10 + 0$	$0.075295 \times 0.98 = 0.073789$
$20 + 0$	$0.002305 \times 0.98 = 0.002258$
$30 + 0$	$0.000032 \times 0.98 = 0.000031$
$40 + 0$	$0 \times 0.98 = 0$
New machine out of service	
$0 + 5 = 5$	$0.922368 \times 0.02 = 0.018447$
$10 + 5 = 15$	$0.075295 \times 0.02 = 0.001506$
$20 + 5 = 25$	$0.002305 \times 0.02 = 0.000047$
$30 + 5 = 35$	$0.000032 \times 0.02 = 0$
$40 + 5 = 45$	$0 \times 0.02 = 0$

The exact state probabilities are combined by adding the probabilities for the mutually exclusive events that have identical outages, the result is shown in Table 8.6.

TABLE 8.6 Table of Combined Probabilities

MW outage x	Exact probability $p(x)$	Cumulative probability $P'_0(x)$
0	0.903921	1.000000
5	0.018447	0.096079
10	0.073789	0.077632
15	0.001506	0.003843
20	0.002259	0.002337
25	0.000047	0.000078
30	0.000031	0.000031
35	0	0
40	0	0
45	0	0

Table 8.6 is the Capacity model for the five-unit system and is usually assumed to be fixed until new machines are added or a machine is retired, or the model is altered to reflect scheduled maintenance outage.

This model was constructed using maximum capacities and calculating capacity outage probability distributions. Similar techniques may be used to construct available capacity distributions. A similar convolution is used in the probabilistic production cost computations developed in the next section. The form of the distribution is different because we are dealing with a scheduling problem rather than with the static, long-range planning problem. In the present case we are interested in a distribution of capacity outage probabilities; in the scheduling problem we require a distribution of unserved load probabilities.

8.3.2 Probabilistic Production Cost Computations

Production cost programs that recognize unit forced outages and that compute the statistically expected energy production cost have been developed and used widely. Mathematical methods based on probability methods make use of probabilistic models of both the load to be served and the energy and capacity resources. The models of the generation need to represent the unavailability of basic energy resources (i.e., hydro availability), the random forced outages of units, and the effects of contracts for energy sales and/or purchases. The computation may also include the expected costs of emergency energy over the tie lines sometimes referred to as the *cost of unsupplied energy.*

The basic difficulties that were noted when using deterministic approaches to the calculation of system production costs were

1. The base-load units of a system are loaded in the models nearly 100% of an interval.

2. The midrange, or "cycling," units are loaded for periods that depend on their priority rank and the shape of the load duration curve.

3. For any system with reasonably adequate reserve level, the peaking units have nearly zero capacity factors.

These conditions are, in fact, all violated to a greater or lesser extent whenever random unit forced outages occur on a real system. The remainder of this chapter outlines specific techniques for developing probabilistic production costs.

The unavailability of thermal-generating units due to unexpected, randomly occurring outages is fairly high for large-sized units. Values of 10 to 20% are common for full forced outages. That is, for a full forced outage rate of q, per unit, the particular generating unit is completely unavailable for $100\,q$ percent of the time it is supposed to be available. Generating units also suffer partial outages where the units must be derated (i.e., run at less than full capacity) for some period of time due to the forced outage of some system component (e.g., a boiler feed pump or a fan motor). These partial forced outages may reach very significant levels. It is not uncommon to see data reflecting a 25% forced reduction in maximum generating unit capability for 20% of the time it is supposed to be available.

Data on unit outage rates were collected and processed in the past in the United States by the Edison Electric Institute, an association of investor-owned utilities. This important function is now done by the National Electric Reliability Council. The collection and processing of these data is an important and difficult task. Performance data of this nature are essential if rational projections of component and system unavailability are to be made.

In the developments that follow it is assumed that data are available describing generating units in the following format.

Maximum power output available (MW)	Probability unit is available to load to this power (per unit)	Cost of generating maximum available (R/h)
$C(1) = 0$	$p(1)$	$F(1) = $ minimum cost
$C(2)$	$p(2)$	$F(2)$
$C(3)$	$p(3)$	$F(3)$
$\vdots$	$\vdots$	$\vdots$
$C(n) = $ Maximum	$p(n)$	$F(n)$

Pictorially the unit characteristics needed are shown in Figure 8.8.

The probabilistic production cost procedure uses thermal unit heat rate characteristics (i.e., heat input rate versus electric power output) that are linear segments. This type of heat rate characteristic is essential to the development of an efficient probabilistic computational algorithm since it results in stepped incremental cost curves. This simplifies the economic scheduling algorithm since any segment is fully loaded before the next is required. These unit input-output characteristics may have any number of segments so that a unit may be represented with as much detail as is desired. Unit thermal data are converted to cost per hour using fuel costs and other operating costs as is the case with any economic dispatching technique.

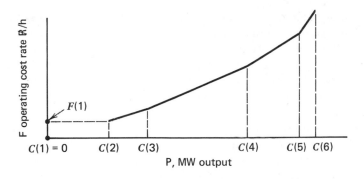

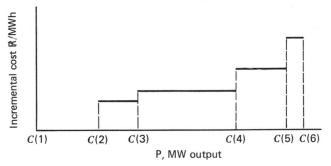

FIG. 8.8 Unit characteristics.

8.3.3. Simulating Economic Scheduling

The probabilistic production cost model simulates economic loading procedures and constraints. Fuel budgeting and planning studies utilize suitable approximations in order to permit the probabilistic computation of expected future costs. For instance, unit commitment will usually be approximated using a priority order. The priority list might be computed on the basis of average cost per megawatt-hour at full load with units grouped in blocks by minimum downtime requirements, taking the longest first. Within each block of units with similar downtimes, units could be ordered economically by average cost per megawatt-hour at full load.

With unit commitment order established, the various available loading segments can be placed in sequence in order of increasing incremental costs. The loading of units in this fashion is identical to using equal incremental cost scheduling where input-output curves are made up of straight-line segments. Finally, emergency sources (i.e., tie lines or pseudo tie lines) are placed last on the loading order list. The essential differences between the results of the probabilistic procedure and the usual economic dispatch computations is that all the units will be required if the forced outages are considered.

"Must-run" units are usually designated in these computations by requiring minimum downtimes equal to or greater than a week (i.e., $7 \times 24 = 168$ h or more). These base-load units are committed first. After the must-run units are committed, they must supply their minimum power. The next lowest cost block of capacity may be either a subsequent loading segment on a committed unit or a new unit to be committed. (Remember that units must be committed before they are loaded further.) Following this or a similar procedure results in a list of unit loading segments, arranged in economic loading order, which is then convenient and efficient to use in the probabilistic production cost calculations and to modify for each scheduling interval.

Storage hydro units and system sales/purchase contracts for interconnected systems must also be simulated in production cost programs. The exact treatment of each depends on the constraints and costs involved. For example, a monthly load model might be modified to account for storage hydro by *peak shaving*. In the peak-shaving approach the hydro unit production is scheduled to serve the peak load levels ignoring hydraulic constraints (but not the capacity limit) and assuming a single incremental cost curve for the thermal system for the entire scheduling interval. This can be done taking into account both hydro-unit forced outages and hydro-energy availability (i.e., amount of interval energy available versus the probability of its being available). System purchases and sales are often simulated as if they were stored energy systems. Sales (or purchases) from specific units are more difficult to model, and the modeling depends on the details of the contract. For instance, a "pure" unit transaction is made only when the unit is available. Other "less pure" contracts might be made where the transaction might still take place using energy produced by other units under specified conditions.

8.3.4 Scheduling Procedures

In the probabilistic production cost approach the load is modeled as it was in the previously illustrated load-duration curve approach as a probability distribution expressed in terms of hours that the load is expected to equal or exceed the value on the horizontal axis. This is a monotonically decreasing function with increasing load and could be converted to a "pure" probability distribution by dividing by the number of hours in the load interval being modeled. This model is illustrated in Figures 8.2, 8.3, 8.5, and 8.6. Therefore, each load-duration curve is treated either as a cumulative probability distribution,

$$P_n(x) \text{ versus } x$$

where $P_n(x)$ = probability of needing x MW, or more, or when expressed in hours, it is $T P_n(x)$, where T is the duration of the particular time interval. Also,

$$P_n(x) = 1 \qquad \text{for } x \leq 0$$

The load distribution is usually expressed in a table, $T P_n(x)$, which may be fairly short. The table needs to be only as long as the maximum load divided by the uniform MW interval size used in constructing the table. In applying this approach to a digital computer, it is both convenient and computationally efficient to think in terms of regular discrete steps and recursive algorithms. Various load-duration curves for the entire interval to be studied are arranged in the sequence to be used in the scheduling logic. There is no requirement that a single distribution $P_n(x)$ be used for all time periods. In developing the unit commitment schedule it is necessary to verify not only that the maximum load plus spinning reserve is equal to or less than the sum of the capacities of the committed units but also that the sum of the minimum loading levels of the committed units is not greater than the minimum load to be served.

A number of different descriptions in the literature are used to explain this probabilistic procedure of thermal unit scheduling. The one following has been found to be the easiest to grasp by someone unfamiliar with this procedure and is theoretically sound. If there is a segment of capacity with a total of C MW available for scheduling, and if we denote:

> q = the probability that C MW are unavailable
> (i.e., its unavailability)

and

> $p = 1 - q$
> = the probability or "availability" of its segment

then after this segment has been scheduled the probability of needing x MW or more is now $P'_n(x)$. Since the occurrence of loads and unexpected unit outages are statistically independent events, the new probability distribution is a combination of mutually exclusive events with the same measure of need for additional capacity. That is,

$$P'_n(x) = q\,P_n(x) + p\,P_n(x + C) \tag{8.5}$$

In words, $q\,P_n(x)$ is the probability new capacity C is unavailable times the probability of needing x, or more, MW, and $p\,P_n(x + C)$ is the probability C is available times the probability $(x + C)$, or more, is needed. These two terms represent two mutually exclusive events, each representing combined events where x MW, or more, remain to be served by the generation system.

This is a recursive computational algorithm similar to the one used to develop the capacity outage distribution in Section 8.3.1 and will be used in sequence to convolve each unit or loading segment with the distribution of load not served. It should be recognized that the argument of the probability distribution can be negative after load has been supplied and that $P_n(x)$ is zero for x greater than the peak load. Initially, when only the load distribution is used to develop $T\,P_n(x)$, $P_n(x) = 1$ for all $x \leq 0$.

8.3.5 Scheduling Algorithm for Probabilistic Production Cost Computations

Example 8B provides an introduction to the complexities involved in an enumerative approach to the problem at hand. By extending some of the ideas presented briefly in the previous portions of this chapter, a recursive technique (i.e., algorithm) may be developed to organize the probabilistic production cost calculations.

First we note that the generation requirements for any generating segment are determined by the knowledge of the distribution $TP_n(x)$ that exists prior to the dispatch (i.e., scheduling) of the particular generating segment. That is, the value of $TP_n(0)$ determines the required hours of operation of a new unit and the area under the distribution $TP_n(x)$ for x between zero and the rating of the unit loading segment determines the requirements for energy production. Assuming the particular generation segment being dispatched is not perfectly reliable, it is unavailable for some fraction of the time it is required, and therefore there will be a residual distribution of demands that cannot be served by this particular segment because of its forced outage.

Let us represent the forced outage (i.e., unavailability) rate for a generation segment of C MW, and $TP_n(x)$, the distribution of unserved load prior to scheduling the unit. Assume the unit segment to be scheduled is a complete generating unit with an input-output cost characteristic.

$$F = F_0 + F_1 P, \, \text{R/h}$$

for $0 \leq P \leq C$ MW. The unit will be required $TP_n(0)$ hours, but on the average it will be available only $(1 - q)TP_n(0)$ hours. The energy required by the load distribution that could be served by the unit is

$$E = T \int_{x=0}^{x=C} P_n(x)\, dx$$

or,

$$= T \sum_{x=0+}^{x=C} P_n(x)\, \Delta x$$

for discrete distributions. The unit can only generate $(1 - q)E$ because of its expected unavailability.

These data are sufficient to compute the expected production costs. These costs for this period are

$$= F_0 \times (1 - q)TP_n(0) + (1 - q)\, E\, F_1, \, \text{R}$$

Having scheduled the unit, there is a residual of unserved demands due to the forced outages of the unit. The recursive algorithm for the distribution of the probabilities of unserved load presented in the previous section may be used to develop the new distribution of unserved load after the unit is scheduled. That is,

$$TP'_n(x) = q\, TP_n(x) + (1 - q)TP_n(x + C) \tag{8.6}$$

The process may be repeated until all units have been scheduled and a residual distribution remains that gives the final distribution of unserved demand.

Refer to the unit data described in Figure 8.8 and the accompanying text. The minimum load cost, $F(1)$, shown on this figure is associated only with the first loading segment, $C(2)$ to $C(3)$, since the demands on this portion of the unit will determine the maximum hours of operation of the unit.

A general scheduling algorithm may be developed based on these conditions. In this development we temporarily put aside until the next section some of the practical and theoretical problems associated with scheduling units with multiple steps and nonzero minimum load restrictions. The procedure shown in flowchart form on Figure 8.9 is a method for computing the expected production costs for a single time period, T hours in duration.

Besides the terms defined on Figure 8.8 we require the following nomenclature and definitions.

$i = 1, 2, \ldots, i_{max}$ ordered capacity segments to be scheduled

$c(i) = C(i + 1) - C(i)$ capacity of the ith segment, MW

$dF(i) = \dfrac{[F(i + 1) - F(i)]}{c(i)}$ incremental cost rate for the ith segment, R/MWh

$F0(i)$ minimum load cost rate for i^{th} segment of unit R/h

$p(i)$ availability of segment i, per unit

$q(i) = 1 - p(i)$ unavailability of segment i, per unit

$x = 0, 1, 2, \ldots, x_{max}$ equally spaced load levels

MW_{step} uniform interval for representing load distribution, MW

$PRCOST(i)$ production costs for i^{th} segment, R

$E, E', E''' \cdots$ remaining unserved load energy

In this algorithm the energy generated by any particular loading segment of a generator is computed as the difference in unserved energy before (E) and after (E') the segment is scheduled. Since the incremental cost $[dF(i)]$ of any segment is constant, this is sufficient to determine the added costs due to loading of the unit above its minimum. For initial portions of a unit, $T P_n(0)$ determines the number of hours of operation required of the unit and is used to add the minimum load operating costs. We will illustrate the application of this procedure to the system described in Examples 8A and 8B.

EXAMPLE 8D

The computation of the expected production costs using the method shown in Figure 8.9 and the procedures involved can be illustrated with the data in Example 8A. Initially we will ignore the forced outage of the two units and then follow this with an extension to incorporate the inclusion of forced outages.

With zero forced outage rates the analysis of Example 8A is merely repeated in a different format where the load-duration curve is treated as a probability distribution. Figure 8.10 shows the initial load-duration curve in part a; the modified curve after unit 1 is loaded is shown in part b, and the final curve after both units are loaded is shown in part c.

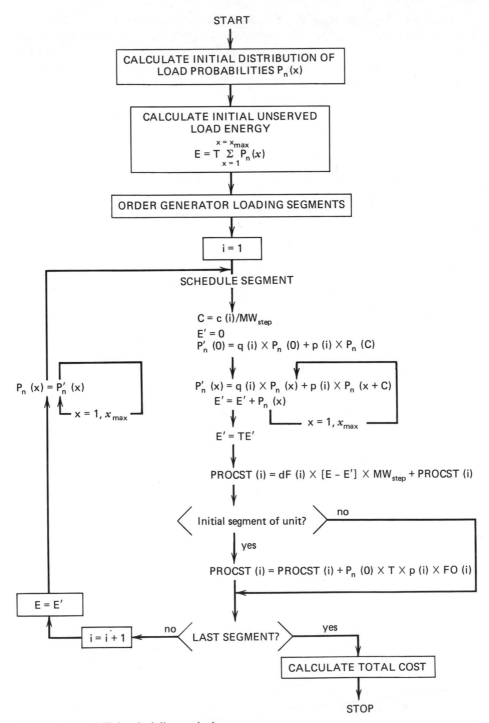

FIG. 8.9 Probabilistic scheduling method.

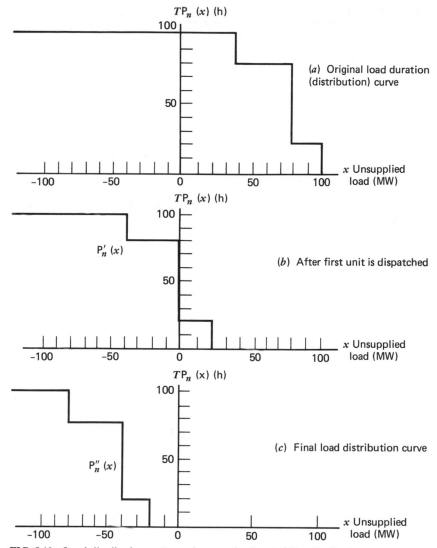

FIG. 8.10 Load-distribution curves redrawn as load probability distributions.

The computations involved in the convolutions may be illustrated in tabular format. In general, in going from the jth distribution to the $(j+1)^{st}$,

$$P_n^{j+1}(x) = q\ P_n^j(x) + p\ P_n^j(x+c)$$

where $p = 1 - q =$ innage rate of unit or segment being loaded

$x + c, x =$ unsupplied load variables (MW)

$c =$ capacity of unit (MW)

$TP_n^j(x) =$ (total duration of interval) $\times$ (probability of needing to supply x or more MW) at j^{th} stage

TABLE 8.7 Load Probability for Unserved Loads After Scheduling Unit 1

x MW	$T\,P_n(x)$ (h)	$T\,P_n'(x) = T\,P_n(x + 80)$ (h)	$T\,P_n''(x) = T\,P_n'(x + 40)$ (h)
0	100	80	0
20	100	20	
40	100	0	
60	80		
80	80		
100	20		
100+	0		
Energy on rhs $= E$		$= E'$	$= E''$

Recall that unit 1 was rated at 80 MW and unit 2 at 40 MW and for Example 8A all $q = 0$ and all $p = 1$.

Table 8.7 shows the load probability for unserved loads of 0 to 100 + MW. The range of valid MW values need not extend beyond the maximum load nor be less than zero. If you wish to consider the distribution extended to show the served load, $TP_n(x)$ may be extended to negative values. Only the energy for the positive portion of this distribution (the right-hand side or rhs) represents real load energy. A negative unsupplied energy is, of course, an energy that has been supplied.

The remaining unsupplied energy levels at each step are denoted on the bottom of each column in Table 8.7 and are computed as follows.

$$E = 100 \times 20 + 80(80 - 20) + 40 \times (100 - 80),\ \text{MWh}$$
$$= 20h \times (100 + 100 + 80 + 80 + 20)\ \text{MW}$$
$$= 7600\ \text{MWh}$$

$$E' = 20 \times (20) = 400\ \text{MWh}$$

$$E'' = 0$$

Unit 1 was on line 100 h and generated $7600 - 400 = 7200$ MWh. Unit 2 was on line 80 h and generated 400 MWh. The unit loadings, loading levels, durations at those levels, fuel consumption, and production costs can easily be determined using these data. The numerical results are the same as shown in Example 8A. You should be able to duplicate those results using the distributions $P_n(x)$, $P_n'(x)$ and $P_n''(x)$.

Next let us consider forced outage rates for each unit. Let

$$q_1 = 0.05\ \text{per unit}$$

and

$$q_2 = 0.10\ \text{per unit}$$

be the forced outage rates of units 1 and 2, respectively. The recursive equation to obtain $P_n'(x)$ from the original load distribution, omitting the common factor T, is now

$$P_n'(x) = 0.05\,P_n(x) + 0.95\,P_n(x + 80)$$

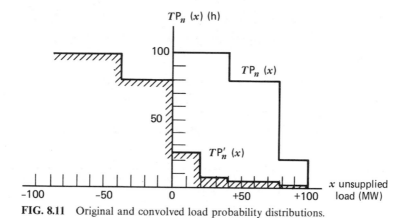

FIG. 8.11 Original and convolved load probability distributions.

The original and resultant unserved load distributions are now as follows. Figure 8.11 shows these distributions.

x (MW)	$T P_n(x)$ (h)	$T P'_n(x)$ (h)
0	100	$76 + 5 = 81$
20	100	$19 + 5 = 24$
40	100	$0 + 5 = 5$
60	80	$0 + 4 = 4$
80	80	$0 + 4 = 4$
100	20	$0 + 1 = 1$
100+	0	0
rhs Energy	7600 MWh	760 MWh

These data may be used to compute the loadings, durations, energy produced, fuel consumption, and production cost for unit 1. Unit 1 may be loaded to 80 MW for 80 h and 40 MW for a maximum of 20 h according to the distribution $TP_n(x)$ shown on Figure 8.11. The unit is available only 95% of the time on the average. The loadings and so forth for unit 1 are as follows and are identical with those from Example 8A.

Unit 1 load (MW)	Duration (h)	Energy (MWh)	Fuel used (10^6 Btu)	Fuel cost (R)
40	$0.95 \times 20 = 19$	760	9,120	9,120
80	$0.95 \times 80 = 76$	6,080	60,800	60,800
		6,840	69,920	69,920

If only production cost and/or fuel consumption are required without detailed loading profiles, the production costs may be computed using the algorithm developed. That is, the production cost of unit 1 is

$$= 160 \text{ R/h} \times 0.95 \times 100 \text{ h} + 8 \text{ R/MWh} \times (7600 - 760) \text{ MWh}$$

$$= 69,920 \text{ R}$$

The detailed loadings and durations for unit 2 may also be computed using the distribution of unserved energy after the unit has been scheduled, $TP_n''(x)$. The unit is required 81 h, is required at zero load for $81 - 24 = 57$ h, may generate 40 MW for 5 h and 20 MW for $24 - 5 = 19$ h. The resulting generation and fuel costs are as follows.

Unit 2 load (MW)	Duration (h)	Energy (MWh)	Fuel used (10^6 Btu)	Fuel cost (R)
0	51.3	0	4,104	8,208
20	17.1	342	4,104	8,208
40	4.5	180	1,800	3,600
	72.9	522	10,008	20,016

However, the fuel consumption and production costs may be easily computed using the scheduling algorithm developed. The convolution of the second unit is done in accord with

$$P_n''(x) = 0.1 \, P_n'(x) + 0.9 \, P_n'(x + 40)$$

where the factor T has again been omitted.

The results are shown in Table 8.8.

TABLE 8.8 Load Probability for Unserved Loads After Scheduling Unit 1 and Unit 2

x (MW)	$T \, P_n'(x)$ (h)	$T \, P_n''(x)$ (h)
0	81	12.6
20	24	6.0
40	5	4.1
60	4	1.3
80	4	0.4
100	1	0.1
100+	0	0
rhs Energy	760 MWh	238 MWh

With these data the production costs for unit 2 are simply

$$= 160 \text{ R/h} \times 0.90 \times 81 \text{ h} + 16 \text{ R/MWh} \times (760 - 238) \text{ MWh}$$

$$= 20,016 \text{ R}$$

The final unserved energy distribution is shown on Figure 8.12. Note that there is still an expected requirement to supply 100 MW. The probability of needing this much capacity is 0.001 per unit (or 0.1%), which is not insignificant.

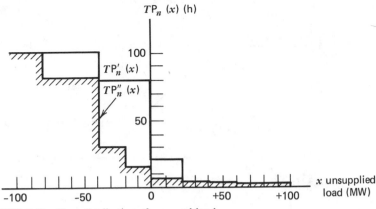

FIG. 8.12 Final distribution of unserved load.

In order to complete the example we may compute the cost of supplying the remaining 238 MWh of unsupplied load energy. This must be based on an estimate of the cost of emergency energy supply or the value of unsupplied energy. For this example let us assume that emergency energy may be purchased (or generated) from a unit with a net heat rate of 12,000 Btu/KWh and a fuel cost of 2 R/MBtu. These are equal to the heat rate and cost associated with unit 2 and are not too far out of line with the costs for energy from the two units previously scheduled. The cost of supplying this 238 MWh is then

$$238 \text{ MWh} \times 12 \text{ MBtu/MWh} \times 2 \text{ R/MBtu} = 5{,}712 \text{ R}$$

In summary, we may compare the results of Example 8A computed with forced outages neglected with the results from Example 8D where they have been included and an allowance has been made for purchasing emergency energy (see Table 8.9). Therefore, ignoring forced outages results in a 1.95% underestimate of fuel consumption, a complete neglect of the need for and costs of emergency energy supplies, and an 8.1% underestimate of the total production costs.

The final unsupplied energy distribution may also be used to provide indexes for the need for additional transmission and/or generation capacity. That is an entire new area, however, and will not be explored here since the primary concern of this text is the operation, scheduling, and cost for power generation.

TABLE 8.9 Results of Examples 8A and 8D Compared

	Fuel used (10^6 Btu)	Fuel cost (R)	Unsupplied energy (MWh)	Cost of emergency energy (R)	Total cost (R)
Example 8A	78,400	83,200	0	0	83,200
Example 8D	79,928	89,936	238	5712	95,648
Difference	1,528	6,736	—	—	12,448
% Difference	1.95%	8.1%	—	—	15%

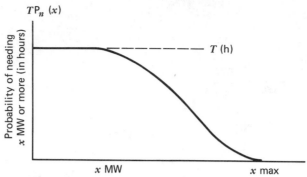

FIG. 8.13 Load probability distribution.

8.3.6 A Discussion of Some Practical Problems

Example 8D illustrated the simplicity of the basic computation of the scheduling technique used in this type of probabilistic production cost program where the load is modeled using a discrete tabular format. There are detailed complications, extensions, exceptions, and such that arise in the practical implementation of any production cost technique. This section will review the procedure used in Example 8D in a general sense and point out some of these considerations. No attempt is made to describe a complete, detailed program. The intent is to point out some of the practical considerations and discuss some of the approaches that may be used.

First, consider Figure 8.13, which shows the *cumulative load distribution* (i.e., a load-duration curve treated as a cumulative probability distribution) for an interval of T hours.

Next, assume an ordered list of loading segments as shown in Table 8.10. Units 3, 1, and 4 are to be committed initially so that the sum of their capacities at full output equals or exceeds the peak load plus capacity required for spinning reserves. If we assume only two segments for each of these three units, this commitment totals 160 MW. Assume such a table includes all the units available that subinterval. The cost data for the first three loading segments are the total costs per hour at the minimum loading levels of 20, 20 and 40 MW, respectively,

TABLE 8.10 Sample Subinterval Loading Data: Segment Data

Unit number (i)	No. (j)	P_{min} MW	P_{max} MW	Cost	Outage rate (q_{ij}) per unit	Innage or availability rate
3	1	0	20	R/h	0.05	0.95
1	1	0	20	R/h	0.02	0.98
4	1	0	40	R/h	0.02	0.98
1	2	20+	60	R/MWh	0.05	0.95
3	2	20+	50	R/MWh	0.05	0.95
4	2	40+	50	R/MWh	0.05	
⋮	⋮	⋮	⋮	⋮	⋮	⋮

and the remaining cost data are the incremental costs in R per MWh for the particular segment. Table 8.10 is the ordered list of loading segments where each segment is loaded, generation and cost are computed, and the cumulative load distribution function is convolved with the segment.

There are two problems presented by these data that have not been discussed previously. First, the minimum loading sections of the initally committed units must be loaded at their minimum load points. For instance, the minimum load for unit 4 is 40 MW, which means it cannot satisfy loads less than 40 MW. Second, each unit has more than one loading segment. The loading of a unit's second loading segment by considering the probability distribution of unserved load after the first segment of a unit has been scheduled would violate the combinatorial probability rules that have been used to develop the scheduling algorithm since the unserved load distribution includes events where the first unit was out of service. That is,the loading of a second or later section is not statistically independent of the availability of the previously scheduled sections of the particular unit. Both these concerns require further exploration in order to avoid the commitment of known errors in the procedure.

The situation with block-loaded units (or a nonzero minimum loading limit) is fairly easily handled. Suppose the unserved load distribution prior to loading such a block-loaded segment is $T P_n(x)$ and the unit data are

$$q = \text{unavailability rate, per unit}$$

$$p = 1 - q$$

$$c = \text{capacity of segment}$$

By *block loading* it is meant that the output of this particular segment is limited to exactly c MW. The nonzero minimum loading limit may be handled in a similar fashion.

The convolution of this segment with $T P_n(x)$ now must be handled in parts. For load demands below the minimum output, c, the unit is completely unavailable. For $x \geq c$, the unit may be loaded to c MW output. The algorithm for combining the mutually exclusive events where x, or more, MW of load remain unserved must now be performed in segments, depending on the load. For load levels, x, such that

$$x \geq c$$

the new unserved load distribution is

$$P_n'(x) = q\,P_n(x) + p\,P_n(x + c) \tag{8.7}$$

where the period length, T, has been omitted. For some loads, $x < c$, the unit cannot operate to supply the load. Let $p_n(x)$ denote the probability density of load x. In discrete form,

$$p_n(x) = P_n(x) - P_n(x + MW_{step}) \tag{8.8}$$

where, MW_{step} = uniform interval in tabulation of $P_n(x)$. For loads equal to or greater than c, the probability of exactly x MW after the unit has been scheduled is

$$p_n'(x) = q\,p_n(x) + p\,p_n(x + c) \tag{8.9}$$

For loads less than c (i.e., $0 \leq x < c$)

$$p_n'(x) = p_n(x) + p\, p_n(x + c) \tag{8.10}$$

For convenience in computation, let

$$p_n(x) = (q + p)\, p_n(x) \tag{8.11}$$

for $0 \leq x < c$. Then for this same load range,

$$p_n'(x) = q\, p_n(x) + p\, p_n(x + c) + p\, p_n(x) \tag{8.12}$$

Next, the new unserved load energy distribution may be found by integration of the density function from the maximum load to the load in question. For discrete representations and for $x \geq c$,

$$P_n'(x) = q\, P_n(x) + p\, P_n(x + c) \tag{8.13}$$

For loads less than c, that is, $0 \leq x \leq c$,

$$P_n'(x) = q\, P_n(x) + p\, P_n(x + c) + p[P_n(x) - P_n(c)] \tag{8.14}$$

The last term represents those events for loads between x and c wherein the unit cannot operate. The term $[P_n(x) - P_n(c)]$ is the probability density of those loads taken as a whole. The first term, $q\, P_n(x)$, resulted from assuming that the unit could supply any load below its maximum.

This format for the block-loaded unit makes it easy to modify the scheduling algorithm presented previously. The effects of restriction to block loading a unit may be illustrated using the data from Example 8D.

EXAMPLE 8E

The two-unit system and load distribution of the previous example will be used with one modification. Instead of allowing the second unit to operate anywhere between 0 and 40 MW output, we will assume its operation is restricted to 40 MW only. The cost of this unit was

$$F_2 = 160 + 16\, P_2,\ \text{R/h}$$

so that for $P_2 = 40$ MW, $F_2 = 800$ R/h.

Recall (see Table 8.8) that after the first 80 MW unit was scheduled, the unserved load distribution was

x (MW)	$T\, P_n'(x)$ (h)
0	81
20	24
40	5
60	4
80	4
100	1

with an unserved load energy of 760 MWh. With a restriction to block loading, the unit is on line only 5 h. The energy it generates is therefore $5 \times 40 \times 0.9 = 180$ MWh. The new distribution of unserved load after the unit is scheduled is as follows.

x (MW)	$T P'_n(x)$ (h)	$q\, T P'(x) + p\, T P'(x + c)$	$p\, T[P_n(x) - P_n(c)]$	$T P''_n(x)$ (h)
0	81	12.6	$0.9[81 - 5]$	81.0
20	24	6.0	$0.9[24 - 5]$	23.1
40	5	4.1	—	4.1
60	4	1.3	—	1.3
80	4	0.4	—	0.4
100	1	0.1	—	0.1

with an unserved load energy of 580 MWh.

The quantitative significance of the precise treatment of block-loaded units has been magnified by the smallness of this example. In studies of practical-sized systems, block-loading restrictions are frequently ignored by removing the restriction on minimum loadings or are treated in some satisfactory, approximate fashion. For long-range studies, these restrictions usually have minor impact on overall production costs.

The analysis of the effects of the statistical dependence of the multiple-loading segments of a unit is somewhat more complicated. In the development that follows the analysis makes use of the fact that the distribution of unserved load probabilities, $T P_n(x)$, at any point in the scheduling algorithm is independent of the order in which various units are scheduled. Only the generation and hours of operation are dependent on the scheduling order. This may easily be verified by a simple numerical example, or it may be deduced from the recursive relationship presented for $T P_n(x)$.

Suppose we have a second section to be incrementally loaded for some machine at a point in the computations where the distribution of unserved load is $T P_n(x)$. The outage of this second incremental loading section is obviously not statistically independent of the outage of the unit as a whole. Therefore the effect of the first section must be removed from $T P_n(x)$ prior to determining the loading of the second segment. This is known as *deconvolution*.

For this illustration of one method for handling multiple segments we will assume:

1. The capacity of the segment extends from C_1 to C_2 where $C_2 > C_1$.
2. The first segment had a capacity of C_1.
3. The outage rates of both segments are equal to q per unit.

In the process of arriving at the distribution $T P_n(x)$ the initial segment of C_1 MW was convolved in the usual fashion. That is,

$$T P_n(x) = q\, T P'_n(x) + p\, T P'_n(x + C_1) \tag{8.15}$$

The distribution $T P_n(x)$ is independent of the order in which segments are convolved. Only the loading of each segment depends on this order.

Therefore, we may consider that $T P'_n(x)$ represents an artificial distribution of load probabilities with the initial segment of the unit removed. This pseudodistribution, $T P'_n(x)$, must be determined in order to evaluate the loading on the segment between C_1 and C_2. Several techniques may be used to recover $T P'_n(x)$ from $T P_n(x)$. The convolution equation may be solved for either $T P'_n(x)$ or $T P'_n(x + c)$. The deconvolution process is started at the maximum load if the equation is solved for $T P'_n(x)$. That is,

$$T P'_n(x) = \frac{1}{q} T P_n(x) - \frac{p}{q} T P'_n(x + c) \tag{8.16}$$

and

$$T P'_n(x) = 0 \text{ for } x > \text{maximum load}$$

We will use this procedure to illustrate the method because the procedures and algorithms discussed have not preserved the distributions for negative values of unserved load (i.e., already served loads). As a practical computational matter it would be better practice to preserve the entire distribution $T P_n(x)$ and solve the convolution equation for $T P_n(x + c)$. That is,

$$T P_n(x + c) = \frac{1}{p} T P_n(x) - \frac{q}{p} T P'_n(x) \tag{8.17}$$

or shifting arguments, by letting $y = x + c$

$$T P_n(y) = \frac{1}{p} T P_n(y - c) - \frac{q}{p} T P'_n(y - c) \tag{8.18}$$

In this case the deconvolution is started at the point at which

$$-y = \text{sum of dispatched generation}$$

since

$$T P_n(y) = T$$

for all $y < -$ sum of dispatched generation.

Even though we will use the first deconvolution equation for illustration, the second should be used in any computer implementation where repeated deconvolutions are to take place. Since $q \ll p$, the factors $1/q$ and p/q in the first formulation will amplify any numerical errors that occur in computing the successive distributions. We use this potentially, numerically treacherous formulation here only as a convenience in illustration.

To return, we obtain the deconvolved distribution $T P'_n(x)$ by removing the effects of the first loading segment. Then the loading of the second segment from C_1 to C_2 is determined using $T P'_n(x)$, and the new, remaining distribution of unserved load is obtained by adding the total unit of C_2 MW to the distribution so that

$$T P''_n(x) = q T P'_n(x) + p T P'_n(x + C_2) \tag{8.19}$$

EXAMPLE 8F

Assume that in our previous examples the first unit had a total capacity of 100 MW instead of 80. This last segment might have an incremental cost rate of 20 R/MWh so that it would not be dispatched until after the second unit had been used. Assume the outage rate of 0.05 per unit applies to the entire unit. Let us determine the loading on this second section and the final distribution of unserved load.

The distribution of unserved load from the previous examples is

x	$T\,P_n''(x)$
(MW)	(h)
0	12.6
20	6.0
40	4.1
60	1.3
80	0.4
100	0.1

The deconvolved distribution may be computed starting at $x = 100$ MW using Eq. 8.16 and working up the table. The table was constructed with $C = 80$ MW for the capacity of this unit. The deconvolved distribution is

$$T\,P_n'(100) = \frac{0.1}{0.05} = 2$$

$$T\,P_n'(80) = \frac{0.4}{0.05} = 8$$

$$\vdots$$

The new distribution, adding the entire 100 MW unit, is determined using $C_2 = 100$ MW and is

$$T\,P_n''(x) = 0.05\,T\,P_n'(x) + 0.95\,T\,P_n'(x + 100)$$

The results are as follows.

x	$T\,P_n(x)$	$T\,P_n'(x)$	$T\,P_n''(x)$
(MW)	(h)	(h)	(h)
0	12.6	100	6.9
20	6.0	82	4.1
40	4.1	82	4.1
60	1.3	26	1.3
80	0.4	8	0.4
100	0.1	2	0.1
Energy	238 MWh		200 MWh

Thus, the second section of the first unit generates 38 MWh.

This computation may be verified by examining the detailed results of Example 8B where the various load and outage combination events were enumerated. At a load of 100 MW the second segment of unit 2 would have been loaded to the extent shown by this example. You should be able to identify two periods where the second section would have reduced previous shortages of 0 and 20 MW. This procedure and the example are theoretically correct but computationally tedious. Furthermore, the repeated deconvolution process may lead to numerical round-off errors unless care is taken in any practical implementation.

Frequently there are approximations made in treating sequential loading segments. These are usually based on the assumption that the subsequent loading sections are independent of the previously loaded segments. That is, that they are equivalent to new, independent units with ratings that are equal to the capacity increment of the segment. When these types of approximations are made, they are justified on the basis of numerical tests. They generally perform more than adequately for larger systems but should be avoided for small systems.

The two extensions discussed here are only examples of the many extensions and modifications that may be made. The computations of expected production costs may be made as a function of the load to be served. These characteristics may be used as pseudogenerators in scheduling hydroelectric plants, pumped-storage units, or units with limited fuel supplies.

There have been further extensions in the theoretical development as well. It is quite feasible to represent the distribution of available capacity by the use of suitable orthogonal polynomials. Gram-Charlier series are frequently used to model probabilistic phenomena. They are most useful with a reasonably uniform set of generator capacities and outage rates. By representing the expected load distribution also as an analytic function it is possible to develop analytical expressions for unserved energy distributions and expected production costs. Care must be exercised in using these approximations when one or two very large generators are added to systems previously composed of a uniform array of capacities. We will not delve further into this area in this text. The remainder of the chapter is devoted to a further example and problems.

8.4 SAMPLE COMPUTATION AND EXERCISE

The discussion of the probabilistic techniques is more difficult than their performance. We will illustrate the method further using a three-unit system. The heart of the technique is the convolution of the probability distribution of needing x, or more, MW, $TP_n(x)$, with the various states of the unit and calculating the energy generation and production cost in the process.

The three generating units each may be loaded from 0 MW to their respective ratings. For ease of computation, we assume linear input-output cost curves and only full forced outage rates (that is, the unit is either completely available or completely unavailable). The unit data are as follows.

Unit no.	Maximum rating (MW)	Input-output cost curve (R/h)	Full forced outage rate (per unit)
1	60	$60 + 3 P_1$	0.2
2	50	$70 + 3.5 P_2$	0.1
3	20	$80 + 4 P_3$	0.1

In these cost curves P_i are in MW. In addition, the system is served over a tie line. Emergency energy is available without limit (MW or MWh) at a cost rate of 5 R/MWh.

The load model is a distribution curve for a 4-week interval (a 672-h period). That is, the expected load is as shown in Table 8.11. The total load energy is 43,680 MWh

8.4.1 No Forced Outages

The economic dispatch of these units for each load level is straightforward and simple. The units are listed in economic order whether measured by average cost at full load or by incremental cost. The commitment problem is ignored for this example only.

The sum of the peak load demand (100 MW) and the total capability (130 MW) is 230 MW. Therefore, the probability table of needing capacity will extend eventually from -130 MW to $+100$ MW. It is convenient in digital computer implementation to work in uniform MW steps. For this example we will use 10 MW.

As each unit is dispatched, the probability distribution of needing x or more MW [i.e., $P_n(x)$] is modified (i.e., convolved) using the following algorithm.

$$TP'_n(x) = TP_n(x + c) \tag{8.20}$$

where $P'_n(x)$ and $P_n(x)$ = new and old distributions, respectively

T = time period, 672 h in this instance

c = capability of unit or segment when it is in state j.

Table 8.12 shows initial distribution in the second column. The load energy to be served is

$$E = 672 \sum_{x=0}^{100} P_n(x) \Delta x = 43,680 \text{ MWh}$$

TABLE 8.11 Load Distribution

Load level (MW)	Hours of existence	Probability	Hours load equals or exceeds	Probability of needing load or more (pu)
30	134.4	0.2	672.0	1.00
50	134.4	0.2	537.6	0.80
70	134.4	0.2	403.2	0.60
80	168.0	0.25	268.8	0.40
100	100.8	0.15	100.8	0.15
	672.0			

TABLE 8.12 Three-Unit Example: Zero Forced Outage Rates

x (MW)	$P_n(x)$ (pu)	$P'_n(x)$ (pu)	$P''_n(x)$ (pu)
−130	↑	↑	↑
−120			
−110			
−100			
−90			
−80			1.0
−70			0.8
−60			0.8
−50			0.6
−40			0.6
−30		1.0	0.4
−20		0.8	0.15
−10		0.8	0.15
0		0.6	0
10		0.6	
20		0.4	
30	1.0	0.15	
40	0.8	0.15	
50	0.8	0	
60	0.6		
70	0.6		
80	0.4		
90	0.15		
100	0.15		
110	0	↓	↓
E/672	65	13	0
MWh	43,680	8,736	0

With zero forced outage rate the 60 MW unit loading results in the $P_n(x)$ distribution shown in the third column. The resultant load energy to be served is now:

$$E' = (0.15 \times 20 + 0.4 \times 10 + 0.6 \times 10) \times 672 = 8736 \text{ MWh}$$

which means unit 1 generated

$$43,680 - 8736 = 34,944 \text{ MWh}$$

The unit was on-line 672 h, and the incremental cost rate was 3 R/h. Therefore, the cost for unit 1 is

$$\text{Total cost} = \sum_T F(P_t) \cdot \Delta t = \sum_T (60 + 3\,P_t)\,\Delta t = \sum_T (60\,\Delta t + 3\,P_t\,\Delta t)$$

$$= 60\,T + 3\,(\text{MWh generated}), \text{ since MWh} = \sum_T P_t\,\Delta t$$

$$= 60 \text{ R/h} \times 672 \text{ h} + 34,944 \text{ MWh} \times 3 \text{ R/MWh} = 145,152 \text{ R}$$

Unit 2 serves the remaining load distribution (third column) and results in the distribution shown in the fourth column. This unit is only on-line 60% of the interval

TABLE 8.13 Summary of Results: Zero Forced Outage Rates

Unit number	Capacity (MW)	Outage rate (pu)	Hours on-line	Energy generated (MWh)	Cost (R)
1	60	0.000	672.0	34,944.0	145,152.0
2	50	0.000	403.0	8,736.0	58,800.0
3	20	0.000	0	0	0
4	100	0.000	0	0	0
Total	230			43,680.0	203,952.0

Average system cost = 4.6692 R/MWh.

so that its cost is

$$0.6 \times 70 \text{ R/h} \times 672 \text{ h} + 8736 \text{ MWh} \times 3.5 \text{ R/MWh} = 58,800 \text{ R}$$

The total system cost is 203, 952 R, and unit 3 is not used at all. These results are summarized in Table 8.13.

8.4.2 Forced Outages Included

When the forced outage is included, the convolution of the probability distribution is accomplished by

$$P'_n(x) = q\, P_n(x) + p\, P_n(x + c)$$

where   q = forced outage rate (pu)

p = 1 − q = "innage" rate

Table 8.14 shows the computations for the first unit in the third and fourth columns.
   The first unit is on-line $0.8 \times 672 = 537.6$ h and generates 27,955.2 MWh. (The initial load demand contains 43,680 MWh; the modified distribution in column 4 contains 15,724.8 MWh.) Therefore, the first unit's cost is

$$60 \text{ R/h} \times 537.6 \text{ h} + 3 \text{ R/MWh} \times 27,955.2 \text{ MWh} = 116,121.6 \text{ R}$$

The distribution of needed capacity is shown *partially* in the sixth column of Table 8.14. Sufficient data are shown to compute the load energy remaining. (*Load energy* is the portion of the distribution, $P_n^j(x)$, for $x \geq 0$). The unserved load energy after scheduling unit 2 is

$$0.576 \times 10 \times 672 = 3,870.72 \text{ MWh}$$

This means unit 2 generated an energy of $15,724.8 - 3,870.72 = 11,854.08$ MWh at an incremental cost of 3.5 R/MWh; or 41,489.28 R. The unit was on-line 411.264 h at a cost rate of 70 R/h. This brings the total cost to 70,277.76 R for unit 2. Note that the operating time (i.e., the "hours on-line") is $0.9 \times 0.68 \times 672$ h.
   The first factor represents the probability that the unit is available, the second the fraction of the time interval that the load requires unit 2 and the 672-h factor is the length of the interval.

TABLE 8.14 Three-Unit Example Including Forced Outage Rates

x (MW)	$P_n(x)$ (pu)	$P_n(x+60)$ (pu)	$P'_n(x)$ (pu)	$P'_n(x+50)$ (pu)	$P''_n(x)$ (pu)	$P'''_n(x)$ (pu)
−130	↑	↑	↑	↑		
−120						
−110						
−100						
− 90						
− 80				1.00		
− 70				0.84		
− 60				0.84		
− 50				0.68		
− 40				0.68		
− 30		1.0	1.0	0.52		
− 20		0.8	0.84	0.32		
− 10		0.8	0.84	0.28		
0		0.6	0.68	0.16	0.212	
10		0.6	0.68	0.12	0.176	
20		0.4	0.52	0.12	0.160	
30	1.0	0.15	0.32	0.08	0.104	
40	0.8	0.15	0.28	0.03	0.055	
50	0.8	0	0.16	0.03	0.043	
60	0.6		0.12	0	0.012	
70	0.6		0.12		0.012	
80	0.4		0.08		0.008	
90	0.15		0.03		0.003	
100	0.15		0.03		0.003	
110	0	↓	0	↓	0	
E/672	65		23.4		5.76	
MWh	43,680		15,724.8		3,870.72	

8.4.3 Results

Table 8.15 shows a summary of the results for this three-unit plus tie line sample exercise when outage rates are included. The third unit and tie line are utilized a substantial amount compared with ignoring forced outages. The total cost for the 4-wk interval increased by almost 5%.

TABLE 8.15 Results

Unit number	Capacity (MW)	Outage rate (pu)	Hours On-line	Energy generated (MWh)	Cost (R)
1	60	0.200	538.0	27,955.0	116,122.0
2	50	0.100	411.0	11,854.0	70,278.0
3	20	0.100	128.0	2,032.0	18,386.0
4	100	0.000	111.0	1,839.0	9,193.0
Total	230			43,680	213,979.0

Average system cost = 4.8589 R/MWh.

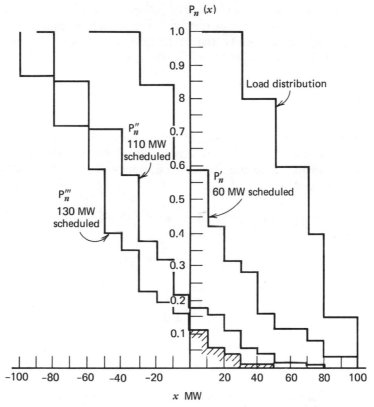

FIG. 8.14 Successive convolutions.

The resulting successive convolutions are shown in Figure 8.14. After the entire 130 MW of generating capacity has been dispatched, the distribution of unserved load is represented by the portion of the lowest curve to the right of the zero MW point (it is shaded).

Table 8.16 shows the distribution of emergency energy delivery over the tie line.

TABLE 8.16 Emergency Energy

Level number	Loading (MW)	Hours
1	10.0	30.71
2	20.0	11.02
3	30.0	22.04
4	40.0	0.81
5	50.0	4.50
6	60.0	3.02
7	70.0	0.27
8	80.0	2.15
9	100.0	0.20
Total		74.73

This chapter has provided only an introduction to this area. Practical schemes exist to handle much more complex unit and load models, to incorporate limited energy and pumped-storage units, and to compute generation reliability indices. They are all based on techniques similar to those introduced here.

PROBLEMS

8.1 Add another unit to Example 8C. The new unit should have a capacity of 10 MW and an availability of 90%. That is, its outage rate is 0.10 per unit. Use the recursive algorithm illustrated in the section. How far must the MW outage table be extended?

8.2 If the probability density function of unsupplied load power for a 1-h interval is $p_n(x)$ and the cumulative distribution is

$$P_n(x) = 1 - \int_0^x P_n(y) \, dy$$

demonstrate using ordinary calculus that the unsupplied energy is

$$\int_0^{x_{max}} P_n(y) \, dy$$

where x_{max} = maximum load in the 1-h interval
   y = dummy variable used to represent the load

Hint: $P_n(x)$ is the probability, or normalized duration, that a load of x MW exists. The energy represented by this load is then $x p_n(x)$. Find the total energy represented by the entire load distribution.

8.3 Complete Table 8.14 for the second unit (i.e., complete the sixth column). Convolve the third unit and determine the data for column 7 $[P_n'''(x)]$ and the energy generation of the third unit and its total cost. Find the distribution of energy to be served over the tie line. If this energy costs 5 R/MWh, what is the cost of this emergency supply and the total cost of production for this 4-wk interval?

8.4 Repeat Example 8D to find the minimum cost dispatch assuming that the fuel for unit 2 has been obtained under a take-or-pay contract and is limited to 4500 MBtu. Emergency energy will be purchased at 50 R/MWh. Find the minimum expected system cost including the cost of emergency energy.

8.5 In some applications it would be desirable to compute the expected operating costs and associated probability of having that much generation available both as functions of the power generated. Take the two-machine example of Example 8D and create a table of expected costs and probabilities for 20 MW intervals from 0 to 100 MW generation. Use these data to check the solution in Example

8D. See if you can devise an algorithm to perform these computations in an orderly and efficient manner.

8.6 Repeat the sample computation of section 8.4 except assume the input/output characteristic of unit 2 with its ratings changed to the following.

Output (MW)	Input/output cost curve (R/h)	Forced outage rate
Sec. 1 0–50	$70 + 3.5\, P_2$	0.1
Sec. 2 50–60	$245 + 4.5(P_2 - 50)$	0.1

Schedule section 2 of unit 2 after unit 3 and before the emergency energy. Use the techniques of Example 8F and deconvolve section 1 of unit 2 prior to determining the loading on section 2. Repeat the analysis ignoring the statistical dependence of section 2 on section 1 (That is, schedule a 10 MW "unit" to represent section 2 without deconvolving section 1.)

FURTHER READING

The literature concerning production cost simulations is profuse. A survey of various types of models is contained in reference 1. References 2–4 describe deterministic models designed for long-range planning. Reference 5 provides an entry into the literature of Monte Carlo simulation methods applied to generation planning and production cost computations.

The two texts referred to in references 6 and 7 provide an introduction to the use of probabilistic models and methods for power-generation planning. Reference 8 illustrates the application to a single area. These methods have been extended to consider the effects of transmission interconnections on generation system reliability in references 9–12.

The original probabilistic production cost technique was presented by E. Jamoulle and his associates in a difficult to locate Belgian publication (13). The basic methodology has been discussed and illustrated in a number of IEEE papers. References 14–16 are examples.

In many of these articles the presentation of the probabilistic methodology is couched in a sometimes confusing manner. Where authors such as R. R. Booth and others discuss an "equivalent load distribution" they are referring to the same distribution, $T\,P_n(x)$, discussed in this chapter. These authors allow the distribution to grow from zero load to some maximum value equal to the sum of the maximum load plus the sum of the capacity on forced outage. We have found this concept difficult to impart and prefer the present presentation. The practical results are identical to those found more commonly in the literature.

The models of approximation using orthogonal expansions to represent capacity distributions have been presented by Stremel and his associates. Reference 17 provides an entry into this literature.

References 15 and 18 lead into the development of the probabilistic production cost method, which is the dual of the method developed in this chapter. In this dual technique the generator parameters are used to develop probability distributions of capacity and expected cost (or expected incremental cost) curves versus the power available from an array of units. These may then be used as pseudogenerator parameters to compute expected production costs and unserved energy with any deterministic or probabilistic form of load model.

This brief introduction to the literature is by no means complete. There are hundreds of papers, reports, and articles on related subjects.

1. Wood, A. J., "Energy Production Cost Models," Symposium on Modeling and Simulation, University of Pittsburgh, April 1972. Published in the Conference Proceedings.

2. Bailey, E. S., Jr., Galloway, C. D., Hawkins, E. S., Wood, A. J., "Generation Planning Programs for Interconnected Systems: Part II, Production Costing Programs," *AIEE Special Supplement*, 1963, pp. 775–788.

3. Brennan, M. K., Galloway, C. D., Kirchmayer, L. K., "Digital Computer Aids Economic-Probabilistic Study of Generation Systems—I," *AIEE Transactions* (Power Apparatus and Systems), Vol. 77, August 1958, pp. 564–571.

4. Galloway, C. D., Kirchmayer, L. K., "Digital Computer Aids Economic-Probabilistic Study of Generation Systems—II," *AIEE Transactions* (Power Apparatus and Systems), Vol. 77, August 1958, pp. 571–577.

5. Dillard, J. K., Sels, H. K., "An Introduction to the Study of System Planning by Operational Gaming Models," *AIEE Transactions* (Power Apparatus and Systems), Vol. 78, Part III, December 1959, pp. 1284–1290.

6. Billinton, R., *Power System Reliability Evaluation*, Gordon and Breach, New York, 1970.

7. Billinton, R., Ringlee, R. J., Wood, A. J., *Power System Reliability Calculations*, MIT Press, Cambridge, Mass., 1973

8. Garver, L. L., "Reserve Planning Using Outage Probabilities and Load Uncertainties," *IEEE Transactions on Power Apparatus and Systems*, PAS-89, April 1970, pp. 514–521.

9. Cook, V. M., Galloway, C. D., Steinberg, M. J., Wood, A. J., "Determination of Reserve Requirements of Two Interconnected Systems," *AIEE Transactions*, Part III (Power Apparatus and Systems), Vol. 82, April 1963, pp. 18–33.

10. Bailey, E. S., Jr., Galloway, C. D., Hawkins, E. S., Wood, A. J., "Generation Planning Programs for Interconnected Systems: Part I, Expansion Programs," *AIEE Special Supplement*, 1963, pp. 761–764.

11. Spears, H. T., Hicks, K. L., Lee, S. T., "Probability of Loss of Load for Three Areas," *IEEE Transactions on Power Apparatus and Systems*, Vol. PAS-89, April 1970, pp. 521–527.

12. Pang, C. K., Wood, A. J., "Multi-Area Generation System Reliability Calculations," *IEEE Transactions on Power Apparatus and Systems*, Vol. PAS-94, March/April 1975, pp. 508–517.

13. Baleriaux, H., Jamoulle, E., Fr. Linard de Guertechin, "Simulation de l'Exploitation d'un Parc de Machines Thermiques de Production d'Electricite Couple a des Station de Pompage," *Review E (Edition SRBE)*, Vol. 5., No. 7, 1967, pp. 3–24.

14. Booth, R. R., "Power System Simulation Model Based on Probability Analysis," *IEEE Transactions on Power Apparatus and Systems*, Vol. PAS-91, January/February 1972, pp. 62–69.

15. Sager, M. A., Ringlee, R. J., Wood, A. J., "A New Generation Production Cost Program to Recognize Forced Outages," *IEEE Transactions on Power Apparatus and Systems*, Vol. PAS-91, September/October 1972, pp. 2114–2124.

16. Sager, M. A., Wood, A. J., "Power System Production Cost Calculations—Sample Studies Recognizing Forced Outages," *IEEE Transactions on Power Apparatus and Systems*, PAS-92, January/February 1973, pp. 154–158.

17. Stremel, J. P., Jenkins, R. T., Babb, R. A., Bayless, W. D., "Production Costing Using the Cumulant Method of Representing the Equivalent Load Curve," *IEEE Transactions on Power Apparatus and Systems*, Vol. PAS-98, September/October 1980, pp. 1947–1956.

18. Sidenblad, K. M., Lee, S. T. Y., "A Probabilistic Production Costing Methodology for Systems with Storage," *IEEE Transactions on Power Apparatus and Systems*, Vol. PAS-100, June 1981, pp. 3116–3124.

chapter 9

Control of Generation

9.1 INTRODUCTION

So far, this text has concentrated on methods of establishing optimum dispatch and scheduling of generating units. It is important to realize, however, that such optimized dispatching would be useless without a method of control over the generator units. Indeed, the control of generator units was the first problem faced in early power-system design. The methods developed for control of individual generators and eventually control of large interconnections play a vital role in modern energy control centers.

A generator driven by a steam turbine can be represented as a large rotating mass with two opposing torques acting on the rotation. As shown in Figure 9.1, T_{mech}, the mechanical torque, acts to increase rotational speed whereas T_{elec}, the electrical torque, acts to slow it down. When T_{mech} and T_{elec} are equal in magnitude, the rotational speed, ω, will be constant. If the electrical load is increased so that T_{elec} is larger than T_{mech}, the entire rotating system will begin to slow down. Since it would be damaging to let the equipment slow down too far, something must be done to increase the mechanical torque T_{mech} to restore equilibrium, that is, to bring the

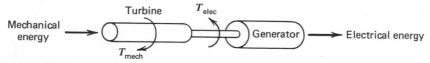

FIG. 9.1 Mechanical and electrical torques in a generating unit.

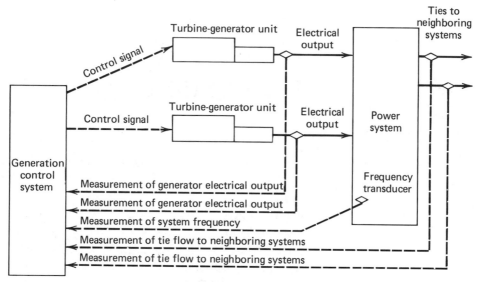

FIG. 9.2 Overview of generation control problem.

rotational speed back to an acceptable value and the torques to equality so that the speed is again held constant.

This process must be repeated constantly on a power system because the loads change constantly. Furthermore, because there are many generators supplying power into the transmission system, some means must be provided to allocate the load changes to the generators. To accomplish this, a series of control systems are connected to the generator units. A governor on each unit maintains its speed while supplementary control, usually originating at a remote control center, acts to allocate generation. Figure 9.2 shows an overview of the generation control problem.

9.2 GENERATOR MODEL

Before starting, it will be useful for us to define our terms.

ω = rotational speed (rad/sec)

α = rotational acceleration

δ = phase angle of a rotating machine

T_{net} = net accelerating torque in a machine

T_{mech} = mechanical torque exerted on the machine by the turbine

T_{elec} = electrical torque exerted on the machine by the generator

P_{net} = net accelerating power

P_{mech} = mechanical power input

P_{elec} = electrical power output

I = moment of inertia for the machine

M = angular momentum of the machine

Where all quantities (except phase angle) will be in per unit on the machine base, or in the case of ω on the standard system frequency base. Thus for example, M is in per unit power/per unit frequency/sec.

In the development to follow, we are interested in deviations of quantities about steady-state values. All steady-state or nominal values will have a "0" subscript (e.g., ω_0, T_{net0}), and all deviations from nominal will be designated by a "Δ" (e.g., $\Delta\omega$, ΔT_{net}). Some basic relationships are

$$I\alpha = T_{net} \tag{9.1}$$

$$M = \omega I \tag{9.2}$$

$$P_{net} = \omega T_{net} = \omega(I\alpha) = M\alpha \tag{9.3}$$

To start, we will focus our attention on a single rotating machine. Assume that the machine has a steady speed of ω_0 and phase angle δ_0. Due to various electrical or mechanical disturbances, the machine will be subjected to differences in mechanical and electrical torque causing it to accelerate or decelerate. We are chiefly interested in the deviations of speed, $\Delta\omega$, and deviations in phase angle, $\Delta\delta$, from nominal.

The phase angle deviation, $\Delta\delta$, is equal to the difference in phase angle between the machine as subjected to an acceleration of α and a reference axis rotating at exactly ω_0. If the speed of the machine under acceleration is

$$\omega = \omega_0 + \alpha t \tag{9.4}$$

Then

$$\Delta\delta = \underbrace{\int (\omega_0 + \alpha t)\, dt}_{\substack{\text{Machine absolute} \\ \text{phase angle}}} - \underbrace{\int \omega_0 dt}_{\substack{\text{Phase angle of} \\ \text{reference axis}}}$$

$$= \omega_0 t + \frac{1}{2}\alpha t^2 - \omega_0 t$$

$$= \frac{1}{2}\alpha t^2 \tag{9.5}$$

The deviation from nominal speed, $\Delta\omega$, may then be expressed as

$$\Delta\omega = \alpha t = \frac{d}{dt}(\Delta\delta)$$ (9.6)

The relationship between phase angle deviation, speed deviation, and net accelerating torque is

$$T_{net} = I\alpha = I\frac{d}{dt}(\Delta\omega) = I\frac{d^2}{dt^2}(\Delta\delta)$$ (9.7)

Next, we will relate the deviations in mechanical and electrical power to the deviations in rotating speed and mechanical torques. The relationship between net accelerating power and the electrical and mechanical powers is

$$P_{net} = P_{mech} - P_{elec}$$ (9.8)

which is written as the sum of the steady-state value and the deviation term,

$$P_{net} = P_{net_0} + \Delta P_{net}$$ (9.9)

where $P_{net_0} = P_{mech_0} - P_{elec_0}$
$\qquad \Delta P_{net} = \Delta P_{mech} - \Delta P_{elec}$

Then

$$P_{net} = (P_{mech_0} - P_{elec_0}) + (\Delta P_{mech} - \Delta P_{elec})$$ (9.10)

Similarly for torques,

$$T_{net} = (T_{mech_0} - T_{elec_0}) + (\Delta T_{mech} - \Delta T_{elec})$$ (9.11)

Using Eq. 9.3, we can see that

$$P_{net} = P_{net_0} + \Delta P_{net} = (\omega_0 + \Delta\omega)(T_{net_0} + \Delta T_{net})$$ (9.12)

substituting Eqs. 9.10 and 9.11, we obtain

$$(P_{mech_0} - P_{elec_0}) + (\Delta P_{mech} - \Delta P_{elec}) = (\omega_0 + \Delta\omega)[(T_{mech_0} - T_{elec_0})$$
$$+ (\Delta T_{mech} - \Delta T_{elec})]$$ (9.13)

Assume that the steady-state quantities can be factored out since

$$P_{mech_0} = P_{elec_0}$$

and

$$T_{mech_0} = T_{elec_0}$$

and further assume that the second-order terms involving products of $\Delta\omega$ with ΔT_{mech} and ΔT_{elec} can be neglected. Then

$$\Delta P_{mech} - \Delta P_{elec} = \omega_0(\Delta T_{mech} - \Delta T_{elec})$$ (9.14)

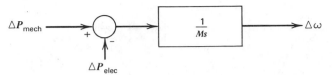

FIG. 9.3 Relationship between mechanical and electrical power and speed change.

As shown in Eq. 9.7, the net torque is related to the speed change as follows.

$$(T_{mech_0} - T_{elec_0}) + (\Delta T_{mech} - \Delta T_{elec}) = I\frac{d}{dt}(\Delta\omega) \tag{9.15}$$

then since $T_{mech_0} = T_{elec_0}$, we can combine Eqs. 9.14 and 9.15 to get

$$\Delta P_{mech} - \Delta P_{elec} = \omega_0 I \frac{d}{dt}(\Delta\omega)$$

$$= M\frac{d}{dt}(\Delta\omega) \tag{9.16}$$

This can be expressed in LaPlace transform operator notation as

$$\Delta P_{mech} - \Delta P_{elec} = Ms\,\Delta\omega \tag{9.17}$$

This is shown in block diagram form in Figure 9.3.

The units for M are watts per radian per second per second. We will always use per unit power over per unit speed per second where the per unit refers to the machine rating as the base (see Example 9A).

9.3 LOAD MODEL

The loads on a power system consist of a variety of electrical devices. Some of them are purely resistive, some are motor loads with variable power-frequency characteristics, and others exhibit quite different characteristics. Since motor loads are a dominant part of the electrical load, there is a need to model the effect of a change in frequency on the net load drawn by the system. The relationship between the change in load due to the change in frequency is given by

$$\Delta P_{L(freq)} = D\,\Delta\omega \qquad \text{or} \qquad D = \frac{\Delta P_{L(freq)}}{\Delta\omega}$$

where D is expressed as percent change in load divided by percent change in frequency. For example, if load changed by 1.5% for a 1% change in frequency, then D would equal 1.5. However, the value of D used in solving for system dynamic response must be changed if the system base MVA is different from the nominal value of load. Suppose the D referred to here was for a net connected load of 1200 MVA and the entire dynamics problem were to be set up for 1000 MVA system base. Note that $D = 1.5$ tells us that the load would change by 1.5 pu for a 1 pu change in

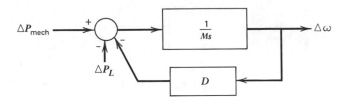

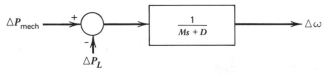

FIG. 9.4 Block diagram of rotating mass and load as seen by prime-mover output.

frequency. That is, the load would change by 1.5 × 1200 MVA or 1800 MVA for a 1 pu change in frequency. When expressed on a 1000 MVA base, D becomes

$$D_{1000 \text{ MVA Base}} = 1.5 \times \left(\frac{1200}{1000}\right) = 1.8$$

The net change in P_{elec} in Figure 9.3 (Eq. 9.15) is

$$\Delta P_{\text{elec}} = \underbrace{\Delta P_L}_{\substack{\text{Nonfrequency-} \\ \text{sensitive load} \\ \text{change}}} + \underbrace{D \, \Delta\omega}_{\substack{\text{Frequency-sensitive} \\ \text{load change}}} \qquad (9.18)$$

Including this in the block diagram results in the new block diagram shown in Figure 9.4.

EXAMPLE 9A

We are given an isolated power system with a 600 MVA generating unit having an M of 7.6 pu MW/pu frequency/sec on machine base. The unit is supplying a load of 400 MVA. The load changes by 2% for a 1% change in frequency. First, we will set up the block diagram of the equivalent generator load system. Everything will be referenced to a 1000 MVA base.

$$M = 7.6 \times \frac{600}{1000} = 4.56 \text{ on 1000 MVA base}$$

$$D = 2 \times \frac{400}{1000} = 0.8 \text{ on 1000 MVA base}$$

Then the block diagram is as shown in Figure 9.5.

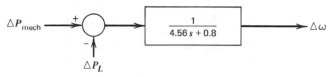

FIG. 9.5 Block diagram for system in Example 9A.

Suppose the load suddenly increases by 10 MVA (or 0.01 pu), that is,

$$\Delta P_L(s) = \frac{0.01}{s}$$

Then

$$\Delta\omega(s) = -\frac{0.01}{s}\left(\frac{1}{4.56\,s + 0.8}\right)$$

or taking the inverse LaPlace transform,

$$\Delta\omega(t) = (0.01/0.8)e^{-0.8/4.56\,t} - 0.01/0.8$$
$$= 0.0125\,e^{-0.1754\,t} - 0.0125$$

The final value of $\Delta\omega$ is -0.0125 pu, which is a drop of 0.75 Hz on a 60 Hz system.

When two or more generators are connected to a transmission system network, we must take account of the phase angle difference across the network in analyzing frequency changes. However, for the sake of governing analysis, which we are interested in here, we can assume that frequency will be constant over those parts of the network that are tightly interconnected. When making such an assumption, we can then lump the rotating mass of the turbine generators together into an equivalent that is driven by the sum of the individual turbine mechanical outputs. This is illustrated in Figure 9.6 where all turbine generators were lumped into a single equivalent rotating mass, M_{equiv}. Similarly, all individual system loads were lumped into an equivalent load with damping coefficient, D_{equiv}.

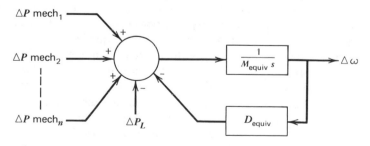

FIG. 9.6 Multi-turbine-generator system equivalent.

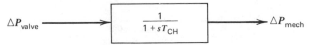

FIG. 9.7 Prime-mover model.

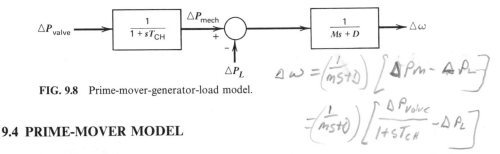

FIG. 9.8 Prime-mover-generator-load model.

$$\Delta \omega = \left(\frac{1}{ms+D}\right)\left[\Delta P_M - \Delta P_L\right]$$

$$= \left(\frac{1}{ms+D}\right)\left[\frac{\Delta P_{valve}}{1+sT_{cH}} - \Delta P_L\right]$$

9.4 PRIME-MOVER MODEL

The prime mover driving a generator unit may be a steam turbine or a hydro turbine. The models for the prime mover must take account of the steam supply and boiler control system characteristics in the case of a steam turbine, or the penstock characteristics for a hydro turbine. Throughout the remainder of this chapter, only the simplest prime-mover model, the nonreheat turbine, will be used. The models for other more complex prime movers, including hydro turbines, are developed in the references (see Further Readings).

The model for a nonreheat turbine, shown in Figure 9.7, relates the position of the valve that controls emission of steam into the turbine to the power output of the turbine, where

$$T_{\mathrm{CH}} = \text{"charging time" time constant}$$

$$\Delta P_{\mathrm{valve}} = \text{per unit change in valve position from nominal}$$

The combined prime mover-generator-load model for a single generating unit can be built by combining Figures 9.4 and 9.7, as shown in Figure 9.8.

9.5 GOVERNOR MODEL

Suppose a generating unit is operated with fixed mechanical power output from the turbine. The result of any load change would be a speed change sufficient to cause the frequency sensitive load to exactly compensate for the load change (as in Example 9A). This condition would allow system frequency to drift far outside acceptable limits. This is overcome by adding a governing mechanism that senses the machine speed and adjusts the input valve to change the mechanical power output to compensate for load changes and to restore frequency to nominal value. The earliest such mechanism used rotating "flyballs" to sense speed and to provide mechanical motion in response to speed changes. Modern governors use electronic means to sense speed changes and often use a combination of electronic, mechanical, and hydraulic means to affect the required valve position changes. The simplest

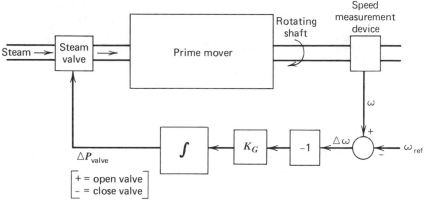

FIG. 9.9 Isochronous governor.

governor, called the *isochronous governor*, adjusts the input valve to a point that brings frequency back to nominal value. If we simply connect the output of the speed-sensing mechanism to the valve through a direct linkage, it would never bring the frequency to nominal. To force the frequency error to zero, one must provide what control engineers call reset action. Reset action is accomplished by integrating the frequency (or speed) error, which is the difference between actual speed and desired or reference speed.

We will illustrate such a speed-governing mechanism with the diagram shown in Figure 9.9. The speed measurement device's output, ω, is compared with a reference, ω_{ref}, to produce an error signal, $\Delta\omega$. The error, $\Delta\omega$, is negated and then amplified by a gain K_G and integrated to produce a control signal, ΔP_{valve}, which causes the main steam supply valve to open (ΔP_{valve} positive) when $\Delta\omega$ is negative. If, for example, the machine is running at reference speed and the electrical load increases, ω will fall below ω_{ref} and $\Delta\omega$ will be negative. The action of the gain and integrator will be to open the steam valve, causing the turbine to increase its mechanical output, thereby increasing the electrical output of the generator and increasing the speed ω. When ω exactly equals ω_{ref}, the steam valve stays at the new position (further opened) to allow the turbine-generator to meet the increased electrical load.

The isochronous (constant speed) governor of Figure 9.9 cannot be used if two or more generators are electrically connected to the same system since each generator would have to have precisely the same speed setting or they would "fight" each other, each trying to pull the system's speed (or frequency) to its own setting. To be able to run two or more generating units in parallel on a generating system, the governors are provided with a feedback signal that causes the speed error to go to zero at different values of generator output.

This can be accomplished by adding a feedback loop around the integrator as shown in Figure 9.10. Note that we have also inserted a new input, called the *load reference*, that we will discuss shortly. The block diagram for this governor is shown in Figure 9.11 where the governor now has a net gain of $1/R$ and a time constant T_G.

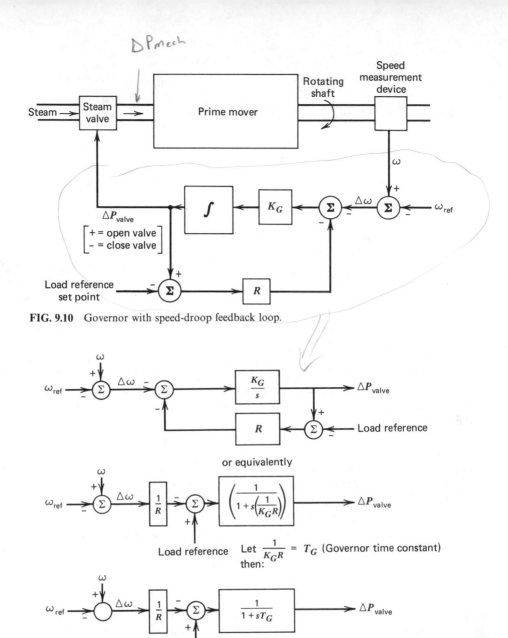

ΔP_{mech}

ΔP_{valve}

$\begin{bmatrix} + = \text{open valve} \\ - = \text{close valve} \end{bmatrix}$

FIG. 9.10 Governor with speed-droop feedback loop.

or equivalently

Let $\dfrac{1}{K_G R} = T_G$ (Governor time constant)
then:

FIG. 9.11 Block diagram of governor with droop.

The result of adding the feedback loop with gain R is a governor characteristic as shown in Figure 9.12. The value of R determines the slope of the characteristic. That is, R determines the change in the unit's output for a given change in frequency. Common practice is to set R on each generating unit so that a change from 0 to 100% (i.e., rated) output will result in the same frequency change for each unit. As a result, a

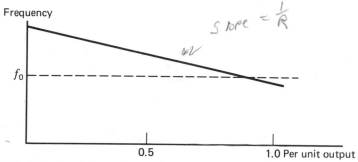

FIG. 9.12 Speed-droop characteristic.

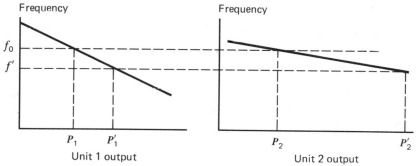

FIG. 9.13 Allocation of unit outputs with governor droop.

change in electrical load on a system will be compensated by generator unit output changes proportional to each unit's rated output.

If two generators with drooping governor characteristics are connected to a power system, there will always be a unique frequency at which they will share a load change between them. This is illustrated in Figure 9.13 showing two units with drooping characteristics connected to a common load.

As shown in Figure 9.13, the two units start at a nominal frequency of f_0. When a load increase, ΔP_L, causes the units to slow down, the governors increase output until the units seek a new common operating frequency, f'. The amount of load pickup on each unit is proportional to the slope of its droop characteristic. Unit 1 increases its output from P_1 to P'_1, unit 2 increases its output from P_2 to P'_2 such that the net generation increase, $P'_1 - P_1 + P'_2 - P_2$, is equal to ΔP_L. Note that the actual frequency sought also depends on the load's frequency characteristic as well.

Figure 9.10 shows an input labeled "load reference set point." By changing the load reference, the generator's governor characteristic can be set to give reference frequency at any desired unit output. This is illustrated in Figure 9.14. *The basic control input to a generating unit as far as generation control is concerned is the* **load reference set point**. By adjusting this set point on each unit, a desired unit dispatch can be maintained while holding system frequency close to the desired nominal value.

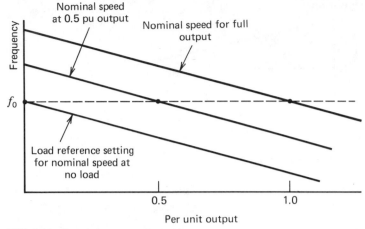

FIG. 9.14 Speed-changer settings.

Note that a steady-state change in ΔP_{valve} of 1.0 pu requires a value of R pu change in frequency, $\Delta\omega$. One often hears unit regulation referred to in percent. For instance, a 3% regulation for a unit would indicate that a 100% (1.0 pu) change in valve position (or equivalently a 100% change in unit output) requires a 3% change in frequency. Therefore, R is equal to pu change in frequency divided by pu change in unit output. That is,

$$R = \frac{\Delta\omega}{\Delta P} \text{ pu}$$

At this point, we can construct a block diagram of a governor-prime mover-rotating mass/load model as shown in Figure 9.15. Suppose that this generator experiences a step increase in load.

$$\Delta P_L(s) = \frac{\Delta P_L}{s} \tag{9.19}$$

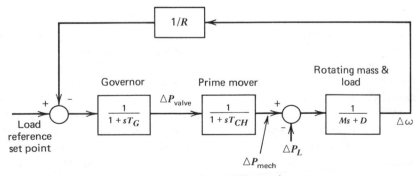

FIG. 9.15 Block diagram of governor, prime mover, rotating mass.

The transfer function relating the load change, ΔP_L, to the frequency change, $\Delta\omega$, is

$$\Delta\omega(s) = \Delta P_L(s) \left[\frac{\dfrac{-1}{Ms + D}}{1 + \dfrac{1}{R}\left(\dfrac{1}{1 + sT_G}\right)\left(\dfrac{1}{1 + sT_{CH}}\right)\left(\dfrac{1}{Ms + D}\right)} \right] \qquad (9.20)$$

The steady-state value of $\Delta\omega(s)$ may be found by

$$\Delta\omega \text{ steady state} = \lim_{s \to 0} \left[s\,\Delta\omega(s)\right]$$

$$= \frac{-\Delta P_L\left(\dfrac{1}{D}\right)}{1 + \left(\dfrac{1}{R}\right)\left(\dfrac{1}{D}\right)} = \frac{-\Delta P_L}{\dfrac{1}{R} + D} \qquad (9.21)$$

Note that if D were zero, the change in speed would simply be

$$\Delta\omega = -R\,\Delta P_L \qquad (9.22)$$

If several generators (each having its own governor and prime mover) were connected to the system, the frequency change would be

$$\Delta\omega = \frac{-\Delta P_L}{\dfrac{1}{R_1} + \dfrac{1}{R_2} + \cdots + \dfrac{1}{R_n} + D} \qquad (9.23)$$

9.6 TIE-LINE MODEL

The power flowing across a transmission line can be modeled using the DC load flow method shown in Chapter 4, Eq. 4.23.

$$P_{\text{tie flow}} = \frac{1}{X_{\text{tie}}}(\theta_1 - \theta_2) \qquad (9.24)$$

This tie flow is a steady-state quantity. For purposes of analysis here we will perturb Eq. 9.24 to obtain deviations from nominal flow as a function of deviations in phase angle from nominal.

$$P_{\text{tie flow}} + \Delta P_{\text{tie flow}} = \frac{1}{X_{\text{tie}}}\left[(\theta_1 + \Delta\theta_1) - (\theta_2 + \Delta\theta_2)\right]$$

$$= \frac{1}{X_{\text{tie}}}(\theta_1 - \theta_2) + \frac{1}{X_{\text{tie}}}(\Delta\theta_1 - \Delta\theta_2) \qquad (9.25)$$

Then

$$\Delta P_{\text{tie flow}} = \frac{1}{X_{\text{tie}}}(\Delta\theta_1 - \Delta\theta_2) \qquad (9.26)$$

where $\Delta\theta_1$ and $\Delta\theta_2$ are equivalent to $\Delta\delta_1$ and $\Delta\delta_2$ as defined in Eq. 9.6. Then, using the relationship of Eq. 9.6,

$$\Delta P_{\text{tie flow}} = \frac{T}{s}(\Delta\omega_1 - \Delta\omega_2)$$

(9.27)

where $T = 377 \times 1/X_{\text{tie}}$ (for a 60 Hz system).

Note that $\Delta\theta$ must be in radians for ΔP_{tie} to be in per unit megawatts, but $\Delta\omega$ is in per unit speed change. Therefore, we must multiply $\Delta\omega$ by 377 rad/sec (the base frequency in rad/sec at 60 Hz). T may be thought of as the "tie-line stiffness" coefficient.

Suppose now that we have an interconnected power system broken into two areas each having one generator. The areas are connected by a single transmission line. The power flow over the transmission line will appear as a positive load to one area and an equal but negative load to the other, or vice versa, depending on the direction of flow. The direction of flow will be dictated by the relative phase angle between the areas, which is determined by the relative speed deviations in the areas. A block diagram representing this interconnection can be drawn as in Figure 9.16. Note that the tie power flow was defined as going from area 1 to area 2, therefore, the flow appears as a load to area 1 and a power source (negative load) to area 2. If one

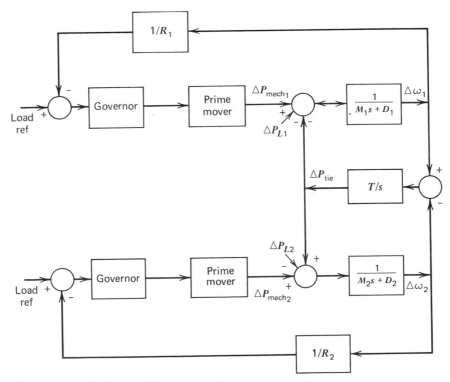

FIG. 9.16 Block diagram of interconnected areas.

No ACE in this model

assumes that mechanical powers are constant, the rotating masses and tie line exhibit damped oscillatory characteristics known as synchronizing oscillations. (See Problem 3 at the end of this chapter.)

It is quite important to analyze the steady-state frequency deviation, tie-flow deviation, and generator outputs for an interconnected area after a load change occurs. Let there be a load change ΔP_{L_1} in area 1. In the steady state, after all synchronizing oscillations have damped out, the frequency will be constant and equal to the same value in both areas. Then

$$\Delta \omega_1 = \Delta \omega_2 = \Delta \omega \quad \text{and} \quad \frac{d(\Delta \omega_1)}{dt} = \frac{d(\Delta \omega_2)}{dt} = 0 \tag{9.28}$$

and

$$\Delta P_{\text{mech}_1} - \Delta P_{\text{tie}} - \Delta P_{L_1} = \Delta \omega D_1$$

$$\Delta P_{\text{mech}_2} + \Delta P_{\text{tie}} = \Delta \omega D_2$$

$$\Delta P_{\text{mech}_1} = \frac{-\Delta \omega}{R_1} \tag{9.29}$$

$$\Delta P_{\text{mech}_2} = \frac{-\Delta \omega}{R_2}$$

by making appropriate substitutions in Eq. 9.29,

$$-\Delta P_{\text{tie}} - \Delta P_{L_1} = \Delta \omega \left(\frac{1}{R_1} + D_1 \right)$$

$$+\Delta P_{\text{tie}} = \Delta \omega \left(\frac{1}{R_2} + D_2 \right) \tag{9.30}$$

or, finally,

$$\Delta \omega = \frac{-\Delta P_{L_1}}{\dfrac{1}{R_1} + \dfrac{1}{R_2} + D_1 + D_2} \tag{9.31}$$

from which we can derive the change in tie flow:

$$\Delta P_{\text{tie}} = \frac{-\Delta P_{L_1} \left(\dfrac{1}{R_2} + D_2 \right)}{\dfrac{1}{R_1} + \dfrac{1}{R_2} + D_1 + D_2} \tag{9.32}$$

Note that the conditions described in Eqs. 9.28 through 9.32 are for the new steady-state conditions after the load change. The new tie flow is determined by the net change in load and generation in each area. We do not need to know the tie stiffness to determine this new tie flow, although the tie stiffness will determine how much difference in phase angle across the tie will result from the new tie flow.

EXAMPLE 9B

You are given two system areas connected by a tie line with the following characteristics.

Area 1	Area 2
$R = 0.01$ pu	$R = 0.02$ pu
$D = 0.8$ pu	$D = 1.0$ pu
Base MVA = 500	Base MVA = 500

A load change of 100 MW (0.2 pu) occurs in area 1. What is the new steady-state frequency and what is the change in tie flow? Assume both areas were at nominal frequency (60 Hz) to begin.

$$\Delta\omega = \frac{-\Delta P_{L_1}}{\dfrac{1}{R_1} + \dfrac{1}{R_2} + D_1 + D_2} = \frac{-.2}{\dfrac{1}{0.01} + \dfrac{1}{0.02} + 0.8 + 1} = -0.00131752 \text{ pu}$$

$$f_{new} = 60 - 0.00132(60) = 59.92 \text{ Hz}$$

$$\Delta P_{tie} = \Delta\omega\left(\frac{1}{R_2} + D_2\right) = -0.00131752\left(\frac{1}{0.02} + 1\right) = -0.06719368 \text{ pu}$$

$$= -33.6 \text{ MW}$$

The change in prime mover powers would be

$$\Delta P_{mech_1} = \frac{-\Delta\omega}{R_1} = -\left(\frac{-0.00131752}{0.01}\right) = 0.13175231 \text{ pu} = 65.876 \text{ MW}$$

$$\Delta P_{mech_2} = \frac{-\Delta\omega}{R_2} = -\left(\frac{-0.00131752}{0.02}\right) = 0.06587615 \text{ pu} = 32.938 \text{ MW}$$

$$= 98.814 \text{ MW}$$

The total change in generation is 98.814, which is 1.186 MW short of the 100 MW load change. The change in total area load due to frequency drop would be

$$\text{For Area 1} = \Delta\omega D_1 = -0.0010540 \text{ pu} = -0.527 \text{ MW}$$

$$\text{For Area 2} = \Delta\omega D_2 = -0.00131752 \text{ pu} = -0.6588 \text{ MW}$$

Therefore, the total load change is $= 1.186$ MW, which accounts for the difference in total generation change and total load change. (See Problem 9.2 for further variations on this problem.)

If we were to analyze the dynamics of the two-area systems, we would find that a step change in load would always result in a frequency error. This is illustrated in Figure 9.17, which shows the frequency response of the system to a step-load change. Note that Figure 9.17 only shows the average frequency (omitting any high-frequency oscillations).

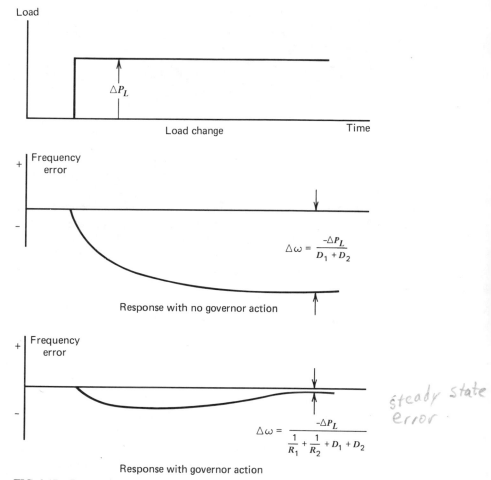

$$\Delta\omega = \frac{-\Delta P_L}{D_1 + D_2}$$

Response with no governor action

$$\Delta\omega = \frac{-\Delta P_L}{\dfrac{1}{R_1} + \dfrac{1}{R_2} + D_1 + D_2}$$

steady state error

Response with governor action

FIG. 9.17 Frequency response to load change.

9.7 GENERATION CONTROL

Automatic generation control (AGC) is the name given to a control system having three major objectives:

1. To hold system frequency at or very close to a specified nominal value (e.g., 60 Hz).
2. To maintain the correct value of interchange power between control areas.
3. To maintain each unit's generation at the most economic value.

9.7.1 Supplementary Control Action

To understand each of the three objectives just listed, we may start out assuming that we are studying a single generating unit supplying load to an isolated power system.

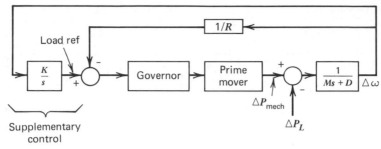

FIG. 9.18 Supplementary control added to generating unit.

As shown in Section 9.5, a load change will produce a frequency change with a magnitude that depends on the droop characteristics of the governor and the frequency characteristics of the system load. Once a load change has occurred, a supplementary control must act to restore the frequency to nominal value. This can be accomplished by adding a reset (integral) control to the governor as shown in Figure 9.18.

The reset control action of the supplementary control will force the frequency deviation to zero by adjustment of the speed reference set point. For example, the error shown in the bottom diagram of Figure 9.17 would be forced to 0.

9.7.2 Tie-Line Control

When two utilities interconnect their systems, they do so for several reasons. One is to be able to buy and sell power with neighboring systems whose operating costs make such transactions profitable. Further, even if no power is being transmitted over ties to neighboring systems, if one system has a sudden loss of a generating unit, the units throughout all the interconnection will experience a frequency change and can help in restoring frequency.

Interconnections present a very interesting control problem with respect to allocation of generation to meet load. The hypothetical situation in Figure 9.19 will be used to illustrate this problem. Assume both systems in Figure 9.19 have equal generation and load characteristics ($R_1 = R_2, D_1 = D_2$) and, further, assume system 1 was sending 100 MW to system 2 under an interchange agreement made between the operators of each system. Now let system 2 experience a sudden load increase of 30 MW. Since both units have equal generation characteristics, they will both experience a 15 MW increase, and the tie line will experience an increase in flow from 100 MW to 115 MW. Thus, the 30 MW load increase in system 2 will have been satisfied by a 15 MW increase in generation in system 2 plus a 15 MW increase in tie flow into system 2. This would be fine except that system 1 contracted to sell only 100 MW, not 115 MW, and its generating costs have just gone up without anyone to bill the extra cost to. What is needed at this point is a control scheme that recognizes the fact that the 30 MW load increase occurred in system 2 and, therefore, would increase generation in system 2 by 30 MW while restoring frequency to nominal

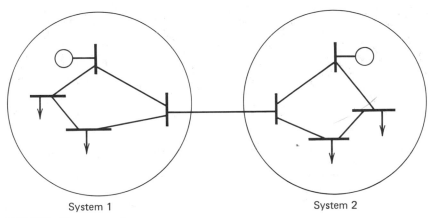

System 1   System 2

FIG. 9.19 Two-area system.

value. It would also restore generation in system 1 to its output before the load increase occurred.

Such a control system must use two pieces of information: the system frequency and the net power flowing in or out over the tie lines.

Such a control scheme would of necessity have to recognize the following.

1. If frequency decreased and net interchange power leaving the system increased, a load increase has occurred outside the system.

2. If frequency decreased and net interchange power leaving the system decreased, a load increase has occurred inside the system.

This can be extended to cases where frequency increases. We will make the following definitions.

$$P_{\text{net int}} = \text{total actual net interchange}$$
$$\text{(+ for power leaving the system; − for power entering)}$$

$$P_{\text{net int sched}} = \text{scheduled or desired value of interchange} \tag{9.33}$$

$$\Delta P_{\text{net int}} = P_{\text{net int}} - P_{\text{net int sched}}$$

Then a summary of the tie line-frequency control scheme can be given as in the table in Figure 9.20.

We define a *control area* to be a part of an interconnected system within which the load and generation will be controlled as per the rules in Figure 9.20. The control area's boundary is simply the tie-line points where power flow is metered. All tie lines crossing the boundary must be metered so that total control area net interchange power can be calculated.

The rules set forth in Figure 9.20 can be implemented by a control mechanism that weighs frequency deviation, $\Delta\omega$, and net interchange power, $\Delta P_{\text{net int}}$. The frequency response and tie flows resulting from a load change, ΔP_{L_1}, in the two area

$\Delta\omega$	$\Delta P_{\text{net int}}$	Load change	Resulting control action
$-$	$-$	ΔP_{L_1} $+$ ΔP_{L_2} 0	Increase P_{gen} in system 1
$+$	$+$	ΔP_{L_1} $-$ ΔP_{L_2} 0	Decrease P_{gen} in system 1
$-$	$+$	ΔP_{L_1} 0 ΔP_{L_2} $+$	Increase P_{gen} in system 2
$+$	$-$	ΔP_{L_1} 0 ΔP_{L_2} $-$	Decrease P_{gen} in system 2

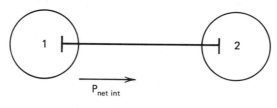

ΔP_{L_1} = Load change in area 1

ΔP_{L_2} = Load change in area 2

FIG. 9.20 Tie-line frequency control actions for two-area system.

system of Figure 9.16 are derived in Eqs. 9.28 through 9.32. These results are repeated here.

Load change	Frequency change	Change in net interchange
ΔP_{L_1}	$\Delta\omega = \dfrac{-\Delta P_{L_1}}{\dfrac{1}{R_1} + \dfrac{1}{R_2} + D_1 + D_2}$	$\Delta P_{\text{net int}_1} = \dfrac{-\Delta P_{L_1}\left(\dfrac{1}{R_2} + D_2\right)}{\dfrac{1}{R_1} + \dfrac{1}{R_2} + D_1 + D_2}$ (9.34)

This corresponds to the first row of the table in Figure 9.20; we would therefore require that

$$\Delta P_{\text{gen}_1} = \Delta P_{L_1}$$

$$\Delta P_{\text{gen}_2} = 0$$

The required change in generation, historically called the *area control error* or ACE, represents the shift in the area's generation required to restore frequency and net interchange to their desired values. The equation for ACE for each area is

$$\text{ACE}_1 = -\Delta P_{\text{net int}_1} - B_1\,\Delta\omega$$

$$\text{ACE}_2 = -\Delta P_{\text{net int}_2} - B_2\,\Delta\omega$$

(9.35)

where B_1 and B_2 are called *frequency bias factors*. We can see from Eq. 9.34 that setting bias factors as follows:

$$B_1 = \left(\frac{1}{R_1} + D_1\right)$$

$$B_2 = \left(\frac{1}{R_2} + D_2\right)$$

(9.36)

results in

$$\text{ACE}_1 = \left(\frac{+\Delta P_{L_1}\left(\dfrac{1}{R_2} + D_2\right)}{\dfrac{1}{R_1} + \dfrac{1}{R_2} + D_1 + D_2}\right) - \left(\frac{1}{R_1} + D_1\right)\left(\frac{-\Delta P_{L_1}}{\dfrac{1}{R_1} + \dfrac{1}{R_2} + D_1 + D_2}\right) = \Delta P_{L_1}$$

$$\text{ACE}_2 = \left(\frac{-\Delta P_{L_1}\left(\dfrac{1}{R_2} + D_2\right)}{\dfrac{1}{R_1} + \dfrac{1}{R_2} + D_1 + D_2}\right) - \left(\frac{1}{R_2} + D_2\right)\left(\frac{-\Delta P_{L_1}}{\dfrac{1}{R_1} + \dfrac{1}{R_2} + D_1 + D_2}\right) = 0$$

This control can be carried out using the scheme outlined in Figure 9.21. Note that the values of B_1 and B_2 would have to change each time a unit was committed or

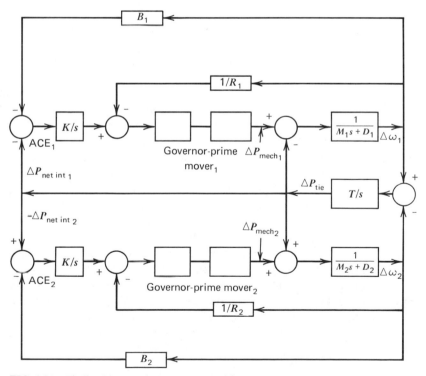

FIG. 9.21   Tie-line bias supplementary control for two areas.

decommitted in order to have the exact values as given in Eq. 9.36. Actually, the integral action of the supplementary controller will guarantee a reset of ACE to zero even when B_1 and B_2 are in error.

9.7.3 Generation Allocation

If each control area in an interconnected system had a single generating unit, the control system of Figure 9.21 would suffice to provide stable frequency and tie-line interchange. However, power systems consist of control areas with many generating units with outputs that must be set according to economics. That is, we must couple an economic dispatch calculation to the control mechanism so it will know how much of each area's total generation is required from each individual unit.

One must remember that a particular total generation value will not usually exist for a very long time since the load on a power system varies continually as people and industries use individual electric loads. Therefore, it is impossible to simply specify a total generation, calculate the economic dispatch for each unit, and then give the control mechanism the values of megawatt output for each unit—unless such a calculation can be made very quickly. Until the widespread use of digital computer based control systems, it was common practice to construct control mechanisms such as we have been describing using analog computers. Although analog computers are not generally proposed for new control-center installations today, there are many in active use. An analog computer can provide the economic dispatch and allocation of generation in an area on an instantaneous basis through the use of function generators set to equal the units' incremental heat rate curves. B matrix loss formulas were also incorporated into analog schemes by setting the matrix coefficients on precision potentiometers.

When using digital computers, it is desirable to be able to carry out the economic-dispatch calculations at intervals of one to several minutes. Either the output of the economic-dispatch calculation is fed to an analog computer (i.e., a "digitally directed analog" control system) or the output is fed to another program in the computer that executes the control functions (i.e., a "direct digital" control system). Whether the control is analog or digital, the allocation of generation must be made instantly when the required area total generation changes. Since the economic-dispatch calculation is to be executed every few minutes, a means must be provided to indicate how the generation is to be allocated for values of total generation other than that used in the economic-dispatch calculation.

The allocation of individual generator output over a range of total generation values is accomplished using base points and participation factors. The economic dispatch calculation is executed with a total generation equal to the sum of the present values of unit generation as measured. The result of this calculation is a set of base-point generations, $P_{i_{base}}$, which is equal to the most economic output for each generator unit. The rate of change of each unit's output with respect to a change in total generation is called the unit's *participation factor, pf* (see Section 3.6 and Example 3G in Chapter 3). The base point and participation factors are used as follows

$$P_{i_{des}} = P_{i_{base}} + pf_i \cdot \Delta P_{total} \qquad (9.37)$$

where $\quad \Delta P_{total} = P_{new\ total} - \sum_{\substack{all \\ gen}} P_{i_{base}} \qquad (9.38)$

and $\quad P_{i_{des}}$ = new desired outout from unit i

$\qquad P_{i_{base}}$ = base point generation for unit i

$\qquad pf_i$ = participation factor for unit i

$\qquad \Delta P_{total}$ = change in total generation

$\qquad P_{new\ total}$ = new total generation

Note that by definition (e.g., see Eq. 3.24) the participation factors must sum to unity. In a direct digital control scheme the generation allocation would be made by running a computer code that was programmed to execute according to Eqs. 9.37 and 9.38.

9.7.4 Automatic Generation Control (AGC) Implementation

Modern implementation of automatic generation control (AGC) schemes usually consist of a central location where information pertaining to the system is telemetered. Control actions arc determined in a digital computer and then transmitted to the generation units via the same telemetry channels. To implement an AGC system, one would require the following information at the control center.

1. Unit megawatt output for each committed unit.

2. Megawatt flow over each tie line to neighboring systems.

3. System frequency.

The output of the execution of an AGC program must be transmitted to each of the generating units. Present practice is to transmit raise or lower pulses of varying lengths to the unit. Control equipment then changes the unit's load reference set point up or down in proportion to the pulse length. The "length" of the control pulse may be encoded in the bits of a digital word that is transmitted over a digital telemetry channel. The use of digital telemetry is becoming commonplace in modern systems wherein supervisory control (opening and closing substation breakers), telemetry information (measurements of MW, MVAR, MVA voltage, etc.) and control information (unit raise/lower) is all sent via the same channels.

The basic reset control loop for a unit consists of an integrator with gain K as shown in Figure 9.22. The control loop is implemented as shown in Figure 9.23. The P_{des} control input used in Figures 9.22 and 9.23 is a function of system frequency

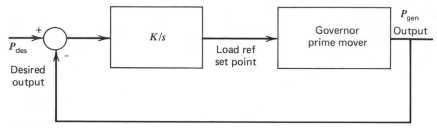

FIG. 9.22 Basic generation control loop.

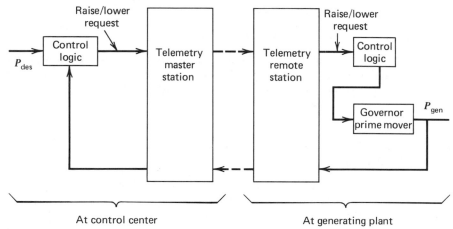

FIG. 9.23 Basic generation control loop via telemetry.

deviation, net interchange error, and each unit's deviation from its scheduled economic output.

The overall control scheme we are going to develop starts with ACE, which is a measure of the error in total generation from total desired generation. ACE is calculated according to Figure 9.24. ACE serves to indicate when total generation must be raised or lowered in a control area. However, ACE is not the only error signal that must "drive" our controller. The individual units may deviate from the economic output as determined by the base point and participation-factor calculation.

The AGC control logic must also be driven by the errors in unit output so as to force the units to obey the economic dispatch. To do this, the sum of the unit output errors is added to ACE to form a composite error signal that drives the entire control system. Such a control system is shown schematically in Figure 9.25 where we have combined the ACE calculation, the generation allocation calculation, and the unit control loop.

Investigation of Figure 9.25 shows an overall control system that will try to drive ACE to zero as well as driving each unit's output to its required economic value. Readers are cautioned that there are many variations to the control execution

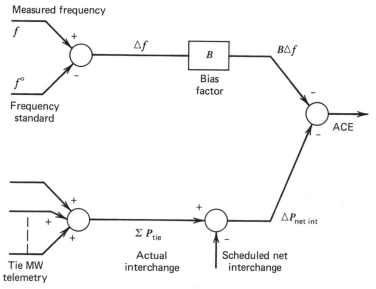

FIG. 9.24 ACE calculation.

shown in Figure 9.25. This is especially true of direct digital implementations of AGC where great sophistication can be programmed into an AGC computer code.

Often the question is asked as to what constitutes "good" AGC design. This is difficult to answer other than in a general way since what is "good" for one system may be different in another. Three general criteria can be given.

1. The ACE signal should ideally be kept from becoming too large. Since ACE is directly influenced by random load variations, this criterion can be treated statistically by saying that the standard deviation of ACE should be small.

2. ACE should not be allowed to "drift." This means that the integral of ACE over an appropriate time should be small. "Drift" in ACE has the effect of creating system time errors or what are termed *inadvertent interchange errors*.

3. The amount of control action called for by the AGC should be kept to a minimum. Many of the errors in ACE, for example, are simply random load changes that need not cause control action. Trying to "chase" these random load variations will only wear out the unit speed changing hardware.

To achieve the objectives of good AGC, many features are added as described briefly in the next section.

9.7.5 AGC Features

This section will serve as a simple catalog of some of the features that can be found on most AGC systems.

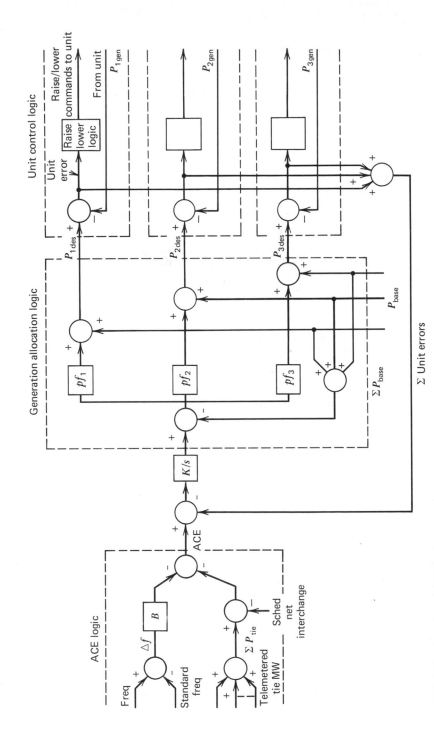

FIG. 9.25 Overview of AGC logic.

Assist Action: Often the incremental heat rate curves for generating units will give trouble to an AGC when an excessive ACE occurs. If one unit's participation factor is dominant, it will take most of the control action and the other units will remain relatively fixed. Although it is the proper thing to do as far as economics are concerned, the one unit that is taking all the action will not be able to change its output fast enough when a large ACE calls for a large change in generation. The assist logic then comes into action by moving more of the units to correct ACE. When the ACE is corrected, the AGC then restores the units back to economic output.

Filtering of ACE: As indicated earlier, much of the change in ACE may be random noise that need not be "chased" by the generating units. Most AGC programs use elaborate, adaptive nonlinear filtering schemes to try to filter out random noise from true ACE deviations that need control action.

Telemetry Failure Logic: Logic must be provided to ensure that the AGC will not take wrong action when a telemetered value it is using fails. The usual design is to suspend all AGC action when this condition happens.

Unit Control Detection: Sometimes a generating unit will not respond to raise/lower pulses. For the sake of overall control the AGC ought to take this into account. Such logic will detect a unit that is not following raise/lower pulses and suspend control to it, thereby causing the AGC to reallocate control action among the other units on control.

Ramp Control: Special logic allows the AGC to ramp a unit from one output to another at a specified rate of change in output. This is most useful in bringing units on-line and up to full output.

Rate Limiting: All AGC designs must account for the fact that units cannot change their output too rapidly. This is especially true of thermal units where mechanical and thermal stresses are limiting. The AGC must limit the rate of change such units will be called on to undergo during fast load changes.

Unit Control Modes: Many units on power systems are not under full AGC control. Various special control modes must be provided such as manual, base load, base load and regulating. For example, base load and regulating units are held at their base load value—but are allowed to move as assist action dictates and are then restored to base load value.

PROBLEMS

9.1 Suppose that you are given a single area with three generating units as shown in Figure 9.26.

Unit	Rating	Speed droop R (per unit on unit base)
1	100 MVA	0.01
2	500 MVA	0.015
3	500 MVA	0.015

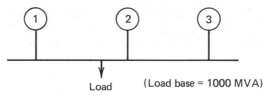

Load   (Load base = 1000 MVA)

FIG. 9.26   Three-generator system for Problem 9.1.

The units are initially loaded as follows.

$$P_1 = \quad 80 \text{ MW}$$

$$P_2 = 300 \text{ MW}$$

$$P_3 = 400 \text{ MW}$$

Assume $D=0$; what is the new generation on each unit for a 50 MW load increase? Repeat with $D = 1.0$ pu (i.e., 1.0 pu on load base). Be careful to convert all quantities to a common base when solving.

9.2 Using the values of R and D in each area, for Example 9B, resolve for the 100 MW load change in area 1 under the following conditions.

Area 1   Base MVA = 2000 MVA

Area 2   Base MVA = 500 MVA

Then solve for a load change of 100 MW occurring in area 2 with Rs and Ds as in Example 9B and base MVA for each area as before.

9.3 Given the block diagram of two interconnected areas shown in Figure 9.27 (consider the prime-mover output to be constant, i.e., a "blocked" governor):

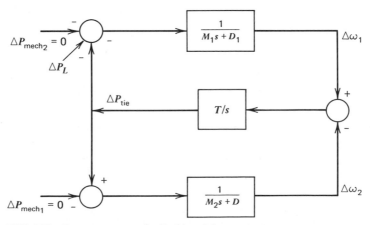

FIG. 9.27   Two-area system for Problem 9.3.

a. Derive the transfer functions that relate $\Delta\omega_1(s)$ and $\Delta\omega_2(s)$ to a load change $\Delta P_L(s)$.

b. For the following data (all quantities refer to a 1000 MVA base),

$$M_1 = 3.5 \text{ pu} \qquad D_1 = 1.00$$
$$M_2 = 4.0 \text{ pu} \qquad D_2 = 0.75$$
$$T = 377 \times 0.02 \text{ pu} = 7.54 \text{ pu}$$

calculate the final frequency for load-step change in area 1 of 0.2 pu (i.e., 200 MW). Assume frequency was at nominal and tie flow was 0 pu.

c. Derive the transfer function relating tie flow, $\Delta P_{tie}(s)$ to $\Delta P_L(s)$. For the data of part (b) calculate the frequency of oscillation of the tie power flow. What happens to this frequency as tie stiffness increases (i.e. $T \rightarrow \infty$)?

9.4 Given two generating units with data as follows.

Unit 1: Fuel cost: $F_1 = 1.0 \text{ R/MBtu}$

$\quad\qquad\qquad H_1(P_1) = 500 + 7\,P_1 + 0.002\,P_1^2 \text{ MBtu/h}$

$\quad\qquad\qquad 150 < P_1 < 600 \qquad$ Rate limit $= 2 \text{ MW/min}$

Unit 2: Fuel cost: $F_2 = 0.98 \text{ R/MBtu}$

$\quad\qquad\qquad H_2(P_2) = 200 + 8\,P_2 + 0.0025\,P_2^2 \text{ MBtu/h}$

$\quad\qquad\qquad 125 \le P_2 \le 500 \text{ MW} \qquad$ Rate limit $= 2 \text{ MW/min}$

a. Calculate the economic base points and participation factors for these two units supplying 500 MW total. Use Eq. 3.24 to calculate participation factors.

b. Assume a load change of 10 MW occurs and that we wish to clear the ACE to 0 in 5 min. Is this possible if the units are to be allocated by base points and participation factors?

c. Assume the same load change as in part (b), but assume that the rate limit on unit 1 was now 0.5 MW/min.

This problem demonstrates the flaw in using Eq. 3.24 to calculate the participation factors. An alternate procedure would generate participation factors as follows.

Let t be the time in minutes between economic-dispatch calculation executions. Each unit will be assigned a range that must be obeyed in performing the economic dispatch.

$$P_i^{max} = P_i^0 + t \times \text{rate limit}_i$$
$$P_i^{max} = P_i^0 - t \times \text{rate limit}_i$$

(P4.1)

The range thus defined is simply the maximum and minimum excursion the unit could undergo within t minutes. If one of the limits described is outside the units' normal economic limits, the economic limit would be used. Participation factors can then be calculated by resolving the economic dispatch at a higher value and enforcing the new limits described previously.

d. Assume $T = 5$ min and that the perturbed economic dispatch is to be resolved for 510 MW. Calculate the new participation factors as

$$pf_i = \frac{P_i^\Delta - P_{i\text{base pt}}}{\Delta P_{\text{total}}}$$

where $P_{i\text{base pt}}$ = base economic solution

$$P_{1\text{base}} + P_{2\text{base}} = 500 \text{ MW}$$

P_i^Δ = perturbed solution

$$P_1^\Delta + P_2^\Delta = 510 \text{ MW}$$

with limits as calculated in Eq. P4.1.

Assume the initial unit generations P_i^0 were the same as the base points found in part (a). And assume the rate limits were as in part (c) (i.e., unit 1 rate lim = 0.5 MW/min, unit 2 rate lim = 2 MW/min). Now check to see if part (c) gives a different result.

FURTHER READING

A great deal of literature in the past 20 years has dealt with control theory and associated technologies. The reader should be familiar with the basics of control theory before attempting to read many of the references cited here. A good introduction to automatic generation control is the book *Control of Generation and Power Flow on Interconnected Systems*, by Nathan Cohn (cited in Chapter 1). Other sources of introductory material are contained in references 1–3.

Descriptions of how steam turbine generators are modeled are found in references 4 and 5; reference 6 shows how hydro units can be modeled. Reference 7 shows the effects to be expected from various prime-mover and governing systems. References 8, 9, and 10 are representative of advances made in AGC techniques through the late 1960s and early 1970s. Other special interests in AGC design include special purpose optimal filters (see references 10 and 11), direct digital control schemes (see references 12–15), and control of jointly owned generating units (see reference 16).

Research in control theory toward "optimal control" techniques was used in several papers presented in the late 1960s and early 1970s. As far as is known to the authors, optimal control techniques have not, as of the writing of this text, been utilized successfully in a working AGC system. Reference 17 is representative of the papers using optimal control theory.

Recent research has included an approach that takes the short-term load forecast, economic dispatch, and AGC problems and approaches them as one overall control problem. References 18 and 19 illustrate this approach.

1. Friedlander, G. D., "Computer-Controlled Power Systems, Part I—Boiler-Turbine Unit Controls," *IEEE Spectrum*, April 1965, pp. 60–81.

2. Friedlander, G. D., "Computer-Controlled Power Systems, Part II—Area Controls and Load Dispatch," *IEEE Spectrum, May* 1965, *pp.* 72–91.

3. Ewart, D. N., "Automatic Generation Control—Performance Under Normal Conditions," *Systems Engineering for Power: Status and Prospects*, U.S. government document CONF-750867, 1975, pp. 1–14.

4. Anderson, P. M., *Modeling Thermal Power Plants for Dynamic Stability Studies*, Cyclone Copy Center, Ames, Iowa, 1974.

5. IEEE Committee Report, "Dynamic Models for Steam and Hydro Turbines in Power System Studies," *IEEE Transactions on Power Apparatus and Systems*, Vol. PAS-92, November/December 1973, pp. 1904–1915.

6. Undrill, J. M., Woodward, J. L., "Nonlinear Hydro Governing Model and Improved Calculation for Determining Temporary Droop," *IEEE Transactions on Power Apparatus and Systems*, Vol. PAS-86, April 1967, pp. 443–453.

7. Concordia, C., Kirchmayer, L. K., de Mello, F. P., Schulz, R. P., "Effect of Prime-Mover Response and Governing Characteristics on System Dynamic Performance," *Proceedings American Power Conference*, 1966.

8. Cohn, N., "Considerations in the Regulation of Interconnected Areas," *IEEE Transactions on Power Apparatus and Systems*, Vol. PAS-86, December 1967, pp. 1527–1538.

9. Cohn, N. "Techniques for Improving the Control of Bulk Power Transfers on Interconnected Systems," *IEEE Transactions on Power Apparatus and Systems*, Vol. PAS-90, November/December 1971, pp. 2409–2419.

10. Cooke, J. L., "Analysis of Power System's Power-Density Spectra," *IEEE Transactions on Power Apparatus and Systems*, Vol. PAS-83, January 1964, pp. 34–41.

11. Ross, C. W., "Error Adaptive Control Computer for Interconnected Power Systems," *IEEE Transactions on Power Apparatus and Systems*, Vol. PAS-85, July 1966, pp. 742–749.

12. Ross, C. W., "A Comprehensive Direct Digital Load-Frequency Controller," *IEEE Power Industry Computer Applications Conference Proceedings*, 1967.

13. Ross, C. W., Green, T. A., "Dynamic Performance Evaluation of a Computer-Controlled Electric Power Systems," *IEEE Transactions on Power Apparatus and Systems*, Vol. PAS-91, May/June 1972, pp. 1158–1165.

14. de Mello, F. P., Mills, R. J., B'Rells, W. F., "Automatic Generation Control—Part I: Process Modeling," *IEEE Transactions on Power Apparatus and Systems*, Vol. PAS-92, March/April 1973, pp. 710–715.

15. de Mello, F. P., Mills, R. J., B'Rells, W. F., "Automatic Generation Control—Part II: Digital Control Techniques," *IEEE Transactions on Power Apparatus and Systems*, Vol. PAS-92, March/April 1973, pp. 716–724.

16. Podmore, R., Gibbard, M. J., Ross, D. W., Anderson, K. R., Page, R. G., Argo, K., Coons, K., "Automatic Generation Control of Jointly Owned Generating Unit," *IEEE Transactions on Power Apparatus and Systems*, Vol. PAS-98, January/February 1979, pp. 207–218.

17. Elgerd, O. I., Fosha, C. E., "The Megawatt-Frequency Control Problem: A New Approach Via Optimal Control Theory," Proceedings, *Power Industry Computer Applications Conference*, 1969.

18. Zaborszky, J., Singh, J., "A Reevaluation of the Normal Operating State Control of the Power Systems Using Computer Control and System Theory: Estimation," *Power Industry Computer Applications Conference Proceedings*, 1979.

19. Mukai, H., Singh, J., Spare, J., Zaborszky, J. "A Reevaluation of the Normal Operating State Control of the Power System Using Computer Control and System Theory—Part II: Dispatch Targeting," *IEEE Transactions on Power Apparatus and Systems*, Vol. PAS-100, January 1981, pp. 309–317.

chapter 10

Interchange Evaluation and Power Pools

10.1 INTRODUCTION

This chapter discusses some of the reasons why electric utility systems interconnect with neighboring systems. Except where geographical or political barriers prevent it, the interconnection of electric systems is almost universal throughout the world. The reasons are quite simple, and always make sense no matter what system we are dealing with.

Basically, electric power systems interconnect because the interconnected system is more reliable, it is a better system to operate, and it may be operated at less cost than if left as separate parts. We saw in Chapter 9 that interconnected systems have better regulating characteristics since a load change in any of the systems is taken care of by all units in the interconnection, not just the units in the control area where the load change occurred. This fact also makes interconnections more reliable since the loss of a generating unit in one of them can be made up from spinning reserve among units throughout the interconnection. Thus, if a unit is lost in one control area, governing action from units in all connected areas will increase generation outputs to make up the deficit until standby units can be brought on-line. If a power system were to run isolated and lose a large unit, the chance of the other units in that isolated system being able to make up the deficit are greatly reduced. Extra units would have to be run as spinning reserve, and this would mean less economic operation. Furthermore, a generation system will generally require a smaller installed generation capacity reserve if it is planned as part of an interconnected system.

One of the most important reasons for interconnecting with neighboring systems centers on the better economics of operation that can be attained when interconnected. This opportunity to improve the operating economics arises any time two power systems are operating with different incremental costs. As Example 10A will show, if there is a sufficient difference in the incremental cost between the systems, it will pay both systems to exchange power at an equitable price. To see how this can happen, one need merely reason as follows. Given the following situation:

- Utility A is generating at a lower incremental cost than utility B.

- If utility B were to buy the next megawatt of power for its load from utility A at a price less than if it generated that megawatt from its own generation, it would save money in supplying that increment of load.

- Utility A would benefit economically from selling power to utility B as long as utility B is willing to pay a price that is greater than utility A's cost of generating that block of power.

The key to achieving a mutually beneficial transaction is in establishing a "fair" price for the economy interchange sale.

There are other, longer-term interchange transactions that are economically advantageous to interconnected utilities. One system may have a surplus of power and energy and may wish to sell it to an interconnected company on a long-term, firm supply basis. It may, in other circumstances, wish to arrange to sell this excess only on a "when, and if available" basis. The purchaser would probably agree to pay more for a firm supply (the first case) than for the interruptable supply of the second case.

In all these transactions the question of a "fair and equitable price" enters into the arrangement. In this text, the economy interchange examples that follow are all based on an equal division of the operating costs that are saved by the utilities involved in the interchange. This is not always the case since "fair and equitable" is a very subjective concept; what is fair and equitable to one party may appear as grossly unfair and inequitable to the other. The 50-50 split of savings in the examples in this chapter should not be taken as advocacy of this particular price schedule. It is used since it is quite common in interchange practices in the U.S. economy and in "normal circumstances" appears to be nondiscriminatory. Pricing arrangements for long-term interchange vary widely and may include "take-or-pay" contracts, split savings, or fixed price schedules.

10.2 ECONOMY INTERCHANGE

Before we look at the pricing of interchange power, we will present an example showing the benefit of interchange power.

EXAMPLE 10A

Two utility operating areas are shown in Figure 10.1. Data giving the heat rates and fuel costs for each unit in both areas are given here.

Unit Data:
$$F_i(P_i) = f_i(a_i + b_iP_i + c_iP_i^2)$$
$$P_i^{\min} \leq P_i \leq P_i^{\max}$$

Unit number	Fuel cost f_i (R/MBtu)	Cost coefficients			Unit limits	
		a_i	b_i	c_i	$P_i^{\min}$ (MW)	$P_i^{\max}$ (MW)
1	2.0	561.	7.92	0.001562	150.	600.
2	2.0	310.	7.85	0.00194	100.	400.
3	2.0	78.	7.97	0.00482	50.	200.
4	1.9	500.	7.06	0.00139	140.	590.
5	1.9	295.	7.46	0.00184	110.	440.
6	1.9	295.	7.46	0.00184	110.	440.

Area 1:

$$\text{Load} = 700 \text{ MW}$$
$$\text{Max total gen} = 1200 \text{ MW}$$
$$\text{Min total gen} = 300 \text{ MW}$$

Area 2:

$$\text{Load} = 1100 \text{ MW}$$
$$\text{Max total gen} = 1470 \text{ MW}$$
$$\text{Min total gen} = 360 \text{ MW}$$

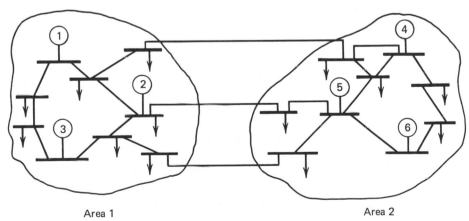

Area 1   Area 2

FIG. 10.1 Interconnected areas for Example 10A.

First, we will assume that each area operates independently, that is, each will supply its own load from its own generation. This will necessitate performing a separate economic dispatch calculation for each area. The results of an independent economic dispatch are given here.

Area 1:
$$P_1 = 322.7 \text{ MW}$$
$$P_2 = 277.9 \text{ MW}$$
$$P_3 = 99.4 \text{ MW}$$
Total gen = 700 MW

$$\lambda = 17.856 \text{ R/MWh}$$

Operating cost area 1 = 13,677.21 R/h

Area 2:
$$P_4 = 524.7 \text{ MW}$$
$$P_5 = 287.7 \text{ MW}$$
$$P_6 = 287.7 \text{ MW}$$
Total gen = 1100 MW

$$\lambda = 16.185 \text{ R/MWh}$$

Operating cost area 2 = 18,569.23 R/h

Total operating cost for both areas = 13,677.21 + 18,569.23
$$= 32,246.44 \text{ R/h}$$

Now suppose the two areas are interconnected by several transmission circuits such that the two areas may be thought of and operated as one system. If we now dispatch them as one system considering the loads in each area to be the same as just shown, we get a different dispatch for the units.

$$P_1 = 184.0 \text{ MW}$$
$$P_2 = 166.2 \text{ MW}$$
$$P_3 = 54.4 \text{ MW}$$
Total gen in area 1 = 404.6 MW

$$P_4 = 590.0 \text{ MW}$$
$$P_5 = 402.7 \text{ MW}$$
$$P_6 = 402.7 \text{ MW}$$
Total gen in area 2 = 1,395.4 MW

Total generation for entire system = 1,800.0 MW

$$\lambda = 16.990 \text{ R/h}$$

Operating cost area 1 = 8,530.93 R/h

Operating cost area 2 = 23,453.89 R/h

Total operating cost = 31,984.82 R/h

Interchange power = 295.4 MW from area 2 to area 1

Note that area 1 is now generating less than when it was isolated and area 2 is generating more. If we ignore losses, we can see that the change in generation in each area corresponds to the net power flow over the interconnecting circuits. This is called the *interchange power*. Note also that the overall cost of operating both systems is now less than the sum of the costs to operate the areas when each supplied its own load.

Example 10A has shown that interconnecting two power systems can have a marked economic advantage when power can be interchanged. If we look at the net change in operating cost for each area, we will discover that area 1 had a decrease in operating cost while area 2 had an increase. Obviously, area 1 should pay area 2 for the power transmitted over the interconnection, but how much should be paid? This question can be, and is, approached differently by each party. Assume the systems were operated with the 295.4 MW of interchange power for 1 h.

Area 1: Can argue that area 2 had a net operating cost increase of 4884.66 R̪ and therefore area 1 ought to pay area 2 4884.66 R̪. Note that if this were agreed to, area 1 would reduce its net operating cost by $13,677.21 - (8530.93 + 4884.66) = 261.62$ R̪ when the cost of the purchase is included.

Area 2: Can argue that area 1 had a net decrease in operating cost of 5146.28 R̪ and therefore area 1 ought to pay area 2 5146.28 R̪. Note that if this were agreed to, area 2 would have a net decrease in its operating costs when the revenues from the sale are included of $18,569.23 - (23,453.89 + 5146.28) = 261.62$ R̪.

The problem with this approach is, of course, that there is no agreement concerning a mutually acceptable "fair" price. In both cases one party to the transaction gets all the economic benefits while the other gains nothing. A common practice in such cases is to price the sale at the cost of generation plus one-half the savings in operating costs of the purchaser. This splits the savings equally between the two operating areas. This means that area 1 would pay area 2 5015.47 R̪ and that each area would have 130.81 R̪ reduction in operating costs.

Such transactions are usually not carried out if the net savings are very small. In such a case the errors in measuring interchange flows might cause the transaction to be uneconomic. The transaction may also appear to be uneconomic to a potential seller if the utility is concerned with conserving its fuel resources to serve its own customers. This, however, is an institutional problem and not one of engineering economics.

10.3 ECONOMY INTERCHANGE EVALUATION

In Example 10A we saw how two power systems could operate interconnected for less money than if they operated separately. We obtained a dispatch of the interconnected systems by assuming that we had all the information necessary (input-output curves, fuel costs, unit limits, on-line status, etc.) in one location and could calculate the overall dispatch as if the areas were part of the same system. However, unless the two power systems have formed a power pool or transmit this information to each other or a third party who will arrange the transaction (see Section 10.9), this assumption is incorrect. The most common situation involves system operations personnel located in offices within each of the control areas who can talk to each other by telephone. We can assume that each office has the data and computation equipment needed to perform an economic dispatch calculation for its own power system and that all information about the neighboring system must come over the telephone (or perhaps by teletype). How should the two operations offices coordinate their operations to obtain best economic operation of both systems?

The simplest way to coordinate the operations of the two power systems is to note that if someone were performing an economic dispatch for both systems combined, the most economic way to operate would require the incremental cost to be the same at each generating plant assuming losses are ignored. The two operations offices can achieve the same result by taking the following steps.

1. Assume there is no interchange power being transmitted between the two systems.

2. Each system operations office runs an economic dispatch calculation for its own system.

3. By talking over the telephone, the offices can determine which system has the lower incremental cost. The operations office in the system with lower incremental cost then runs a series of economic dispatch calculations each one having a greater total demand (that is, the total load is increased at each step). Similarly, the operations office in the system having higher incremental cost runs a series of economic dispatch calculations, each having a lower total demand.

4. Each increase in total demand on the system with lower incremental cost will tend to raise its incremental cost, and each decrease in demand on the high incremental cost system will tend to lower its incremental cost. By running the economic dispatch steps and conversing over the telephone, the two operations offices can determine the level of interchange energy that will bring the two systems toward most economic operation.

Under idealized "free market" conditions where both utilities are attempting to minimize their respective operating costs, and assuming no physical limitations on the transfer, their power negotiations (or bartering) will lead to the same economic results as a pool dispatch performed on a single area basis. These assumptions, however, are critical. In many practical situations there are both physical and institutional constraints that prevent interconnected utility systems from achieving optimum economic dispatch.

EXAMPLE 10B

Starting from the "no interchange" conditions of Example 10A, we will find the most economic operation by carrying out the steps outlined earlier. Since area 2 has a lower incremental cost before the transaction, we will run a series of economic dispatch calculations with increasing load steps of 50 MW and an identical series on area 1 with decreasing load steps of 50 MW.

Area 1:

Step	Demand (MW)	Area 1 incremental cost (R/MWh)	Assumed Interchange from area 2 (MW)
1	700	17.856	0
2	650	17.710	50
3	600	17.563	100
4	550	17.416	150
5	500	17.270	200
6	450	17.123	250
7	400	16.976	300
8	350	16.816	350

Area 2:

Step	Demand (MW)	Area 1 incremental cost (R/MWh)	Assumed interchange to area 1 (MW)
1	1100	16.185	0
2	1150	16.291	50
3	1200	16.395	100
4	1250	16.501	150
5	1300	16.656	200
6	1350	16.831	250
7	1400	17.006	300
8	1450	17.181	350

Note that at step 6, area 1's incremental cost is just slightly above area 2's incremental cost, but that the relationship then changes at step 7. Thus, for minimum total operating costs, the two systems ought to be interchanging between 250 and 300 MW interchange.

This procedure can be repeated with smaller steps between 250 and 300 MW if desired.

10.4 INTERCHANGE EVALUATION WITH UNIT COMMITMENT

In Examples 10A and 10B there was an implicit assumption that conditions remained constant on the two power systems as the interchange was evaluated. Usually this assumption is a good one if the interchange is to take place for a period of up to 1 h. However, there may be good economic reasons to transmit interchange power for periods extending from several hours to several days. Obviously, when studying such extended periods, we will have to take into account many more factors than just the relative incremental costs of the two systems.

Extended interchange transactions require that a model of the load to be served in each system (i.e., the expected load levels as a function of time) be included as well as the unit commitment schedule for each. The procedure for studying interchange of power over extended periods of time is as follows.

1. Each system must run a base-unit commitment study extending over the length of the period in question. These base-unit commitment studies are run without the interchange, each system serving its own load as given by a load forecast extending over the entire time period.

2. Each system then runs another unit commitment, one system having an increase in load, the other a decrease in load over the time the interchange is to take place.

3. Each system then calculates a total production cost for the base unit commitment and for the unit commitment reflecting the effect of the interchange. The difference in cost for each system represents the cost of the interchange power (a positive change in cost for the selling system and a negative change in cost for the buying system). The price for the interchange can then be negotiated. If the agreed on pricing policy is to "split the savings," the price will be set by splitting the savings of the purchaser and adding the change in the cost for the selling system. If the savings are negative, it obviously would not pay to carry out the interchange contract.

The unit commitment calculation allows the system to adjust for the start-up and shut-down times to take more effective advantage of the interchange power. It may pay for one system to leave an uneconomical unit off-line entirely during a peak in load and buy the necessary interchange power instead.

10.5 MULTIPLE INTERCHANGE CONTRACTS

Most power systems are interconnected with all their immediate neighboring systems. This may mean that one system will have interchange power being bought and sold simultaneously with several neighbors. In this case, the price for the interchange must be set while taking account of the other interchange. For example, if one system were to sell interchange power to two neighboring systems in sequence, it would probably quote a higher price for the second sale since the first sale would have raised its incremental cost. On the other hand, if the selling utility was a member of a power pool, the sale price might be set by the power and energy pricing portions of the pool agreement to be at a level such that the seller receives the cost of the generation for the sale plus one-half the total savings of all the purchasers. In this case, assuming that a pool control center exists, the sale prices would be computed by this center and would differ from the prices under multiple interchange contracts. As we will see in the next section, the order in which the interchange transaction agreements are made is very important in costing the interchange where there is no central pool dispatching office.

Another phenomenon that can take place with multiple neighbors occurs when a system's transmission system is simply being used to transmit power from one neighbor, through an intermediate system, to a third system. The intermediate system's AGC will keep net interchange to a specified value regardless of the power being passed through it. The power being passed through will change the transmission losses incurred in the intermediate system. When the losses are increased, this can represent an unfair burden on the intermediate system, since if it is not part of the interchange agreement, the increased losses will be supplied by the intermediate system's generation. As a result, systems often assess a "wheeling" charge for such power passed through its transmission network. Three-party transactions are frequently made in the United States by negotiations between pairs of systems. "A" locates power and energy in "C" and makes arrangement with an intervening system "B" for transmission. Then "C" sells to "B" and "B" sells to "A." The price level to "A" may be set as the cost of "C"'s generation plus the wheeling charges of "B" plus one-half "A"'s savings. It may also be set at "B"'s net costs plus one-half "A"'s savings. Price is a matter of negotiation in this type of transaction when prior agreements on pricing policies are absent. This type of multiparty exchange is known as *displacement*.

10.6 AFTER-THE-FACT PRODUCTION COSTING

Often utility companies will enter into interchange agreements that give the amount and schedule of the interchange power but leave the final price out. Instead of agreeing on the price, the contract specifies that the systems will operate with the interchange and then decide on its cost after it has taken place. By doing so, the systems can use the actual load on the systems and the actual unit commitment schedules rather than the predicted load and commitment schedules. Even when the price has been negotiated prior to the interchange, utilities will many times wish to verify the economic gains projected by performing after-the-fact production costs.

As described in Section 10.5, power systems are often interconnected with many neighboring systems and interchange may be carried out with each one. When carrying out the after-the-fact production costs, the operations offices must be careful to duplicate the order of the interchange agreements. This is illustrated in Example 10C.

EXAMPLE 10C

Suppose area 1 of Example 10A was interconnected with a third system here designated area 3 and that interchange agreements were entered into as follows.

Interchange agreement A: Area 1 buys 300 MW from Area 2

Interchange agreement B: Area 1 sells 100 MW to Area 3

Data for area 1 and area 2 will be the same as in Example 10A. For this example, we assume that Area 3 will not reduce its own generation below 450 MW for reasons that might include unit commitment or spinning-reserve requirements. The Area 3 cost characteristics are as follows.

Total demand (MW)	Area 3 incremental cost (ℝ/MWh)	Area 3 total production cost (ℝ/h)
450	18.125	8,220.00
550	18.400	10,042.00

First, let us see what the cost would be under a split-savings pricing policy if the interchange agreements were made with agreement A first then agreement B.

	Area 1 gen (MW)	Area 1 cost (ℝ/h)	Area 2 gen (MW)	Area 2 cost (ℝ/h)	Area 3 gen (MW)	Area 3 cost (ℝ/h)
Start	700	13,677.21	1100	18,569.23	550	10,042.00
After agreement A	400	8,452.27	1400	23,532.25	550	10,042.00
After agreement B	500	10,164.57	1400	23,532.25	450	8,220.00

Agreement A: Saves area 1   5,224.94 ℝ
Costs area 2   4,963.02 ℝ

After splitting savings, area 1 pays area 2
5,093.98 ℝ

Agreement B: Costs area 1   1,712.30 ℝ
Saves area 3   1,822.00 ℝ

After splitting savings, area 3 pays area 1
1,767.15 ℝ

Summary of payments:

Area 1 pays a net   3,326.83 ℝ
Area 2 receives   5,093.98 ℝ
Area 3 pays   1,767.15 ℝ

Now let the transactions be costed assuming the same split-savings pricing policy but with the interchange agreements made with agreement B first, then agreement A.

	Area 1 gen (MW)	Area 1 cost (R/h)	Area 2 gen (MW)	Area 2 cost (R/h)	Area 3 gen (MW)	Area 3 cost (R/h)
Start	700	13,677.21	1100	18,569.23	550	10,042.00
After agreement B:	800	15,477.55	1100	18,569.23	450	8,220.00
After agreement A:	500	10,164.57	1400	23,532.25	450	8,220.00

Agreement B: Costs area 1   1,800.34 R
                 Saves area 3   1,822.00 R

                 After splitting savings, area 3 pays area 1
                       1,811.17 R

Agreement A: Saves area 1   5,312.98 R
                 Costs area 2   4,963.02 R

                 After splitting savings, area 1 pays area 2
                       5,138.00 R

Summary of payments:
                 Area 1 pays a net   3,326.83 R
                 Area 2 receives   5,138.00 R
                 Area 3 pays   1,811.17 R

Except for area 1, the payments for the interchanged power are different depending on the order in which the agreements were carried out. If agreement A were carried out first, area 2 would be selling power to area 1 at a lower incremental cost than if agreement B were carried out first. Obviously, it would be to a seller's (area 2 in this case) advantage to sell when the buyer's (area 1) incremental cost is high, and, conversely, it is to a buyer's (area 3) advantage to buy from a seller (area 1) whose incremental cost is low.

When several two-party interchange agreements are made, the pricing must follow the proper sequence. In this example, the utility supplying the energy receives more than its incremental production costs no matter which transaction is costed initially. The rate that the other two areas pay per MWh are different and depend on the order of evaluation. These differences may be summarized as follows in terms of R/MWh.

	Cost rates (R/MWh)	
Area	A costed first	B costed first
1 pays	16.634	16.634
2 receives	16.980	17.127
3 pays	17.672	18.112

As we will see in Section 10.9.3, the central dispatch of a pool can avoid this problem by developing a single cost rate for every transaction that takes place in a given interval.

10.7 TRANSMISSION LOSSES IN TRANSACTION EVALUATION

Up to now we have assumed that transmission losses could be neglected in interchange transaction analysis. This may not be true in the operation of an actual power system. It might seem quite profitable to transmit power to a neighboring system if transmission losses are ignored. If, however, some of the transmitted power is lost in transmission losses, it would be necessary to generate extra power to make the actual interchanged power equal to what was agreed with the buyer. The extra power generated would cost something and might wipe out the savings calculated by ignoring losses. For this reason, in systems where losses are of significance, interchanged power is usually thought of as being delivered at the boundary of the system. The price of the power to be interchanged must also therefore be priced assuming delivery at the boundary.

To correctly price the interchange at the boundary of the system, we must perform all economic dispatch calculations using penalty factors calculated using either a loss matrix or load flow. To start, we can assume that each tie line crossing the boundary of a system connects to the system at a unique bus. The interchange of power can then be thought of as power flowing into or out of the network from these buses. When the system is selling power, the "tie buses" appear to have loads on them, and, when buying power, they appear to have generators on them. Let us assume then that there are N generator buses in the system and M tie buses. Then we can construct vectors of interchange quantities as follows.

$$\mathbf{P_{gen}} = \begin{bmatrix} P_{g1} \\ P_{g2} \\ \vdots \\ P_{gN} \end{bmatrix} \qquad \text{Vector of } N \text{ generator buses} \qquad (10.1)$$

$$\mathbf{P_{tie}} = \begin{bmatrix} P_{T1} \\ P_{T2} \\ \vdots \\ P_{TM} \end{bmatrix} \qquad \text{Vector of } M \text{ tie buses} \qquad (10.2)$$

We can then construct a loss matrix that has rows and columns corresponding to each of the generator buses and to each of the tie buses. The system losses can then be calculated using Eq. 10.3 where the **P** vector of Chapter 4 is made up of two sets of quantities as shown.

$$P_{\text{loss}} = [\mathbf{P_{gen}^T} \mid \mathbf{P_{Tie}^T}][B_{ij}] \begin{bmatrix} \mathbf{P_{gen}} \\ \mathbf{P_{Tie}} \end{bmatrix} + [\mathbf{P_{gen}^T} \mid \mathbf{P_{Tie}^T}][B_{i0}] + B_{00} \qquad (10.3)$$

If we were able to predict the power flow over the tie lines, the elements in $\mathbf{P_{tie}}$ could be used along with the individual generation MW from the system's generators to calculate an economic dispatch with interchange. However, the value of the tie flows themselves are very difficult to predict. What is known is the net MW interchange with each neighbor. When the interchange is implemented by offsetting each control area's ACE calculation, the power flow over the ties will be determined by the impedances of both networks and how the generating units were loaded during the interchange.

A useful approximation is to assume that each tie flow is a linear function of all the interchange agreements now taking place. This can be pictured easily if we think of the interchange power from a particular neighbor as always distributing in the same manner over the ties. Mathematically, we can say that each tie flow is equal to a "distribution factor" times the net interchange MW with a particular neighboring utility. The distribution factor will be different for each neighbor.

If we say that interchange is to take place with K neighbors, then the interchange powers can be expressed as a K dimensional vector.

$$\mathbf{P}_{\text{int}} = \begin{bmatrix} P_{\text{int}_1} \\ P_{\text{int}_2} \\ \vdots \\ P_{\text{int}_K} \end{bmatrix} \quad \text{Vector of interchanges with } K \text{ neighboring systems} \tag{10.4}$$

The tie flow can then be calculated from the interchange vector using the distribution factors here arranged to form the tie-distribution matrix.

$$\mathbf{P}_{\text{tie}} = [TD]\mathbf{P}_{\text{int}} \tag{10.5}$$

where $[TD] = K \times M$ tie-distribution matrix

A more useful expression allows us to calculate the combined vector of generator powers and tie powers given the generator powers and the net interchange power with each neighboring utility.

$$\begin{bmatrix} \mathbf{P}_{\text{gen}} \\ \hline \mathbf{P}_{\text{tie}} \end{bmatrix} = \begin{bmatrix} I & 0 \\ \hline 0 & TD \end{bmatrix} \begin{bmatrix} \mathbf{P}_{\text{gen}} \\ \hline \mathbf{P}_{\text{int}} \end{bmatrix} = [TD'] \begin{bmatrix} \mathbf{P}_{\text{gen}} \\ \hline \mathbf{P}_{\text{int}} \end{bmatrix} \tag{10.6}$$

$$I = \text{identity matrix}$$

Finally, the network losses can be expressed as a function of the generator powers and the interchange powers.

$$P_{\text{loss}} \doteq [\mathbf{P}_{\text{gen}}^{\mathbf{T}} \mid \mathbf{P}_{\text{int}}^{\mathbf{T}}] [B'_{ij}] \begin{bmatrix} \mathbf{P}_{\text{gen}} \\ \hline \mathbf{P}_{\text{int}} \end{bmatrix} + [\mathbf{P}_{\text{gen}}^{\mathbf{T}} \mid \mathbf{P}_{\text{int}}^{\mathbf{T}}] [B'_{i0}] + B_{00} \tag{10.7}$$

where:

$$[B'_{ij}] = [TD']^T [B_{ij}][TD']$$

and

$$[B'_{i0}] = [TD']^T [B_{i0}]$$

Any calculations of production cost during interchange agreements should be made with this formula to price the power at the true boundary delivery points correctly. This can be illustrated by a fairly simple procedure as follows. In this illustration let us represent the system generators by P_i. The internal system load at an equivalent

---

load bus is P_L. The interchange points are considered as previously to be equivalent generators. A representative interchange would therefore be P_{int_k}.

In this illustrative development, we will use scalar equations to show how one computes the incremental cost of power at a typical interchange bus. The Lagrangian functions for a real generator at i and a hypothetical interchange generator are

$$F_i - \lambda(P_i + P_{int_k} - P_L - P_{loss}) = 0$$

for $i = 1, 2, 3$, etc. and

$$F_{int_k} - \lambda(P_i + P_{int_k} - P_L - P_{loss}) = 0$$

Therefore,

$$\frac{dF_i}{dP_i} = \lambda\left(1 - \frac{\partial P_{loss}}{\partial P_i}\right) \tag{10.8}$$

and

$$\frac{dF_{int_k}}{dP_{int_k}} = \lambda\left(1 - \frac{\partial P_{loss}}{\partial P_{int_k}}\right) \tag{10.9}$$

The incremental cost of the interchange bus is therefore,

$$\frac{dF_{int_k}}{dP_{int_k}} = \frac{dF_i}{dP_i}\left[\frac{1 - \dfrac{\partial P_{loss}}{\partial P_{int_k}}}{1 - \dfrac{\partial P_{loss}}{\partial P_i}}\right] = \frac{Pf_i}{Pf_{int_k}}\frac{dF_i}{dP_i} \tag{10.10}$$

for each generator in the system.

However, this is an inconvenient format. A better and more practical format involves the following.

1. Solve the coordination equations for the value of λ at the equivalent load bus. From Eq. (10.8),

$$\lambda = Pf_i\frac{dF_i}{dP_i} \tag{10.11}$$

2. Use this result in Eq. (10.10) to find the incremental cost of power delivered to (or from) the interchange bus. That is,

$$\frac{dF_{int_k}}{dP_{int_k}} = \frac{\lambda}{Pf_{int_k}}$$

For a block of interchange power, ΔP_{int_k}, the delivered cost is therefore,

$$\frac{dF_{int_k}}{dP_{int_k}}\Delta P_{int_k} = \frac{\lambda \Delta P_{int_k}}{Pf_{int_k}} \tag{10.12}$$

10.8 OTHER TYPES OF INTERCHANGE

There are other reasons for interchanging power than simply obtaining economic benefits. Arrangements are usually made between power companies to interconnect for a variety of reasons. Ultimately, of course, economics plays the dominant role, as we will see.

10.8.1 Capacity Interchange

Normally, a power system will add generation to make sure that the available capacity of the units it has equals its predicted peak load plus a reserve to cover unit outages. If for some reason this criterion cannot be met, the system may enter into a capacity agreement with a neighboring system provided that neighboring system has surplus capacity beyond what it needs to supply its peak load and maintain its own reserves. In selling capacity, the system that has a surplus agrees to cover the reserve needs of the other system. This may require running an extra unit during certain hours, which represents a cost to the selling system. The advantage of such agreements is to let each system schedule generation additions at longer intervals by buying capacity when it is short and selling capacity when a large unit has just been brought on-line and it has a surplus. Pure capacity reserve interchange agreements do not entitle the purchaser to any energy other than emergency energy requirements.

10.8.2 Diversity Interchange

Daily diversity interchange arrangements may be made between two large systems covering operating areas that span different time zones. Under such circumstance, one system may experience its peak load at a different time of the day than the other system simply because the second system is 1 h behind. If the two systems experience such a phenomenon, they can help each other by interchanging power during the peak. The system that peaked first would buy power from the other and then pay it back when the other system reached its peak load.

This type of interchange can also occur between systems that peak at different seasons of the year. Typically, one system will peak in the summer due to air-conditioning load and the other will peak in winter due to heating load. The winter peaking system would buy power during the winter months from the summer peaking system whose system load is presumably lower at that time of year. Then in the summer, the situation is reversed and the summer peaking system buys power from the winter peaking system.

10.8.3 Energy Banking

Energy-banking agreements usually occur when a predominantly hydro system is interconnected to a predominantly thermal system. During high water runoff periods, the hydro system may have energy to spare and will sell it to the thermal system. Conversely, the hydro system may also need to import energy during periods of low runoff. The prices for such arrangements are usually set by negotiations between the specific systems involved in the agreement.

Instead of accounting for the interchange and charging each other for the transactions on the basis of hour-by-hour operating costs, it is common practice in some areas for utilities to agree to a banking arrangement whereby one of the systems acts as a bank and the other acts as a depositor. The depositor would "deposit" energy whenever it had a surplus and only the MWh "deposited" would be accounted for. Then whenever the depositor needed energy, it would simply withdraw the energy up to the MWh it had in the account with the other system. Which system is "banker" or "depositor" depends on the exchange contract. It may be that the roles are reversed as a function of the time of year.

10.8.4 Emergency Power Interchange

It is very likely that at some future time a power system will have a series of generation failures that require it to import power or shed load. Under such emergencies it is useful to have agreements with neighboring systems that commit them to supply power so that there will be time to shed load. This may occur at times that are not convenient or economical from an incremental cost point of view. Therefore, such agreements often stipulate that emergency power be priced very high.

10.8.5 Inadvertant Power Exchange

The AGC systems of utilities are not perfect devices with the result that there are regularly occurring instances where the error in controlling interchange results in a significant, accumulated amount of energy. This is known as *inadvertant interchange*. Under normal circumstances, system operators will "pay back" the accumulated inadvertant interchange energy megawatt-hour for megawatt-hour, usually during similar time periods in the next week. Differences in cost rates are ignored.

Occasionally utilities will suffer prolonged shortages of fuel or water, and the inadvertant interchange energy may grow beyond normal practice. If done deliberately, this is known as "leaning on the ties." When this occurs, systems will normally agree to pay back the inadvertant energy at the same time of day that the errors occurred. This tends to equalize the economic transfer. In severe fuel shortage situations, interconnected utilities may agree to compensate each other by paying for the inadvertant interchange at price levels that reflect the real cost of generating the exchange energy.

10.9 POWER POOLS

Interchange of power between systems is economically advantageous, as has been demonstrated previously. However, when a system is interconnected with many neighbors, the process of setting up one transaction at a time with each neighbor can become very time consuming and will rarely result in the best overall economic interchange. To overcome this burden, several utilities may form a power pool that incorporates a central dispatch office. The power pool is administered from a central location that has responsibility for setting up interchange between members as well as other administrative tasks. The pool members relinquish certain responsibilities to the pool operating office in return for greater economies in operation.

The agreement the pool members sign is usually very complex. The complexity arises because the members of the pool are attempting to gain greater benefits from the pool operation and to allocate these benefits equitably among the members. In addition to maximizing the economic benefits of interchange between the pool members, pools help member companies by coordinating unit commitment and maintenance scheduling, providing a centralized assessment of system security at the pool office, calculating better hydro schedules for member companies, and so forth. Pools provide increased reliability by allowing members to draw energy from the pool transmission grid during emergencies as well as covering each others' reserves when units are down for maintenance or on forced outage.

Some of the difficulties in setting up a power pool involving nonaffiliated companies or systems arise because the member companies are independently owned and for the most part independently operated. Therefore, one cannot just make the assumption that the pool is exactly the same entity as a system under one ownership. If one member's transmission system is heavily loaded with power flows that chiefly benefit that member's neighbors, then that system is entitled to a reimbursement for the use of the transmission facilities. If one member is directed to commit a unit to cover a reserve deficiency in a neighboring system, that system is also likewise entitled to a reimbursement.

These reimbursement arrangements are built into the agreement that the members sign when forming the pool. The more the members try to push for maximum economic operation, the more complicated such agreements become. Nevertheless, the savings obtainable are quite significant and have led many interconnected utility systems throughout the world to form pools when feasible.

A list of operating advantages to power pools ordered by greatest expected economic advantage might look as follows.

1. Minimize operating costs (maximize operating efficiency).
2. Perform a system-wide unit commitment.
3. Minimize the reserves being carried throughout the system.
4. Coordinate maintenance scheduling to minimize costs and maximize reliability by sharing reserves during maintenance periods.
5. Maximize the benefits of emergency procedures.

There are disadvantages that must be weighed against these operating and economic advantages. Although it is generally true that power pools with centralized dispatch offices will reduce overall operating costs, some of the individual utilities may perceive the pool requirements and disciplines as disadvantageous. Factors that have been cited include.

1. The complexity of the pool agreement and the continuing costs of supporting the interutility structure required to manage and administer the pool.

2. The operating and investment costs associated with the central dispatch office and the needed communication and computation facilities.

3. The relinquishing of the right to engage in independent transactions outside of the pool by the individual companies to the pool office and the requirement that any outside transactions be priced on a split-saving basis.

4. The additional complexity that may result in dealing with regulatory agencies if the pool operates in more than one state.

5. The feeling on the part of the management of some utilities that the pool structure is displacing some of the individual system's management responsibilities and restricting some of the freedom of independent action possible to serve the needs of its own customers.

Power pools without central dispatch control centers can be administered through a central office that simply acts as a brokerage house to arrange transactions among members. In the opposite extreme, the pool can have a fully staffed central office with real-time data telemetered to central computers that calculate the best pool-wide economic dispatch and provide control signals to the member companies.

By far the most difficult task of pool operation is to decide who will pay what to whom for all the economic transactions and special reimbursements built into the pool agreement. There are several ways to solve this problem, and some will be illustrated in Section 10.9.3.

10.9.1 The Energy-Broker System

As with sales and purchases of various commodities or financial issues (i.e., stock), it is often advantageous for interconnected power systems to deal through a broker who sets up sales and purchases of energy instead of dealing directly with each other. The advantage of this arrangement is that the broker can observe all the buy and sell offers at one time and achieve better economy of operation. When utilities negotiate exchanges of power and energy in pairs, the "market place" is somewhat haphazard like a bazaar. The introduction of a central broker to accept quotations to sell and quotations to purchase creates an orderly marketplace where supply, demand, and prices are known simultaneously.

In one power broker scheme in use, the companies that are members of the broker system send hourly buy and sell offers for energy to the broker who matches them according to certain rules. Hourly, each member transmits an incremental cost and the number of megawatt-hours it is willing to sell or its decremental cost and the number of megawatt-hours it will buy. The broker sets up the transactions by matching the lowest cost seller with the highest cost buyer, proceeding in this manner until all offers are processed. The matched buyers and sellers will price the transaction on the basis of rules established in setting up the power broker scheme. A common arrangement is to compensate the seller for his incremental generation costs and split the savings of the buyer equally with the seller. The pricing formula for this arrangement is as follows. Let

$$F'_s = \text{incremental cost of the selling utility, R/MWh}$$

$$F'_b = \text{decremental cost of the buying utility, R/MWh}$$

$$F_c = \text{cost rate of the transaction, R/MWh}$$

Then,

$$F_c = F'_s + \frac{1}{2}(F'_b - F'_s)$$

$$= \frac{1}{2}(F'_s + F'_b)$$

In words, the transaction's cost rate is the average of the seller's incremental cost and the purchaser's decremental cost. In this text decremental cost is the reduction in incremental operating cost when the generation is reduced a small amount. Example 10D illustrates the power brokerage process.

EXAMPLE 10D

In this example four power systems have sent their buy/sell offers to the broker. In the table that follows, these are tabulated and the maximum pool savings possible is calculated.

Utilities selling energy	Incremental cost	MWh for sale	Seller's total increase in cost
A	25 R/MWh	100	2500 R
B	30 R/MWh	100	3000 R

Utilities buying energy	Decremental cost	MWh for purchase	Buyer's total decrease in cost
C	35 R/MWh	50	1750 R
D	45 R/MWh	150	6750 R

$$\text{Net pool savings} = (1750\ R + 6750\ R) - (2500\ R + 3000\ R)$$
$$= 8500\ R - 5500\ R = 3000\ R$$

The broker sets up transactions as shown in the following table.

Transaction	Savings computation		Total transaction savings
1. A sells 100 MWh to D	100 MWh (45 − 25) R/MWh	=	2000 R
2. B sells 50 MWh to D	50 MWh (45 − 30) R/MWh	=	750 R
3. B sells 50 MWh to C	50 MWh (35 − 30) R/MWh	=	250 R
Total			3000 R

The rates and total payments are easily computed under the split savings arrangement. These are

Transaction	Price (R/MWh)	Total cost (R)
1. A sells 100 MWh to D	35.0	3500
2. B sells 50 MWh to D	37.5	1875
3. B sells 50 MWh to C	32.5	1625
Total		7000

A receives 3500 R from D; B receives 3500 R from D and C. Note that each participant benefits: A receives 1000 R above its costs; B receives 500 R above its costs; C saves 125 R; and D saves 1375 R;

The chief advantage of a broker system is its simplicity. All that is required to get a broker system into operation is a telephone circuit to each member's operations office and some means of setting up the transactions. The transactions can be set up manually or, in the case of more modern brokerage arrangements, by a computer program that is given all the buy/sell offers and automatically sets up the trans-actions. The quoting systems are normally only informed of the "match" suggested by the broker and are free to enter into the transaction or not as each sees fit.

It appears quite feasible and economic to extend power broker schemes to handle long-term economy interchange and to arrange capacity sales. This would enable brokers to assist in minimizing costs for spinning reserves and coordinate unit commitments in interconnected systems.

10.9.2 Centralized Economic Dispatch of a Power Pool

The greatest savings from pool operations come when the members of the pool agree to set up the necessary equipment to effect a centralized economic dispatch. There are several ways to calculate the pool-wide economic dispatch. To some extent the methods historically were dependent on the available technology when first designed.

If the power pool has the ability to set up a real-time model (see Chapter 12 for a definition of "real-time model") of the entire pool transmission network, then penalty factors can be calculated as shown in Chapter 4, Section 4.2.4. Once penalty factors for the entire network are calculated, a conventional economic dispatch using coordination equations can be run and, if necessary, the penalty factors can be recalculated by iterating the economic dispatch with a load flow.

However, this is not the only method of pool dispatch now in use. One could propose that a pool-wide loss formula be developed to calculate the penalty factors. However, this would necessitate making the assumption that loads throughout the pool all conformed with each other (see Chapter 4, Section 4.2.3). This assumption probably would introduce significant errors for a large pool. What is needed is a method wherein each member system's incremental losses are obtained from a loss formula for that system and a method of coordinating the economic dispatch of the member systems in such a way that the overall dispatch is optimum for the pool as a whole.

To illustrate how this is done, we will show three systems as members of a pool with ties between them and between the members and an external system. Figure 10.2 illustrates the three-area pool and external system.

The following definitions will be made for the areas. Let

$$P_{gA}, P_{gB}, P_{gC} = \text{generation in areas A, B, C, respectively}$$

$$P_{\text{loss A}}, P_{\text{loss B}}, P_{\text{loss C}} = \text{transmission losses in areas A, B, C, respectively}$$

$$P_{\text{int A}}, P_{\text{int B}}, P_{\text{int C}} = \text{net MW interchange for areas A, B, C, respectively}$$

$$P_{\text{load A}}, P_{\text{load B}}, P_{\text{load C}} = \text{load in areas A, B, C, respectively}$$

$$F(P_{gA}), F(P_{gB}), F(P_{gC}) = \text{total operating cost in areas A, B, C, respectively}$$

$$P_{\text{int P}} = \text{net pool interchange}$$

$$P_{\text{tie } i} = \text{flow on tie } i$$

$$\{i_A\}, \{i_B\}, \{i_C\} = \text{set of ties connected to areas A, B, C, respectively}$$

For each area, the sum of the load, losses, and interchange out of the area must be equal to the generation within the area. In addition, the sum of the net interchange

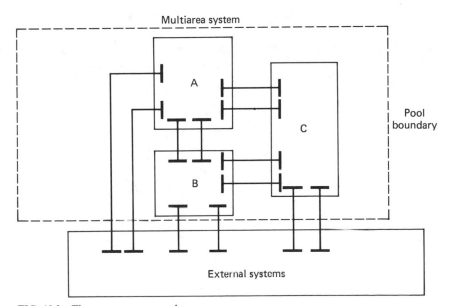

FIG. 10.2   Three-area power pool.

for all the pool members will equal the net pool export. These relationships are set up as equations as follows.

$$P_{\text{load A}} + P_{\text{loss A}} + P_{\text{int A}} - P_{g\text{A}} = 0$$

$$P_{\text{load B}} + P_{\text{loss B}} + P_{\text{int B}} - P_{g\text{B}} = 0$$

$$P_{\text{load C}} + P_{\text{loss C}} + P_{\text{int C}} - P_{g\text{C}} = 0$$

$$P_{\text{int A}} + P_{\text{int B}} + P_{\text{int C}} = P_{\text{int P}}$$

In order to calculate the incremental losses for each area, a matrix relating the tie flows to the area generations and interchange must be calculated. This matrix is called the *interarea matrix* and contains sensitivities similar to the matrix set up in Section 10.7.

$$
\begin{bmatrix} \Delta P_{\text{tie 1}} \\ \Delta P_{\text{tie 2}} \\ \vdots \end{bmatrix} =
\begin{bmatrix} \text{Interarea} \\ \text{matrix} \\ \text{sensitivity} \\ \text{coefficients} \end{bmatrix}
\begin{bmatrix} \Delta P_{g\text{A}} \\ \Delta P_{g\text{B}} \\ \Delta P_{g\text{C}} \\ \Delta P_{\text{int A}} \\ \Delta P_{\text{int B}} \\ \Delta P_{\text{int C}} \end{bmatrix}
$$

For small changes in the generation and interconnected powers,

$$\frac{\partial P_{\text{tie }i}}{\partial P_{g\text{A}}} = \text{term in row } i \text{ of the interarea matrix corresponding to column } P_{g\text{A}}$$

$$\frac{\partial P_{\text{tie }i}}{\partial P_{\text{int A}}} = \text{term in row } i \text{ of the interarea matrix corresponding to column } P_{\text{int A}}$$

In solving for the overall economic dispatch, we first set up a Lagrange function that includes each area constraint and the pool export constraint.

$$
\begin{aligned}
\mathscr{L} = {} & F(P_{g\text{A}}) + F(P_{g\text{B}}) + F(P_{g\text{C}}) \\
& + \lambda_{\text{A}}(P_{\text{load A}} + P_{\text{loss A}} + P_{\text{int A}} - P_{g\text{A}}) \\
& + \lambda_{\text{B}}(P_{\text{load B}} + P_{\text{loss B}} + P_{\text{int B}} - P_{g\text{B}}) \\
& + \lambda_{\text{C}}(P_{\text{load C}} + P_{\text{loss C}} + P_{\text{int C}} - P_{g\text{C}}) \\
& + \lambda_{\text{p}}(-P_{\text{int A}} - P_{\text{int B}} - P_{\text{int C}} + P_{\text{int P}})
\end{aligned}
$$

First, the Lagrange function is differentiated with respect to each generator in each area.

$$\frac{\partial \mathscr{L}}{\partial P_{g\text{A}}} = 0$$

This gives

$$\frac{\partial F(P_{g\text{A}})}{\partial P_{g\text{A}}} = \lambda_{\text{A}} - \left[\lambda_{\text{A}} \frac{\partial P_{\text{loss A}}}{\partial P_{g\text{A}}} - \sum_{i=\{i\text{B}\}} \lambda_{\text{B}} \frac{\partial P_{\text{loss B}}}{\partial P_{\text{tie }i}} \frac{\partial P_{\text{tie }i}}{\partial P_{g\text{A}}} - \sum_{i=\{i\text{C}\}} \lambda_{\text{C}} \frac{\partial P_{\text{loss C}}}{\partial P_{\text{tie }i}} \frac{\partial P_{\text{tie }i}}{\partial P_{g\text{A}}} \right]$$

$$= \lambda_{\text{A}} - [\text{total incremental loss}]$$

Similar equations exist for $\partial\mathcal{L}/\partial P_{gB}$ and $\partial\mathcal{L}/\partial P_{gC}$. Note that the $\partial P_{tie\,i}/\partial P_{gA}$ terms come from the interarea matrix.

The terms in brackets are all incremental loss terms. The first incremental loss term gives the change in losses in area A with respect to a shift of power from a generator in area A to the area A load center. The remaining terms give the change in losses in other areas due to the change in tie flows that would occur when the area A generation shift took place. These terms are often neglected. In the flowchart in Figure 10.3, we refer to all the incremental losses as the total incremental loss.

Next, the Lagrange function is differentiated with respect to each area's interchange. The resulting equations are arranged in a matrix expression that relates

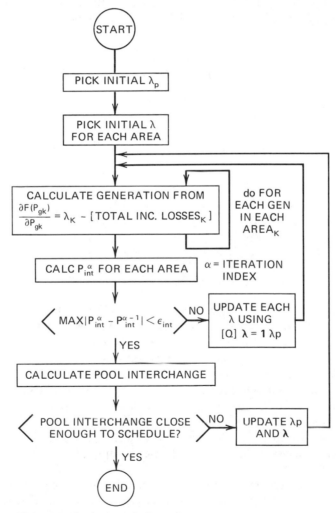

FIG. 10.3 Pool economic dispatch.

each area's lambda to the pool lambda, λ_p:

$$\frac{\partial \mathcal{L}}{\partial P_{\text{int A}}} = 0$$

This gives

$$\lambda_A + \sum_{i=\{i_A\}} \lambda_A \frac{\partial P_{\text{loss A}}}{\partial P_{\text{tie }i}} \frac{\partial P_{\text{tie }i}}{\partial P_{\text{int A}}} + \sum_{i=\{i_B\}} \lambda_B \frac{\partial P_{\text{loss B}}}{\partial P_{\text{tie }i}} \frac{P_{\text{tie }i}}{P_{\text{int A}}}$$

$$+ \sum_{i=\{i_C\}} \lambda_C \frac{\partial P_{\text{loss C}}}{\partial P_{\text{tie }i}} \frac{\partial P_{\text{tie }i}}{\partial P_{\text{int A}}} = \lambda_p$$

Similarly for $\partial \mathcal{L}/\partial P_{\text{int B}}$ and $\partial \mathcal{L}/\partial P_{\text{int C}}$. Note that again the interarea matrix yields the terms $\partial P_{\text{tie }i}/\partial P_{\text{int A}}$, $\partial P_{\text{tie }i}/\partial P_{\text{int B}}$ and $\partial P_{\text{tie }i}/\partial P_{\text{int C}}$. These can be arranged in the following form.

$$[Q]\begin{bmatrix} \lambda_A \\ \lambda_B \\ \lambda_C \end{bmatrix} = \begin{bmatrix} 1 \\ 1 \\ 1 \end{bmatrix}\lambda_p$$

where

$$Q_{MM} = 1 + \sum_{i=\{i_M\}} \lambda_M \frac{\partial P_{\text{loss M}}}{\partial P_{\text{tie }i}} \frac{\partial P_{\text{tie }i}}{\partial P_{\text{int M}}}$$

$$Q_{MN} = \sum_{i=\{i_M\}} \lambda_M \frac{\partial P_{\text{loss M}}}{\partial P_{\text{tie }i}} \frac{\partial P_{\text{tie }i}}{\partial P_{\text{int N}}}$$

and

$$M = \text{A, B, or C}$$

$$N = \text{A, B, or C}$$

The solution of the pool economic dispatch is carried out as shown in the flowchart in Figure 10.3. The solution to the pool economic dispatch shown in Figure 10.3 uses two basic loops, an area lambda loop and a pool lambda loop. For given values of the pool lambda and area lambdas, the generation at each generator bus in each area is calculated. From the generation values obtained, the net interchange in each area is calculated. The net interchange values are compared to the values obtained on the previous iteration. If sufficient change is found, the area lambdas are updated using the "Q" equation. When the area net interchanges converge, the net pool interchange is calculated and compared to the scheduled value for pool interchange. If the pool net interchange is in error, the pool lambda itself is adjusted and the area loop is then reconverged.

Once the central economic dispatch has been calculated, the problem remains to apply this dispatch to the generation units in the member companies. This could be done by executing the generation control functions from the pool control center and would necessitate telemetry channels from each member's generators to the pool control center. A more likely control system to be found in present usage would require the pool control center to transmit information concerning the pool economic dispatch to the member companies' control centers where each member is then responsible to control its units to match the pool economic dispatch. The information sent to the member's control centers can consist of net interchange values or the area incremental cost that each member must match in executing its own economic dispatch.

10.9.3 Allocating Pool Savings

All methods of allocating the savings achieved by a central pool dispatch are based on the premise that no pool member should have higher generation production expenses than it could achieve by dispatching its own generation to meet its own load.

We saw previously in the pool broker system that one of the ways to allocate pool savings is simply to split them in proportion to each system's net interchange during the interval. In the broker method of matching buyers and sellers, calculations of savings are relatively easy to make since the agreed incremental costs and amounts of energy must be transmitted to the broker at the start. When a central economic dispatch is used, it is easier to act as if the power were sold to the pool by the selling systems and then bought from the pool by the buying systems. In addition, allowances may be made for the fact that one system's transmission system is being used more than others in carrying out the pool transactions.

There are two general types of allocations schemes in use at U.S. pool control centers. One, illustrated in Example 10E, may be performed in a real-time mode with cost and savings allocations made periodically using the incremental and decremental costs of the systems. In this scheme, power is sold to and purchased from the pool and participants' accounts are updated currently. In the other approach, illustrated in Example 10F, the allocation of costs and savings is done after the fact using total production costs. Example 10E shows a scheme using incremental costs similar to one used by an Eastern U.S. pool made up of several member systems.

EXAMPLE 10E

Assume that the same four systems as given in Example 10D were scheduled to transact energy by a central dispatching scheme. Also assume that 10% of the gross system's savings was to be set aside to compensate those systems that provided transmission facilities to the pool. The first table shows the calculation of the net system savings.

Utilities selling energy	Incremental cost	MWh for sale	Seller's total increase in cost
A	25 R/MWh	100	2500 R
B	30 R/MWh	100	3000 R

Utilities buying energy	Decremental cost	MWh for purchase	Buyer's total decrease in cost
C	35 R/MWh	50	1750 R
D	45 R/MWh	150	6750 R
Pool savings			3000 R
Savings withheld for transmission compensation[a]			300 R
Net savings			2700 R

[a] 10% savings withheld for transmission compensation.

Next, the weighted average incremental costs for selling and buying power are calculated.

Seller's weighted average incremental cost

$$= \left[\frac{(25 \text{R/MWh} \times 100 \text{ MWh}) + (30 \text{R/MWh} \times 100 \text{ MWh})}{100 \text{ MWh} + 100 \text{ MWh}} \right] = 27.50 \text{ R/MWh}$$

Buyer's weighted average decremental cost

$$= \left[\frac{(35 \text{R/MWh} \times 50 \text{ MWh}) + (45 \text{ R/MWh} \times 150 \text{ MWh})}{50 \text{ MWh} + 150 \text{ MWh}} \right] = 42.50 \text{ R/MWh}$$

Finally, the individual transactions savings are calculated.

1. A sells 100 MWh to pool $\qquad 100 \text{ MWh} \dfrac{42.50 - 25 \text{ R/MWh}}{2} \times 0.9$

$$= 787.50 \text{ R}$$

2. B sells 100 MWh to pool $\qquad 100 \text{ MWh} \dfrac{42.50 - 30 \text{ R/MWh}}{2} \times 0.9$

$$= 562.50 \text{ R}$$

3. C buys 50 MWh from pool $\qquad 50 \text{ MWh} \dfrac{35 - 27.50 \text{ R/MWh}}{2} \times 0.9$

$$= 168.75 \text{ R}$$

4. D buys 150 MWh from pool $\qquad 150 \text{ MWh} \dfrac{45 - 27.50 \text{ R/MWh}}{2} \times 0.9$

$$= 1181.25 \text{ R}$$

Net Savings $\qquad\qquad\qquad\qquad 2700.00 \text{ R}$

The total transfers for this hour are then:

$$C \text{ buys } 50 \text{ MWh for } 42.5 \times 50 - 168.75 = 1956.25 \text{ R}$$
$$D \text{ buys } 150 \text{ MWh for } 42.5 \times 150 - 1181.25 = 5193.75 \text{ R}$$
$$\text{Total} \qquad 7150.00 \text{ R}$$

$$A \text{ sells } 100 \text{ MWh for } 27.5 \times 100 + 787.5 = 3537.50 \text{ R}$$
$$B \text{ sells } 100 \text{ MWh for } 27.5 \times 100 + 562.5 = 3312.50 \text{ R}$$
$$6850.00 \text{ R}$$
$$\text{Total Transmission charge} \qquad 300.00 \text{ R}$$
$$\text{Total} \qquad 7150.00 \text{ R}$$

The 300 R that was set aside for transmission compensation would be split up among the four systems according to some agreed rule reflecting each system's contribution to the pool transmission network.

The second type of savings allocation method is based on after-the-fact computations of individual pool member costs as if each were operating strictly so as to serve their own individual load. In this type of calculation, the unit commitment, hydro schedules, and economic dispatch of each individual pool member are recomputed for an interval after the pool load has been served. This "own load dispatch" is performed with each individual system's generating capacity, including any portions of jointly owned units, to achieve maximum operating economy for the individual system.

The costs for these computed individual production costs are then summed and the total pool savings are computed as the difference between this cost and the actual cost determined by the central pool dispatch.

These savings are then allocated among the members of the pool according to the specific rules established in the pool agreement. One method could be based on rules similar to those illustrated previously. That is, any interval for which savings are being distributed, buyers and sellers will split the savings equally.

Specific computational procedures may vary from pool to pool. Those members of the pool supplying energy in excess of the needs of their own loads will be compensated for their increased production expenses and receive a portion of the overall savings due to a pool-wide dispatch. The process is complicated because of the need to perform individual system production cost calculations. Pool agreements may contain provisions for compensation to members supplying capacity reserves as well as energy to the pool. A logical question that requires resolution by the pool members involves the fairness of comparing an after-the-fact production cost analysis that utilizes a known load pattern with a pool dispatch that was forced to use load forecasts. With the load pattern known with certainty, the internal unit commitment may be optimized to a greater extent that was feasible by the pool control center. Example 10F illustrates this type of procedure for the three systems of Example 10C for 1 period. In this example, only the effects of the economic dispatch are shown since the unit commitment process is not involved.

EXAMPLE 10F

The three areas and load levels are identical to those in Example 10C. (Generation data are in Examples 10A and 10B as well.) In this case the three areas are assumed to be members of a centrally dispatched power pool. The pool's rules for pricing pool interchange are as follows.

1. Each area delivering power and energy to the pool in excess of its own load will receive compensation for its increased production costs.

2. The total pool savings will be computed as the difference between the sum of the production costs of the individual areas (computed on the basis that it supplied its own load) and the pool-wide production cost.

3. These savings will be split equally between the suppliers of pool capacity or energy and the areas receiving pool-supplied capacity or energy.

4. In each interval where savings are allocated (usually a week, but in this example only 1 h), the cost rate for pricing the interchange will be one-half the sum of the total pool savings plus the cost of generating the pool energy divided by the total pool energy. The total pool energy is the sum of the energies in the interval supplied by all areas generating energy in excess of its own load.

The pool production costs are as follows.

Area	Area load (MW or MWh)	Own-load production cost (R/h)
1	700	13,677.21
2	1100	18,569.23
3	550	10,042.00
Total	2350	42,288.44 R/h

Under the pool dispatch, areas 1 and 2 are dispatched at an incremental cost of 17.149 R/MWh to generate a total of 1900 MW. Area 3 is limited to supplying 450 MW of its own load at an incremental cost of 18.125 R/MWh. The generation and costs of the three areas and the pool under pool dispatch are given in the following table.

Area	Area generation (MW or MWh)	Production cost (R/h)	Incremental cost (R/MWh)
1	458.9	9,458.74	17.149
2	1441.1	24,232.66	17.149
3	450.0	8,220.00	18.125
Pool	2350.0	41,911.40	17.149

Therefore, the total savings due to the pool dispatch for this 1 h are

$$42,288.44 \text{ R} - 41,911.40 \text{ R} = 377.04 \text{ R}$$

In this example, area 2 is supplying 341.1 MWh in excess of its own load to the pool. This is the total pool energy. Therefore, the price rate for allocating savings is computed as follows.

Cost of Pool Energy: Cost of energy supplied to the pool by area 2

$$= 24{,}232.66 \; \mathrm{R} - 18{,}569.23 \; \mathrm{R} = 5663.43 \; \mathrm{R}$$

$$+ \; 1/2 \text{ pool savings} = \underline{188.52 \; \mathrm{R}}$$

$$\text{Total} \qquad 5851.95 \; \mathrm{R}$$

$$\text{Interchange price rate} = \frac{5851.95}{341.1} = 17.156 \; \mathrm{R/MWh}$$

The final outcome for each area is

Area	Pool energy received (MWh)	Interchange cost (R)	Production cost (R)	Net cost (R)
1	+241.1	4,136.34	9,458.74	13,595.08
2	−341.1	−5,851.95	24,232.66	18,380.71
3	+100	1,715.61	8,220.00	9,935.61
Pool		0	41,911.40	41,911.40

Note that each area's net production costs are reduced as compared with what they would have been under isolated dispatch. Furthermore, the ambiguity involved in pricing different transactions in alternative sequences has been avoided.

Example 10F is based on only a single load level so that after-the-fact unit commitment and production costing is not required. It could have been done on a real-time basis in fact. This example also illustrates the complete transaction allocation that must be done for savings allocation schemes.

Complete own-load dispatch computation for cost and savings allocations are usually performed for a weekly period. The implementation may be complex since hourly loads and unit status data are required. An on-line, real-time allocation scheme avoids these complications.

PROBLEMS

10.1 Four areas are interconnected as shown in Figure 10.4. Each area has a total generation capacity of 700 MW currently on-line. The minimum loading of these units is 140 MW in each area. Area loads for a given hour are as shown in Figure 10.4. The transmission lines are each sufficient to transfer any amount of power required.

The composite input-output production cost characteristics of each area are as follows:

$$F_1 = 200 + 2\,P_1 + 0.005\,P_1^2, \qquad \mathrm{R/h}$$

$$F_2 = 325 + 3\,P_2 + 0.002143\,P_2^2, \quad \mathrm{R/h}$$

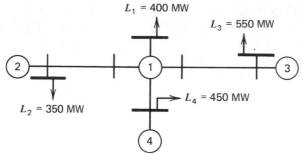

FIG. 10.4   Four-area system for Problem 10.1.

$$F_3 = 275 + 2.6\,P_3 + 0.003091\,P_3^2, \ \text{R/h}$$

$$F_4 = 190 + 3.1\,P_4 + 0.00233\,P_4^2, \ \text{R/h}$$

in all cases, $140 \le P_i \le 700$ MW. Find the cost of each area if each independently supplies its own load and the total cost for all four areas.

10.2 Assume that Area 1 of Problem 10.1 engages in two transactions

a. Area 1 buys 190 MW from area 2.

b. Area 1 sells 120 MW to area 3.

For each of these transactions, the price is based upon a 50-50 split-savings agreement. Find the price of each transaction, the net generation costs for each area including the sum it pays or receives under the split-savings agreement with the order of the transactions being as follows.

1. a then b

2. b then a

In both instances, find the total cost for the four-area pool.

10.3 Assume that the four areas of Problem 10.1 are centrally dispatched by a pool control center.

a. Find the generation and production cost in each area.

b. Assume a split-savings pool agreement such that each area exporting receives its increased costs of production plus its proportionate share of 50% of the pool savings. Find the cost per MWh of transfer energy (i.e., "pool energy") and the net production cost of each area.

10.4 Assume that the four areas of Problem 10.1 are members of a "power broker." Previous to the hour shown in Problem 10.1, each area submits quotations to the broker to sell successive blocks of 25 or 50 MW and bids to purchase blocks of 25 or 50 MW. In furnishing these data to the broker, assume that the prices quoted are the average incremental costs for the block. The broker matching rules are as follows.

Rule 1. Quotations to sell and bids to buy are matched only wherever there is a direct connection between the quoting and bidding company.

Rule 2. Transactions are arranged in a priority order where the lowest remaining incremental cost for the quoting area is matched with the highest decremental cost for the bidding areas. [That is, lowest available incremental cost energy available for sale is matched with the area with the greatest available potential incremental cost savings (= decremental cost).]

Rule 3. "Matches" may be made for all or part of a block. The remainder of the block must be used, if possible, before the next block is utilized. Matching will cease when the absolute value of the difference between the incremental and decremental cost drops below 0.33 R/MWh.

Rule 4. No area may be both a buyer and a seller in any given hour.

Rule 5. The price per MWh for each matched transaction is one-half the sum of the absolute values of the incremental and decremental costs.

For this problem, assume that quotes and bids are supplied to the broker by each area as follows.

Area	Quotes to sell	Quotes to buy
1	100 MW in 25 MW blocks	100 MW in 25 MW blocks
2	200 MW in 50 MW blocks	None
3	None	200 MW in 50 MW blocks
4	25 MW	25 MW

a. Set up the power broker matching system and establish the transactions that can take place and the price of each.

b. Assume that all feasible transactions take place and find the net production cost to each area and the pool.

10.5 Repeat Problem 10.4 with the following assumptions simultaneously taken in place of those in Problem 10.4.

a. Each area is interconnected with every other area and transfers may take place directly between all pairs of areas.

b. The matched transactions will proceed until the difference between decremental costs are zero instead of 0.33 R/MWh.

FURTHER READING

References 1–3 provide a good historical look at the techniques that have gone into power pooling. Reference 4 is an excellent summary of the state-of-the-art (1980) of power brokering and pooling and reviews the practices of most major U.S. power pools. References 5–7 provide the theoretical basis for the interarea matrix and its application to pool economic dispatch calculations.

1. Mochon, H. H. Jr. "Practices of the New England Power Exchange," *Proceedings of the American Power Conference*, Vol. 34, 1972, pp. 911–925.

2. Happ, H. H. "Multi-Computer Configurations & Diakoptics: Dispatch of Real Power in Power Pools," 1967 *Power Industry Computer Applications Conference Proceedings*, pp. 95–107.

3. Roth, J. E., Ambrose, Z. C., Schappin, L. A., Gassert, J. D., Hunt, D. M., Williams, D. D., Wood, W., Matrin, L. W., "Economic Dispatch of Pennsylvania-New Jersey-Maryland Interconnection System Generation on a Multi-Area Basis," 1967 *Power Industry Computer Applications Conference Proceedings*, pp. 109–116.

4. "Power Pooling: Issues and Approaches," DOE/ERA/6385-1, U.S. Department of Energy, 1980.

5. Happ, H. H., "The Interarea Matrix: A Tie Line Flow Model for Power Pools," *IEEE Transactions on Power Apparatus and Systems*, Vol. PAS-90, (January/February 1971, pp. 36–45.

6. Cameron, D. E., Koehler, J. E., Ringlee, R. J. "A Study Mode Multi-Area Economic Dispatch Program," Presented at the IEEE Power Engineering Society Winter Power Meeting, 1974, Paper C 74 157-4.

7. Happ, H. H. *Piecewise Methods and Applications to Power Systems*, 1980, John Wiley, New York.

chapter 11

Power System Security

11.1 INTRODUCTION

Up until now we have been mainly concerned with minimizing the cost of operating a power system. An overriding factor in the operation of a power system is the desire to maintain system security. System security involves practices designed to keep the system operating when components fail. For example, a generating unit may have to be taken off-line because of auxiliary equipment failure. By maintaining proper amounts of spinning reserve, the remaining units on the system can make up the deficit without too low a frequency drop or need to shed any load. Similarly, a transmission line may be damaged by a storm and taken out by automatic relaying. If in commiting and dispatching generation proper regard for transmission flows is maintained, the remaining transmission lines can take the increased loading and still remain within limit.

Because the specific times at which initiating events that cause components to fail are unpredictable, the system must be operated at all times in such a way that the system will not be left in dangerous condition should any credible initiating event occur. Since power system equipment is designed to be operated within certain limits, most pieces of equipment are protected by automatic devices that can cause equipment to be switched out of the system if these limits are violated. If any event occurs on a system that leaves it operating with limits violated, the event may be followed by a series of further actions that switch other equipment out of service. If this process continues, the entire system or large parts of it may completely collapse. This is usually referred to as a *system blackout*.

An example of the type of event sequence that can cause a blackout might start with a single line being opened due to an insulation failure; the remaining transmission circuits in the system will take up the flow that was flowing on the now opened line. If one of the remaining lines is now too heavily loaded, it may open due to relay action, thereby causing even more load on the remaining lines. This type of process is often termed a *cascading outage*. Most power systems are operated such that any single initial failure event will not leave other components overloaded, specifically to avoid cascading failures.

Most large power systems install equipment to allow operations personnel to monitor and operate the system in a reliable manner. This chapter will deal with the techniques and equipment used in these systems. We will lump these under the commonly used title *system security*.

System security can be broken down into three major functions that are carried out in an operations control center.

1. System monitoring.

2. Contingency analysis.

3. Corrective action analysis.

System monitoring provides the operators of the power system with pertinent up-to-date information on the conditions on the power system. Generally speaking, it is the most important function of the three. From the time that utilities went beyond systems of one unit supplying a group of loads, effective operation of the system required that critical quantities be measured and the values of the measurements be transmitted to a central location. Such systems of measurement and data transmission, called *telemetry systems*, have evolved to schemes that can monitor voltages, currents, power flows, and the status of circuit breakers and switches in every substation in a power system transmission network. In addition, other critical information such as frequency, generator unit outputs and transformer tap positions can also be telemetered. With so much information telemetered simultaneously, no human operator could hope to check all of it in a reasonable time frame. For this reason, digital computers are usually installed in operations control centers to gather the telemetered data, process them, and place them in a data base from which operators can display information on large display monitors. More importantly, the computer can check incoming information against prestored limits and alarm the operators in the event of an overload or out-of-limit voltage.

State estimation is often used in such systems to combine telemetered system data with system models to produce the best estimate (in a statistical sense) of the current power system conditions or "state." We will discuss some of the highlights of these techniques in Chapter 12.

Such systems are usually combined with supervisory control systems that allow operators to control circuit breakers and disconnect switches and transformer taps remotely. Together, these systems are often referred to as *SCADA systems*, standing for supervisory control and data acquisition system. The SCADA system allows a few operators to monitor the generation and high voltage transmission systems and to take action to correct overloads or out-of-limit voltages.

The second major security function is contingency analysis. The results of this type of analysis allow systems to be operated defensively. Many of the problems that occur on a power system can cause serious trouble within such a quick time period that the operator could not take action fast enough. This is often the case with cascading failures. Because of this aspect of system operation, modern operations computers are equipped with contingency analysis programs that model possible system troubles before they arise. These programs are based on a model of the power system and are used to study outage events and alarm the operators to any potential overloads or out-of-limit voltages. For example, the simplest form of contingency analysis can be put together with a standard load-flow program such as described in Chapter 4, together with procedures to set up the load-flow data for each outage to be studied by the load-flow program. This permits the system operators to establish defensive operating states where no single contingency event (i.e., a single failure) will cause overloads and/or out-of-limit voltages. This analysis effectively develops operating constraints that may be used in the economic dispatch and unit commitment program. Several variations on this type of contingency analysis scheme involve fast solution methods, automatic contingency event selection, and automatic initializing of the contingency load flows using actual system data and state estimation procedures.

The third major security function, corrective action analysis, allows operating personnel to alter the operation of the power system in the event of an overload or in the event that a contingency analysis program predicts a serious problem should a certain outage occur. A simple type of corrective action involves shifting generation from one generating unit to another. Such shifts can cause power flows to change and thus can alter loading on overloaded lines.

Together, the functions of system monitoring, contingency analysis, and corrective action analysis comprise a very complex set of tools that can aid in the secure operation of a power system. This chapter introduces basic analytical foundations needed in order to understand this topic.

11.2 FACTORS AFFECTING POWER SYSTEM SECURITY

As a consequence of several widespread blackouts in interconnected power systems, the priorities for operation of modern power systems have evolved to the following:

- Operate the system in such a way that power is delivered reliably.
- Within the constraints placed on the system operation by reliability considerations, the system will be operated most economically.

The greater part of this book is devoted to developing methods to operate a power system to gain maximum economy. But what factors affect its operation from a reliability standpoint? We will assume that the engineering groups who have designed the power system's transmission and generation systems have done so with reliability in mind. This means that adequate generation has been installed to meet the load and that adequate transmission has been installed to deliver the generated power to the load. If the operation of the system went on without sudden failures or without experiencing unanticipated operating states, we would probably have no reliability problems. However, any piece of equipment in the system can fail either due to internal causes or due to external causes such as lightning strikes, objects hitting transmission towers, or human errors in setting relays. It is highly uneconomical, if not impossible, to build a power system with so much redundancy (i.e., extra transmission lines, reserve generation, etc.) that failures never cause load to be dropped on a system. Rather, systems are designed so that the probability of dropping load is acceptably small. Thus, most power systems are designed to have sufficient redundancy to withstand all major failure events, but this does not guarantee that the system will be 100% reliable.

Within the design and economic limitations, it is the job of the operators to try to maximize the reliability of the system they have at any given time. Usually a power system is never operated with all equipment "in" (i.e., connected) since failures occur or maintenance may require taking equipment out of service. Thus the operators play a considerable role in seeing that the system is reliable.

In this chapter, we will not be concerned with all the events that can cause trouble on a power system. Instead, we will concentrate on the possible consequence and remedial actions required by two major types of failure events—transmission-line outages and generation-unit failures.

Transmission-line failures cause changes in the flows and voltages on the transmission equipment remaining connected to the system. Therefore, the analysis of transmission failures requires methods to predict these flows and voltages so as to be sure they are within their respective limits. Generation failures can also cause flows and voltages to change in the transmission system with the addition of dynamic problems involving system frequency and generator output.

Whether a given failure causes serious problems is both a function of the configuration of the existing system and how it is operated. Therefore, it will also be useful to ask the question: How should the system be operated so that failures do not cause problems?

11.3 CONTINGENCY ANALYSIS: DETECTION OF NETWORK PROBLEMS

We will briefly illustrate the kind of problems we have been describing by use of the six-bus network used in Chapter 4. The base-case load flow results for Example 4A are shown in Figure 11.1 and indicate a flow of 43.8 MW and 60.7 MVAR on the line from bus 3 to bus 6. The limit on this line can be expressed in MW or in MVA. For

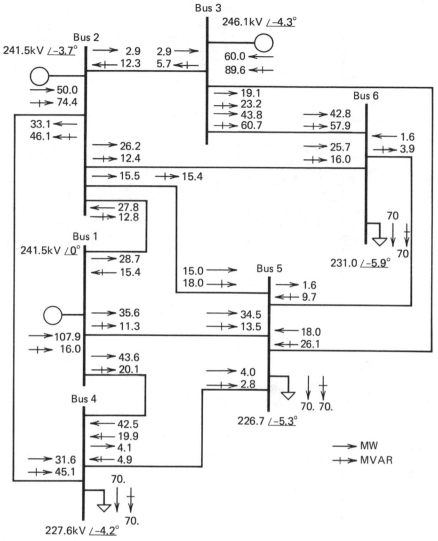

FIG. 11.1 Six-bus network base case AC load flow (see Example 4A).

the purpose of this discussion, assume that we are only interested in the MW loading on the line. Now let us ask what will happen if the transmission line from bus 3 to bus 5 were to open. The resulting flows and voltages are shown in Figure 11.2. Note that the flow on the line from bus 3 to bus 6 has increased to 54.9 MW and that most of the other transmission lines also experienced changes in flow. Note also that the bus voltage magnitudes changed, particularly at bus 5, which is now almost 5% below nominal. Figures 11.3 and 11.4 are examples of generator outages and serve to

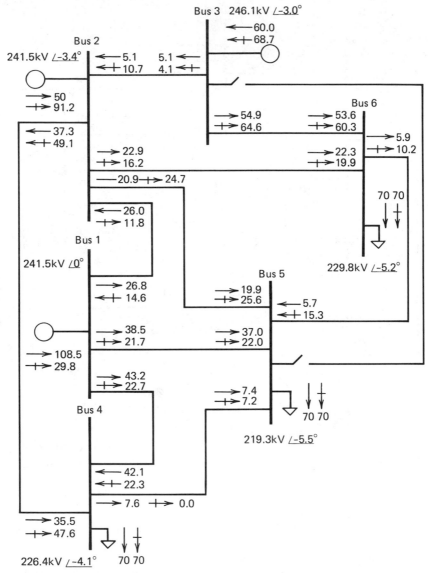

FIG. 11.2 Six-bus network line outage case; line from bus 3 to bus 5 opened.

illustrate the fact that generation outages can also result in changes in flows and voltages on a transmission network. In the example shown in Figure 11.3 all the geration lost from bus 3 is picked up on the generator at bus 1. Figure 11.4 shows the case when the loss of generation on bus 3 is made up by an increase in generation at buses 1 and 2. Clearly, the difference in flows and voltages shows that how the lost generation is picked up by the remaining units is important.

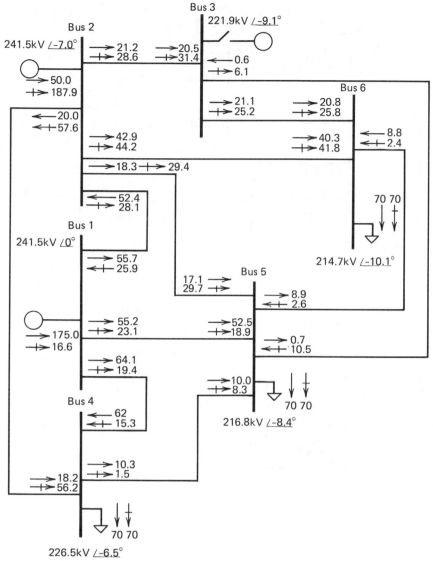

FIG. 11.3 Six-bus network generator outage case. Outage of generator on bus 3; lost generation picked up on generator 1.

If the system being modeled is part of a large interconnected network, the lost generation will be picked up by a large number of generating units outside the system's immediate control area. When this happens, the pickup in generation is seen as an increase in flow over the tie lines to the neighboring systems. To model this we can build a network model of our own system plus an equivalent network of our neighbor's system and place the swing bus or reference bus in the equivalent

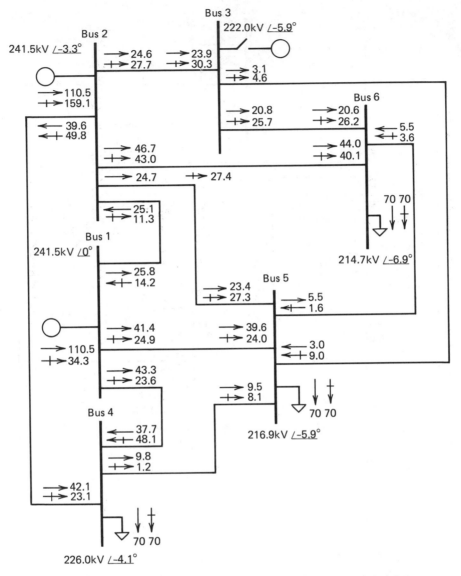

FIG. 11.4 Six-bus network generator outage case. Outage of generator on bus 3; lost generation picked up on generator 1 and generator 2.

system. A generator outage is then modeled so that all lost generation is picked up on the swing bus, which then appears as an increase on the tie flows, thus approximately modeling the generation loss when interconnected. If, however, the system of interest is not interconnected, then the loss of generation must be shown as a pickup in output on the other generation units within the system. An approximate method of doing this is shown in the next section.

Operations personnel must know which line or generation outages will cause flows or voltages to fall outside limits. To predict the effects of outages, contingency analysis techniques are used. Contingency analysis procedures model single failure events (i.e., one line outage or one generator outage) one after another in sequence until "all credible outages" have been studied. For each outage tested, the contingency analysis procedure checks all lines and voltages in the network against their respective limits. The simplest form of such a contingency analysis technique is shown in Figure 11.5.

The most difficult methodological problem to cope with in contingency analysis is the speed of solution of the model used. The most difficult logical problem is the selection of "all credible outages." If each outage case studied were to solve in 1 min and several hundred outages were of concern, it would take hours before all cases could be reported. This would be useful if the system conditions did not change over that period of time. However, power systems are constantly undergoing changes and the operators usually need to know if the present operation of the system is safe without waiting too long for the answer. Contingency analysis execution times of several minutes to a half hour for several hundred outage cases are typical of computer and analytical technology as of 1983.

One way to gain speed of solution in a contingency analysis procedure is to use an approximate model of the power system. For many systems, the use of DC load flow models provides adequate capability. In such systems the voltage magnitudes may not be of great concern and the DC load flow provides sufficient accuracy with respect to the megawatts flows. For other systems voltage is a concern and full AC load flow analysis is required.

11.3.1 Network Sensitivity Methods

Many possible outage conditions could happen to a power system. Thus there is a need to be able to study a large number of them so that operations personnel can be warned ahead of time if one or more outages will cause serious overloads on other equipment.

The problem of studying hundreds of possible outages becomes very difficult to solve if it is desired to present the results quickly so that corrective actions can be taken. One of the easiest ways to provide a quick calculation of possible overloads is to use *network sensitivity factors*. These factors show the approximate change in line flows for changes in generation on the network configuration and are derived from the DC load flow presented in Chapter 4. These factors can be derived in a variety of ways and basically come down to two types.

1. Generation shift factors.

2. Line outage distribution factors.

First, we will describe how these factors are used, and then we will describe how they can be derived from the same network data that were used to calculate the DC load flow.

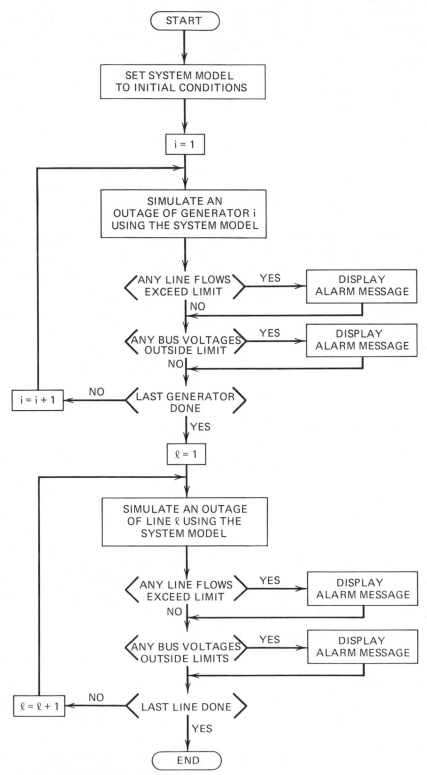

FIG. 11.5 Contingency analysis procedure.

The generation shift factors are designated $a_{\ell i}$ and have the following definition.

$$a_{\ell i} = \frac{\Delta f_\ell}{\Delta P_i} \tag{11.1}$$

where
ℓ = line index

i = bus index

Δf_ℓ = change in megawatt power flow on line ℓ when a change in generation, ΔP_i, occurs at bus i

ΔP_i = change in generation at bus i

It is assumed in this definition that the change in generation, ΔP_i, is exactly compensated by an opposite change in generation at the reference bus and that all other generators remain fixed. The $a_{\ell i}$ factor then represents the sensitivity of the flow on line ℓ to a change in generation at bus i. Suppose one wanted to study the outage of a large generating unit and it was assumed that all the generation lost would be made up by the reference generation (we will deal with the case where the generation is picked up by many machines shortly). If the generator in question was generating P_i^0 MW and it was lost, we would represent ΔP_i as

$$\Delta P_i = -P_i^0 \tag{11.2}$$

and the new power flow on each line in the network could be calculated using a precalculated set of "a" factors as follows.

$$\hat{f}_\ell = f_\ell^0 + a_{\ell i} \Delta P_i \qquad \text{for } \ell = 1, \ldots, L \tag{11.3}$$

where
$\hat{f}_\ell$ = flow on line ℓ after the generator on bus i fails

f_ℓ^0 = flow before the failure

The "outage flow" $\hat{f}_\ell$ on each line can be compared to its limit and those exceeding their limit flagged for alarming. This would tell the operations personnel that the loss of the generator on bus i would result in an overload on line ℓ.

The generation shift sensitivity factors are linear estimates of the change in flow with a change in power at a bus. Therefore, the effects of simultaneous changes on several generating buses can be calculated using superposition. Suppose, for example, that the loss of the generator on bus i were compensated by governor action on machines throughout the interconnected system. One frequently used method assumes that the remaining generators pick up in proportion to their maximum MW rating. Thus, the proportion of generation pickup from unit j ($j \neq i$) would be

$$\gamma_{ji} = \frac{P_j^{\max}}{\sum_{\substack{k \\ k \neq i}} P_k^{\max}} \tag{11.4}$$

where
$P_k^{\max}$ = maximum MW rating for generator k

γ_{ji} = proportionality factor for pickup on generating unit j when unit i fails

Then, to test for the flow on line ℓ under the assumption that all the generators in the interconnection participate in making up the loss, use the following:

$$\hat{f} = f_\ell^0 + \sum_{\substack{j \\ j \neq i}} a_{\ell j} \gamma_{ji} \Delta P_i \qquad (11.5)$$

Note that this assumes that the unit will not actually hit its maximum. If this is apt to be the case, a more detailed generation pickup algorithm that took account of generation limits would be required.

The line outage distribution factors are used in a similar manner, only they apply to the testing for overloads when transmission circuits are lost. By definition, the line outage distribution factor has the following meaning.

$$d_{\ell,k} = \frac{\Delta f_\ell}{f_k^0} \qquad (11.6)$$

where $d_{\ell,k}$ = line outage distribution factor when monitoring line ℓ after an outage on line k

Δf_ℓ = change in MW flow on line ℓ

f_k^0 = original flow on line k before it was outaged (opened)

If one knows the power on line ℓ and line k, the flow on line ℓ with line k out can be determined using "d" factors.

$$\hat{f}_\ell = f_\ell^0 + d_{\ell,k} f_k^0 \qquad (11.7)$$

where f_ℓ^0, f_k^0 = preoutage flows on lines ℓ and k, respectively

$\hat{f}_\ell$ = flow on line ℓ with line k out

By precalculating the line outage distribution factors, a very fast procedure can be set up to test all lines in the network for overload for the outage of a particular line. Furthermore, this procedure can be repeated for the outage of each line in turn with overloads reported to the operations personnel in the form of alarm messages.

Using the generator and line outage procedures described earlier, one can program a digital computer to execute a contingency analysis study of the power system as shown in Figure 11.6. Note that a line flow can be positive or negative so that, as shown in Figure 11.6, we must check f against $-f_\ell^{max}$ as well as f_ℓ^{max}. This figure makes several assumptions, first, it assumes that the generator output for each of the generators in the system is available and that the line flow for each transmission line in the network is also available. Second, it assumes that the sensitivity factors have been calculated and stored, and that they are correct. The assumption that the generation and line flow MWs are available can be satisfied with telemetry systems or with state estimation techniques. The assumption that the sensitivity factors are correct is valid as long as the transmission network has not undergone any significant switching operations that would change its structure. For this reason, control systems that use sensitivity factors must have provision for updating the factors when the network is switched. A third assumption is that all generation pickup will be made on the reference bus. If this is not the case, substitute Eq. 11.5 in the generator outage loop.

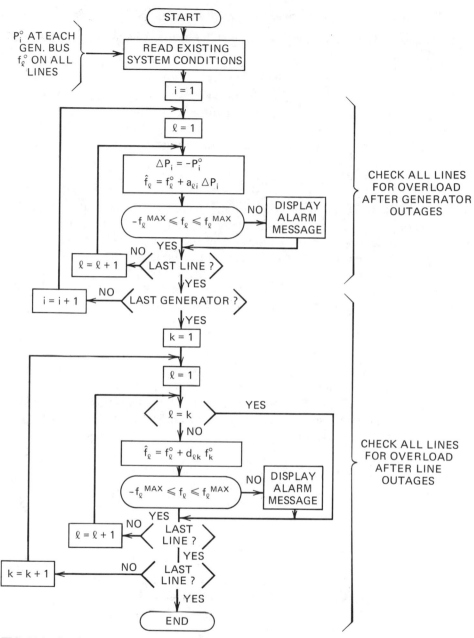

FIG. 11.6 Contingency analysis using sensitivity factors.

11.3.2 Calculation of Network Sensitivity Factors

First, we show how to derive the generation-shift sensitivity factors. To start, repeat Eq. 4.25.

$$\boldsymbol{\theta} = [X]\mathbf{P} \tag{11.8}$$

This is the standard matrix calculation for the DC load flow. Since the DC load-flow model is a linear model, we may calculate perturbations about a given set of system conditions by use of the same model. Thus, if we are interested in the changes in bus phase angles, $\Delta\boldsymbol{\theta}$, for a given set of changes in the bus power injections, $\Delta\mathbf{P}$, we can use the following calculation.

$$\Delta\boldsymbol{\theta} = [X]\Delta\mathbf{P} \tag{11.9}$$

In Eq. 11.8 it is assumed that the power on the swing bus is equal to the sum of the injections of all the other buses. Similarly, the net perturbation of the swing bus in Eq. 11.9 is the sum of the perturbations on all the other buses.

Suppose that we are interested in calculating the generation shift sensitivity factors for the generator on bus i. To do this, we will set the perturbation on bus i to $+1$ and the perturbation on all the other buses to zero. We can then solve for the change in bus phase angles using the matrix calculation in Eq. 11.10.

$$\Delta\boldsymbol{\theta} = [X]\begin{bmatrix} +1 \\ -1 \end{bmatrix}\begin{matrix} \text{-row } i \\ \text{-ref row} \end{matrix} \tag{11.10}$$

The vector of bus power injection perturbations in Eq. 11.10 represents the situation when a 1 pu power increase is made at bus i and is compensated by a 1 pu decrease in power at the reference bus. The $\Delta\theta$'s in Eq. 11.10 are thus equal to the derivative of the bus angles with respect to a change in power injection at bus i. Then the required sensitivity factors are

$$a_{\ell i} = \frac{df_\ell}{dP_i} = \frac{d}{dP_i}\left(\frac{1}{x_\ell}(\theta_n - \theta_m)\right)$$

$$= \frac{1}{x_\ell}\left(\frac{d\theta_n}{dP_i} - \frac{d\theta_m}{dP_i}\right) = \frac{1}{x_\ell}(X_{ni} - X_{mi}) \tag{11.11}$$

where $X_{ni} = \dfrac{d\theta_n}{dP_i} = n^{\text{th}}$ element from the $\Delta\theta$ vector in Eq. 11.10

$X_{mi} = \dfrac{d\theta_m}{dP_i} = m^{\text{th}}$ element from the $\Delta\theta$ vector in Eq. 11.10

$x_\ell = $ line reactance for line ℓ

A line outage may be modeled by adding two power injections to a system, one at each end of the line to be dropped. The line is actually left in the system and the effects of its being dropped are modeled by injections. Suppose line k from bus n to bus m were opened by circuit breakers as shown in Figure 11.7. Note that when the circuit breakers are opened no current flows through them and the line is completely

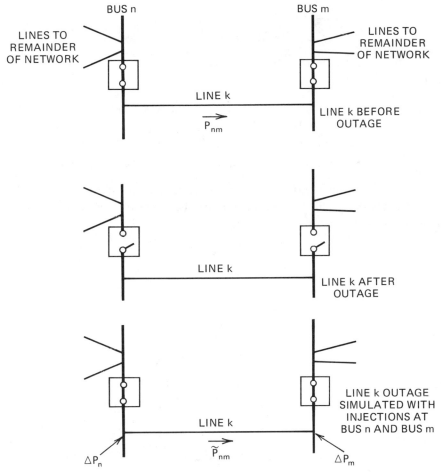

FIG. 11.7 Line outage modeling using injections.

isolated from the remainder of the network. In the bottom part of Figure 11.7 the breakers are still closed but injections ΔP_n and ΔP_m have been added to bus n and bus m, respectively. If $\Delta P_n = \tilde{P}_{nm}$, where $\tilde{P}_{nm}$ is equal to the power flowing over the line, and $\Delta P_m = -\tilde{P}_{nm}$, we will still have no current flowing through the circuit breakers even though they are closed. As far as the remainder of the network is concerned, the line is disconnected.

Using Eq. 11-9 relating to $\Delta\boldsymbol{\theta}$ and $\Delta\mathbf{P}$, we have

$$\Delta\boldsymbol{\theta} = [X]\Delta\mathbf{P}$$

where

$$\Delta\mathbf{P} = \begin{bmatrix} \vdots \\ \Delta P_n \\ \vdots \\ \Delta P_m \end{bmatrix}$$

so that

$$\Delta\theta_n = X_{nn}\Delta P_n + X_{nm}\Delta P_m$$
$$\Delta\theta_m = X_{mn}\Delta P_n + X_{mm}\Delta P_m$$

(11.14)

define

$\theta_n, \theta_m, P_{nm}$ to exist before the outage, where P_{nm} is the flow on line k from bus n to bus m.

$\Delta\theta_n, \Delta\theta_m, \Delta P_{nm}$ to be the incremental changes resulting from the outage.

$\tilde{\theta}_n, \tilde{\theta}_m, \tilde{P}_{nm}$ to exist after the outage.

The outage modeling criteria requires that the incremental injections ΔP_n and ΔP_m equal the power flowing over the outaged line **after** the injections are imposed. Then, if we let the line reactance be x_k

$$\tilde{P}_{nm} = \Delta P_n = -\Delta P_m$$

(11.15)

where

$$\tilde{P}_{nm} = \frac{1}{x_k}(\tilde{\theta}_n - \tilde{\theta}_m)$$

then

$$\Delta\theta_n = (X_{nn} - X_{nm})\Delta P_n$$
$$\Delta\theta_m = (X_{mn} - X_{mm})\Delta P_m$$

(11.16)

and

$$\tilde{\theta}_n = \theta_n + \Delta\theta_n$$
$$\tilde{\theta}_m = \theta_m + \Delta\theta_m$$

(11.17)

giving

$$\tilde{P}_{nm} = \frac{1}{x_k}(\tilde{\theta}_n - \tilde{\theta}_m) = \frac{1}{x_k}(\theta_n - \theta_m) + \frac{1}{x_k}(\Delta\theta_n - \Delta\theta_m)$$

or

(11.18)

$$\tilde{P}_{nm} = P_{nm} + \frac{1}{x_k}(X_{nn} + X_{mm} - 2X_{nm})\Delta P_n$$

then (using the fact that $\tilde{P}_{nm}$ is set to ΔP_n)

$$\Delta P_n = \left[\frac{1}{1 - \frac{1}{x_k}(X_{nn} + X_{mm} - 2X_{nm})}\right]P_{nm}$$

(11.19)

Define a sensitivity factor δ as the ratio of the change in phase angle θ, anywhere in the system, to the original power P_{nm} flowing over a line nm before it was dropped.

That is,

$$\delta_{i,nm} = \frac{\Delta\theta_i}{P_{nm}} \tag{11.20}$$

If neither n or m is the system reference bus, two injections ΔP_n and ΔP_m, are imposed at buses n and m, respectively. This gives a change in phase angle at bus i equal to

$$\Delta\theta_i = X_{in}\Delta P_n + X_{im}\Delta P_m \tag{11.21}$$

Then using the relationship between ΔP_n and ΔP_m, the resulting δ factor is

$$\delta_{i,\,nm} = \frac{(X_{in} - X_{im})x_k}{x_k - (X_{nn} + X_{mm} - 2X_{nm})} \tag{11.22}$$

If either n or m is the reference bus, only one injection is made. The resulting δ factors are

$$\delta_{i,nm} = \frac{X_{in}x_k}{(X_{nm} - X_{nn})} \qquad \text{for } m = \text{ref}$$

$$= \frac{-X_{im}x_k}{(X_{nm} - X_{mm})} \qquad \text{for } n = \text{ref} \tag{11.23}$$

If bus i itself is the reference bus, then $\delta_{i,nm} = 0$ since the reference bus angle is constant.

The expression for $d_{\ell,k}$ is

$$d_{\ell,k} = \frac{\Delta f_\ell}{f_k^0} = \frac{\dfrac{1}{x_\ell}(\Delta\theta_i - \Delta\theta_j)}{f_k^0}$$

$$= \frac{1}{x_\ell}\left(\frac{\Delta\theta_i}{P_{nm}} - \frac{\Delta\theta_j}{P_{nm}}\right)$$

$$= \frac{1}{x_\ell}(\delta_{i,nm} - \delta_{j,nm}) \tag{11.24}$$

if neither i nor j is a reference bus

$$d_{\ell,k} = \frac{1}{x_\ell}\left(\frac{(X_{in} - X_{im})x_k - (X_{jn} - X_{jm})x_k}{x_k - (X_{nn} + X_{mm} - 2X_{nm})}\right)$$

$$= \frac{\dfrac{x_k}{x_\ell}(X_{in} - X_{jn} - X_{im} + X_{jm})}{x_k - (X_{nn} + X_{mm} - 2X_{nm})} \tag{11.25}$$

EXAMPLE 11A

The $[X]$ matrix for our six-bus sample network is shown in Figure 11.8, together with the generation shift distribution factors and the line outage distribution factors.

Generation Shift Factors For Six-Bus Sample System

	Bus 1	Bus 2	Bus 3
$\ell = 1$ (line 1-2)	0	−0.47	−0.40
$\ell = 2$ (line 1-4)	0	−0.31	−0.29
$\ell = 3$ (line 1-5)	0	−0.21	−0.30
$\ell = 4$ (line 2-3)	0	0.05	−0.34
$\ell = 5$ (line 2-4)	0	0.31	0.22
$\ell = 6$ (line 2-5)	0	0.10	−0.03
$\ell = 7$ (line 2-6)	0	0.06	−0.24
$\ell = 8$ (line 3-5)	0	0.06	0.29
$\ell = 9$ (line 3-6)	0	−0.01	0.37
$\ell = 10$ (line 4-5)	0	0	−0.08
$\ell = 11$ (line 5-6)	0	−0.06	−0.13

X Matrix for Six-Bus Sample System (Reference at Bus 1)

$$
\begin{bmatrix}
0 & 0 & 0 & 0 & 0 & 0 \\
0 & 0.09412 & 0.08051 & 0.06298 & 0.06435 & 0.08129 \\
0 & 0.08051 & 0.16590 & 0.05897 & 0.09077 & 0.12895 \\
0 & 0.06298 & 0.05897 & 0.10088 & 0.05422 & 0.05920 \\
0 & 0.06435 & 0.09077 & 0.05422 & 0.12215 & 0.08927 \\
0 & 0.08129 & 0.12895 & 0.05920 & 0.08927 & 0.16328
\end{bmatrix}
$$

Line Outage Distribution Factors for Six-Bus Sample System

	$k=1$ (Line 1-2)	$k=2$ (Line 1-4)	$k=3$ (Line 1-5)	$k=4$ (Line 2-3)	$k=5$ (Line 2-4)	$k=6$ (Line 2-5)	$k=7$ (Line 2-6)	$k=8$ (Line 3-5)	$k=9$ (Line 3-6)	$k=10$ (Line 4-5)	$k=11$ (Line 5-6)
$\ell = 1$ (line 1-2)		0.64	0.54	−0.11	−0.50	−0.21	−0.12	−0.14	0.01	0.01	0.13
$\ell = 2$ (line 1-4)	0.59		0.46	−0.03	0.61	−0.06	−0.04	−0.04	0	−0.33	0.04
$\ell = 3$ (line 1-5)	0.41	0.36		0.15	−0.11	0.27	0.16	0.18	−0.02	0.32	−0.17
$\ell = 4$ (line 2-3)	−0.10	−0.03	0.18		0.12	0.23	0.47	−0.40	−0.53	0.17	0.13
$\ell = 5$ (line 2-4)	−0.59	0.76	−0.17	0.16		0.30	0.17	0.19	−0.02	−0.67	−0.19
$\ell = 6$ (line 2-5)	−0.19	−0.06	0.33	0.22	0.23		0.24	0.27	−0.03	0.31	−0.26
$\ell = 7$ (line 2-6)	−0.12	−0.04	0.21	0.51	0.15	0.27		−0.20	0.58	0.20	0.44
$\ell = 8$ (line 3-5)	−0.12	−0.04	0.20	−0.38	0.14	0.27	−0.17		0.47	0.19	−0.42
$\ell = 9$ (line 3-6)	0.01	0	−0.03	−0.62	−0.02	−0.03	0.64	0.60		−0.02	0.56
$\ell = 10$ (line 4-5)	0.01	−0.24	0.29	0.13	−0.39	0.24	0.14	0.15	−0.02		−0.15
$\ell = 11$ (line 5-6)	0.11	0.03	−0.18	0.12	−0.13	−0.23	0.36	−0.40	0.42	−0.18	

FIG. 11.8 Outage factors for six-bus system.

The generation shift distribution factors that give the fraction of generation shift that is picked up on a transmission line are designated $a_{\ell i}$. The a factor is obtained by finding line ℓ along the rows and then finding the generator to be shifted along the columns. For instance, the shift factor for a change in the flow on line 1-4 when making a shift in generation on bus 3 is found in the second row, third column.

The line outage distribution factors are stored such that each row and column corresponds to one line in the network. The distribution factor $d_{\ell,k}$ is obtained by finding line ℓ along the rows and then finding line k along that row in the appropriate column. For instance, the line outage distribution factor that gives the fraction of flow picked up on line 3-5 for an outage on line 3-6 is found in the eighth row and ninth column. Figure 11.3 shows an outage of the generator on bus 3 with all pickup of lost generation coming on the generator at bus 1. To calculate the flow on line 1-4 after the outage of the generator on bus 3, we need:

$$\text{Base-case flow on line 1-4} = 43.6 \text{ MW}$$

$$\text{Base-case generation on bus 3} = 60 \text{ MW}$$

$$\text{Generation shift distribution factor} = a_{1\text{-}4,\,3} = -0.29$$

Then the flow on line 1-4 after generator outage is $= \text{base-case Flow}_{1\text{-}4} +$ $a_{1\text{-}4,\,3}\,\Delta P_{\text{gen}_3} = 43.6 + (-0.29)(-60 \text{ MW}) = 61 \text{ MW}.$

To show how the line outage and generation shift factors are used, calculate some flows for the outages shown in Figures 11.2 and 11.3. Figure 11.2 shows an outage of line 3-5. If we wish to calculate the power flowing on line 3-6 with line 3-5 opened, we would need the following.

$$\text{Base-case flow on line 3-5} = 19.1 \text{ MW}$$

$$\text{Base-case flow on line 3-6} = 43.8 \text{ MW}$$

$$\text{Line outage distribution factor: } d_{3\text{-}6,3\text{-}5} = 0.60$$

Then the flow on 3-6 after the outage is $= \text{base flow}_{3\text{-}6} + d_{3\text{-}6,\,3\text{-}5} \cdot \text{base flow}_{3\text{-}5} = 43.8 + (0.60) \times (19.1) = 55.26.$

In both outage cases, the flows calculated by the sensitivity methods are reasonably close to the values calculated by the full AC load flows as shown in Figures 11.2 and 11.3.

11.3.3 AC Load Flow Methods

The calculations made by network sensitivity methods are faster than those made by AC load flow methods and therefore find wide use in operations control systems. However, there are many power systems where voltage magnitudes are the critical factor in assessing contingencies. In addition, there are some systems where VAR flows predominate on some circuits, such as underground cables, and an analysis of only the MW flows will not be adequate to indicate overloads. When such situations present themselves, the network sensitivity methods may not be adequate and the operations control system will have to incorporate a full AC load flow for contingency analysis.

When an AC load flow is to be used to study each contingency case, the speed of solution and the number of cases to be studied are critical. To repeat what was said before, if the contingency alarms come too late for operators to act, they are worthless. Most operations control centers that use an AC load flow program for contingency analysis use neither a Newton-Raphson or a method called the *decoupled load flow*. These solution algorithms are used because of their speed of solution and the fact that they are reasonably reliable in convergence when solving difficult cases. The decoupled load flow has the further advantage that a matrix alteration formula can be incorporated into it to simulate the outage of transmission lines without reinverting the system Jacobian matrix at each iteration.

Other techniques that are beginning to be implemented use schemes to limit the number of contingency cases to be run. These "contingency selection" schemes use complex mathematical models of the transmission system to assess the relative severity of one contingency versus another. By ranking the contingencies in relative severity, the algorithms indicate which cases to try first. If after calculating the contingency overloads for several cases it is found that no serious overloads are being encountered, the contingencies that have as yet not been studied can be forgotten since they are ranked lower in severity and will most likely therefore not cause overloads.

11.4 CORRECTING THE GENERATION DISPATCH

Suppose the operators of a power system perform a contingency analysis and discover that if the system experiences a certain transmission line outage there will be some serious overloads on the remaining network. What should the operators do about this situation?

Before answering the question, "What *should* the operators do?" it would be well for us to consider the question, "What *can* the operators do?" Obviously, one option would be simply to do nothing and hope the contingency would never happen. The other options require the operators to take action to prevent the problem should the contingency occur. Note that if the contingency occurs and there is a serious overload the operators then *must* take action or risk damage to equipment or perhaps even worse, a cascading outage resulting in a blackout of the entire system.

The actions that can be taken to relieve power flow overloads are

- Shifting generation to redistribute power flows.
- Adjusting interchanges with neighboring systems.
- Adjusting phase-shift transformers to force power to flow on alternate paths.
- Switching the transmission network.
- Shedding load at selected locations.

Actions that can be taken in the event of voltage problems are

- Adjustment of generator exciters to change the voltage at the generator buses.
- Adjustment of autotransformer taps.
- Switching of reactive sources such as reactors or capacitors.
- Switching the transmission network.

11.4.1 Correcting the Generation Dispatch by Sensitivity Methods

If one has overloads on a transmission network, the following scheme will allow correction of the overloads via generation shifts.

1. For each overloaded line ℓ and each generator i, determine whether to raise or lower generation on bus i from the direction of flow on line ℓ and from the sign of $a_{\ell i}$.
2. From step 1, separate the generators into a "raise" list and a "lower" list.
3. Remove generators that appear on both lists.
4. For each generator on each list, calculate the maximum correction each generator can make on the worst-overloaded line. The maximum correction is the product of the generation shift sensitivity coefficient for the most-overloaded lincand the generator's maximum change. The maximum change is calculated as

   For generators on the "raise" list:

   $$\text{max change} = P_i^{\max} - P_i^0$$

   For generators on the "lower" list:

   $$\text{max change} = P_i^{\min} - P_i^0$$

5. Start with the generator having the greatest correction on the overloaded line and adjust it as much as possible until the overload is eliminated or the generator hits its limit. If the overload cannot be eliminated with the first generator tried, go to the next generator, and so on.

EXAMPLE 11B

Suppose we wished to correct the flow on the line from bus 3 to bus 6 in the base case six-bus system shown in Figure 11.1. The base case flow is 43.8 MW. We will correct it to 40 MW using the procedure outlined in this section. Our desired change in flow is

$$43.8 \text{ MW} + \Delta f_{3\text{-}6} = 40 \text{ MW}$$

so

$$\Delta f_{3\text{-}6} = -3.8 \text{ MW}$$

From Figure 11.8 we find

$$a_{3-6,2} = -0.01$$

$$a_{3-6,3} = 0.37$$

Therefore, if we wish to correct the 3-6 flow downward by 3.8 MW, we must place the generator on bus 2 on the "raise" list and the generator on bus 3 on the "lower" list:

Raise list	Lower list
Gen on bus 2	Gen on bus 3

Next we calculate the maximum change for each generator: the generator min and max generations are taken from Example 4E.

$$50.0 \text{ MW} \leq P_1 \leq 200 \text{ MW}$$

$$37.5 \text{ MW} \leq P_2 \leq 150 \text{ MW}$$

$$45 \text{ MW} \leq P_3 \leq 180 \text{ MW}$$

Then the maximum change in the raise direction for the generator on bus 2 starting at 50 MW (see Figure 11.1) is

$$\text{Max } \Delta P_{g2} \text{ (raise)} = 150 - 50 = +100 \text{ MW}$$

The maximum change in the lower direction for the generator on bus 3 starting at 60 MW is

$$\text{Max } \Delta P_{g3} \text{ (lower)} = 45 - 60 = -15 \text{ MW}$$

Then the maximum corrections to the power flowing on line 3-6 are

$$\text{Max correction for } P_{g2} = +100 \text{ MW} \times (-0.01) = -1 \text{ MW}$$

$$\text{Max correction for } P_{g3} = -15 \text{ MW} \times (0.37) = -5.55 \text{ MW}$$

We will therefore choose to lower generation at bus 3. The amount it should be lowered is

$$\Delta P_{g3} = \frac{\Delta f_{3-6}}{a_{3-6,3}} = \frac{-3.8}{0.37} = -10.27 \text{ MW}$$

11.4.2 Compensated Factors

The fact that the a and d factors are linear models of a power system allow one to use superposition. For example, suppose we wished to know the generation shift sensitivity factor between line ℓ and generator bus i when line k was out. This can be calculated by first assuming that a change in generation on bus i, ΔP_i, has a direct effect on line ℓ and an indirect effect through its influence on the power flowing on

---

line k that then influences line ℓ when line k is out. Then,

$$\Delta f_\ell = a_{\ell i} \Delta P_i + d_{\ell,k} \Delta f_k \qquad (11.26)$$

but

$$\Delta f_k = a_{ki} \Delta P_i$$

so

$$\Delta f_\ell = a_{\ell i} \Delta P_i + d_{\ell,k} a_{ki} \Delta P_i$$
$$= \underbrace{(a_{\ell i} + d_{\ell,k} a_{ki})}_{} \Delta P_i \qquad (11.27)$$
Compensated generation shift sensitivity

The term in brackets in Eq. 11.27 is the new or compensated generation shift sensitivity factor. By using compensated factors, the same generation correction technique shown earlier in this section and in Example 11B can be applied to contingency overloads. That is, if a contingency analysis (as shown in Figure 11.5) indicates that a certain outage will cause overloads, the same corrections can be calculated using compensated generation shift sensitivities as was run with the uncompensated sensitivities. The result of such corrections will be to shift generation in such a way that should the line outage that was studied actually occur, the resulting flows will already have been properly corrected to relieve the overload. It would be proper then to call such corrections *preventive corrections*.

11.4.3 Correcting the Generation Dispatch Using Linear Programming

Correcting the generation dispatch for overloads using sensitivity methods can be quite difficult when the result of correcting for one overload causes another transmission line to overload. Furthermore, if several overloads are present, finding a generation correction that relieves all the overloads and does not create any new ones is almost impossible unless a well-organized and systematic approach is used. Such an approach is readily available using linear programming. Not only can LP be used to relieve the overloaded transmission lines, but it can be set up to so with a minimum amount of generation shifting.

We will start our development of this technique by developing a linear transmission line flow function. We may express the flow on any transmission line ℓ as a Taylor series expansion about an initial flow f_ℓ^0 where the terms in our Taylor series expansion will be the perturbation in generation, ΔP, at each generator bus with only the linear terms retained. Then

$$f_\ell = f_\ell^0 + \sum_{i=1}^{N} \frac{df_\ell}{dP_i} \Delta P_i \qquad (11.28)$$

but we already know that

$$\frac{df_\ell}{dP_i} = a_{\ell i}$$

Then

$$f_\ell = f_\ell^0 + \sum_{i=1}^{N} a_{\ell i} \Delta P_i \tag{11.29}$$

Using Eq. 11.29 we can write an inequality constraint equation that expresses the fact that the flow on line ℓ is to be limited to some maximum flow, $f_\ell^{\max}$

$$-f_\ell^{\max} \leq f_\ell \leq f_\ell^{\max}$$

or

$$-f_\ell^{\max} \leq f_\ell^0 + \sum_{i=1}^{N} a_{\ell i} \Delta P_i \leq f_\ell^{\max}$$

This two-sided inequality can be expressed as two one-sided inequalities.

$$\sum_{i=1}^{N} a_{\ell i} \Delta P_i \leq f_\ell^{\max} - f_\ell^0 \tag{11.30a}$$

and

$$\sum_{i=1}^{N} a_{\ell i} \Delta P_i \geq -f_\ell^{\max} - f_\ell^0 \tag{11.30b}$$

Equations 11-30a and 11-30b are exactly in the form we need to incorporate line overloads into an LP. Note that often the user is aware that either the line flow is positive and is exceeding $f_\ell^{\max}$ or it is negative and exceeding $-f_\ell^{\max}$ and that a correction by the LP is not computed to swing the line from $> f_\ell^{\max}$ to $< -f_\ell^{\max}$ or vice versa. Therefore, often only Eq. 11.30a or Eq. 11.30b will be used, depending on which limit is of interest.

We can state our LP corrective dispatching algorithm as follows.

Minimize: The sum of the generation shifts (11.31)

Subject to: **1.** The generation shifts model the system behavior by maintaining a correct total generation versus total load balance.

2. All known overloads are within limit.

Here we will assume that the generators are already at an economic dispatch. Since we wish to correct for transmission overloads, we will try to do so with a minimum deviation to the economic dispatch schedule. Therefore an objective that minimizes the sum of the generation shifts will often suffice for such a calculation. Note, however, that by representing the generator input/output functions as piecewise linear functions, we could solve the economic dispatch itself with an LP. If we include line overload constraints in such an LP, we would have a "constrained economic dispatch" that gave the best economic dispatch while meeting line overload constraints. For illustration in this chapter we will just assume that the "minimum sum of generation shifts" objective is satisfactory. Our LP variables will be the P_i values for each generator bus. Note that LP algorithms require that all variables be positive but that we do not wish to restrict the corrective generation shifts to be only positive. Therefore we will use the following simple identity.

$$\Delta P_i = \Delta P_i^+ - \Delta P_i^- \tag{11.32}$$

where: $\Delta P_i^+ =$ the net upward shift on bus i

$\Delta P_i^- =$ the net downward shift on bus i

The objective function in Eq. 11.31 for our LP can be expressed as

Minimize:
$$\sum_{i=1}^{N} \left(K \Delta P_i^+ + K \Delta P_i^- \right)$$

Where the value of K can be chosen as any large number. In this case, all generation shifts, either up or down, will be penalized equally.

The first constraint in our LP is placed there to guarantee that the resulting answer models the power system correctly. To see what this means, we must review what we wish to have happening while we are correcting the generation dispatch.

1. The generators are to remain within their limits.

2. The sum of the generator outputs must equal the load (we will ignore losses).

Point (1) requires that all generators be part of the LP algorithm—including the reference bus. Point (2) requires that generation equal load at all times. If we express P_i as

$$P_i = P_i^0 + \Delta P_i = P_i^0 + \Delta P_i^+ - \Delta P_i^- \tag{11.33}$$

and

$$\sum_{i=1}^{N} P_i = P_{\text{load}} \tag{11.34}$$

Equation 11.34 must hold for the precorrection conditions as well as after the corrections, then

$$\sum_{i=1}^{N} P_i^0 = P_{\text{load}} \tag{11.35}$$

Expanding Eq. 11.34 and substituting Eq. 11.35, we get

$$\sum_{i=1}^{N} P_i = \sum_{i=1}^{N} (P_i^0 + \Delta P_i^+ - \Delta P_i^-) = P_{\text{load}}$$

$$= \sum_{i=1}^{N} P_i^0 + \sum_{i=1}^{N} (\Delta P_i^+ - \Delta P_i^-) = P_{\text{load}}$$

or simply

$$\sum_{i=1}^{N} (\Delta P_i^+ - \Delta P_i^-) = 0 \tag{11.36}$$

Equation 11.36 expresses the fact that the sum of the generation shifts over all generators must be zero if the load is to be met.

To solve the LP, we also need to include the limits on the generators. For each generator,

$$P_i^{\min} \leq P_i \leq P_i^{\max} \qquad (11.37)$$

However, we are not dealing with P_i, but rather with ΔP_i^+ and ΔP_i^- as shown in Eqs. 11.32 and 11.33. If the generator is initially generating P_i^0 (as in Eq. 11.33), then we can define limits on ΔP_i^+ and ΔP_i^- as follows.

$$0 \leq \Delta P_i^+ \leq P_i^{\max} - P_i^0$$

and $\qquad\qquad\qquad\qquad\qquad\qquad\qquad\qquad\qquad\qquad\qquad\qquad\qquad$ (11.38)

$$0 \leq \Delta P_i^- \leq P_i^0 - P_i^{\min}$$

Now we can write our final expressions for the LP.

Minimize: $\displaystyle\sum_{i=1}^{N} (K \, \Delta P_i^+ + K \, \Delta P_i^-)$ (Objective)

Subject to: $\displaystyle\sum_{i=1}^{N} (\Delta P_i^+ - \Delta P_i^-) = 0$ (Gen shift equality constraint)

And: $\left.\begin{array}{l} \displaystyle\sum_{i=1}^{N} a_{\ell i}(\Delta P_i^+ - \Delta P_i^-) \\[4pt] \leq f_\ell^{\max} - f_\ell^0 \\[8pt] \displaystyle\sum_{i=1}^{N} a_{\ell i}(\Delta P_i^+ - P_i^-) \\[4pt] \geq -f_\ell^{\max} - f_i^0 \end{array}\right\}$ (Line flow constraint)

And: $\left.\begin{array}{l} 0 \leq \Delta P_i^+ \leq P_i^{\max} - P_i^0 \\ 0 \leq \Delta P_i^- \leq P_i^0 - P_i^{\min} \end{array}\right\}$ (Gen shift limits) (11.39)

$$\text{For } i = 1 \cdots N$$

In solving for the corrections using the LP algorithm, we wish to end with all line flows within limit. This can be accomplished by placing a line flow constraint (see Eq. 11.39) into the LP for each overloaded line. After solving the LP, the generation shifts should be made and a new base case load flow run. Some of the flows for the overloaded lines will have been adjusted until they exactly equal their respective limits while others may be below limit. However, other line flows that were well within limit before the correction was made may now be beyond limit. The safest procedure at this point is to re-execute the LP with all the line flow constraints that were in the first LP included as well as line flow constraints corresponding to the "new" overloads. This procedure can be repeated as often as needed and is termed an *iterative constraint search* since the user is essentially searching for a set of constraints that, when included in the LP, will result in a set of corrections that leave the transmission system network with no overloads.

EXAMPLE 11C

Given the DC load flow solution to the six-bus sample system (see Example 4C) as a starting point, use linear programming to achieve the following corrections to the transmission line flows.

Line	Flow before correction (as calculated by the DC load flow)	Flow limit
1-4	41.6	36.0 MW
3-6	44.9	40.0 MW

The limits on generation are the same as shown in Example 11B, and the initial generation values will be those from Example 4C, that is,

Generator unit	Minimum output (MW)	Initial output (MW)	Maximum output (MW)
1	50	100	200
2	37.5	50	150
3	45	60	180

The sensitivity factors are taken from Figure 11.8.
Using the LP expression in Eq. 11.39 we have

Minimize: $\quad 100\,\Delta P_1^+ + 100\,\Delta P_1^- + 100\,\Delta P_2^+ + 100\,\Delta P_2^- + 100\,\Delta P_3^+$
$$+ 100\,\Delta P_3^-$$

Subject to: $\qquad \Delta P_1^+ - \quad \Delta P_1^- + \quad \Delta P_2^+ - \quad \Delta P_2^- + \quad \Delta P_3^+$
$$-\quad \Delta P_3^- = 0$$

and: (line 1-4) $\qquad\qquad -0.31\,\Delta P_2^+ + 0.31\,\Delta P_2^- - 0.29\,\Delta P_3^+$
$$+ 0.29\,\Delta P_3^- \le 36{-}41.6$$

(line 3-6) $\qquad\qquad -0.01\,\Delta P_2^+ + 0.01\,\Delta P_2^- + 0.37\,\Delta P_3^+$
$$-0.37\,\Delta P_3^- \le 40{-}44.9$$

With variable limits

$$0 \le \Delta P_1^+ \le 100$$
$$0 \le \Delta P_1^- \le 50$$
$$0 \le \Delta P_2^+ \le 100$$
$$0 \le \Delta P_2^- \le 12.5$$
$$0 \le \Delta P_3^+ \le 120$$
$$0 \le \Delta P_3^- \le 15$$

Note that here we are only using the "upper limit" inequality constraints (see Eq. 11.30a) for both line flow corrections.

We can arrange our solution as follows: Using the upper-bound LP algorithm described in the appendix of Chapter 6.

Variable	ΔP_1^+	ΔP_1^-	ΔP_2^+	ΔP_2^-	ΔP_3^+	ΔP_3^-	Slack 1	Slack 2	Z	RHS
Constraint 1:	1	−1	1	−1	1	−1				= 0
Constraint 2:			−0.31	0.31	−0.29	0.29	1			= −5.6
Constraint 3:			−0.01	0.01	0.37	−0.37		1		= −4.9
Objective ftn:	100	100	100	100	100	100			−1	= 0
Variable min:	0	0	0	0	0	0	0	0		
Present value:	0	0	0	0	0	0	−4.5	−4.9	0	
Variable max:	100	50	100	12.5	120	15	∞	∞		

NOTE: The initial basis variables are: ΔP_1^+, Slack 1 and Slack 2

The solution to this LP tableau is

$$\Delta P_1^+ = 0$$

$$\Delta P_1^- = 17.26$$

$$\Delta P_2^+ = 29.70$$

$$\Delta P_2^- = 0$$

$$\Delta P_3^+ = 0$$

$$\Delta P_3^- = 12.44$$

That is,

Generator	Initial generation output (MW)	Shift in generator output (MW)	Final generator output (MW)
1	100	−17.26	92.74
2	50	+29.70	79.70
3	60	−12.44	47.56
	Total 210		Total 210.00

We can check that these shifts do indeed result in the desired corrections using Eq. 11.129.

For line 1-4: $f_{1-4}^{new} = f_{1-4}^0 + \sum_{i=1}^{3} a_{1-4,i} \Delta P_i$

$$= 41.6 + 0(\Delta P_1) - 0.31(\Delta P_2) - 0.29(\Delta P_3)$$
$$= 41.6 - 0.31(29.7) - 0.29(-12.44)$$
$$= 36.0 \text{ MW}$$

For line 3-6: $f_{3-6}^{new} = f_{3-6}^0 + \sum_{i=1}^{3} a_{3-6,i} \Delta P_i$

$$= 44.9 + 0(\Delta P_1) - 0.01(\Delta P_2) + 0.37(\Delta P_3)$$
$$= 44.9 - 0.01(29.7) + 0.37(-12.44)$$
$$= 40.0 \text{ MW}$$

Figure 11.9 shows an LP redispatch algorithm used with a contingency analysis program to calculate a generation redispatch and to calculate a preventive

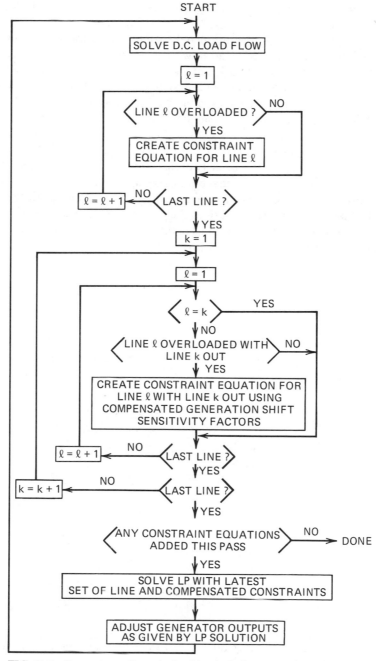

FIG. 11.9 Generation redispatch algorithm including preventive redispatch.

generation redispatch to protect the system from overloads from the next contingency outage.

The linear programming generation redispatch calculation can be used with either an AC load flow or a DC load flow. In Example 11C, the LP was used to correct the power flows for a network that had been solved initially using a DC load flow. Since the LP and the DC load flow are both linear models, the LP's corrections to the generation dispatch gave the exact flow corrections desired. If we had taken the initial flows, f_ℓ^0, from an AC load flow, solved the LP for the generation corrections and then resolved the AC load flow with the generation corrections included, the flows would most likely not be at the values we desired. This is because the AC load flow models the nonlinearity of the power system. The corrected power flows will be close to the limit desired—but not precisely at the limit as when using a DC load flow. To force the power flow to its limit, the new solution to the AC load flow can be taken as the initial conditions and the LP resolved with new values for f_ℓ^0. Usually, within two or three iterations of the AC load flow with the LP the flows can be moved exactly to the desired MW limit.

PROBLEMS

11.1 Figure 11.10 shows a four-bus power system. Also given below are the impedance data for the transmission lines of the system as well as the generation and load values.

Line	Line reactance (pu)
1-2	0.2
1-4	0.25
2-3	0.15
2-4	0.30
3-4	0.40

Bus	Load	Generation
1		150 MW
2		350 MW
3	220 MW	
4	280 MW	

a. Calculate the generation shift sensitivity coefficients for a shift in generation from bus 1 to bus 2.

b. Calculate the line outage sensitivity factors for outages on lines 1-2, 1-4, and 2-3.

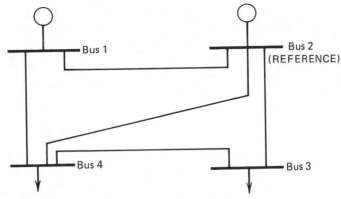

FIG. 11.10   Four-bus network for Problem 11.1.

11.2 In the system shown in Figure 11.11, three generators are serving a load of 1300 MW. The MW flow distribution, bus loads, and generator outputs are as shown. The generators have the following characteristics.

Gen number	P_{min} (MW)	P_{max} (MW)
1	100	600
2	90	400
3	100	500

The circuits have the following limits.

CKT A   600 MW MAX

CKT B   600 MW MAX

CKT C   450 MW MAX

CKT D   350 MW MAX

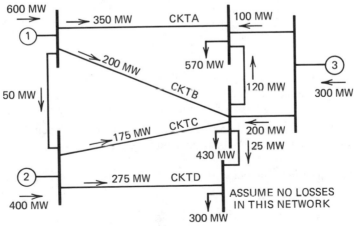

FIG. 11.11   Three-generator system for Problem 11.2.

Throughout this problem we will only be concerned with flows on the circuit labeled A, B, C, and D. The generation shift sensitivity coefficients, $a_{\ell i}$, for circuits, A, B, C, and D are

CKT	Shift on gen 1	Shift on gen 2
A	0.7	0.08
B	0.2	0.02
C	0.06	0.54
D	0.04	0.36

Example:
$$\Delta P_{\text{flow}_\ell} = a_{\ell,i} \cdot \Delta P_i$$

if
$$\ell = C \text{ and } i = 2$$
$$\Delta P_{\text{flow}_c} = (0.54)\Delta P_2$$

Assume a shift on gen 1 or gen 2 will be compensated by an equal (opposite) shift on gen 3. The line outage sensitivity factors $d_{\ell,k}$ are

$$\xrightarrow{\hspace{4cm}} k$$

		A	B	C	D
	A	X	0.8	0.21	0.14
ℓ	B	0.9	X	0.06	0.04
	C	0.06	0.12	X	0.82
	D	0.04	0.08	0.73	X

As an example, suppose the loss of circuit k will increase the loading on circuit ℓ as follows.

$$P_{\text{flow}_\ell} = P_{\text{flow}_\ell} \text{ (before outage)} + d_{\ell,k} \cdot P_{\text{flow}_k} \text{ (before outage)}$$

if
$$\ell = A, \qquad k = B$$

The new flow on ℓ would be

$$P_{\text{flow}_A} = P_{\text{flow}_A} + (0.8)P_{\text{flow}_B}$$

a. Find the contingency (outage) flow distribution on circuits A, B, C, and D for an outage on circuit A. Repeat for an outage on B, then on C, then on D. (Only one circuit is lost at one time.) Are there any overloads?

b. Can you shift generation from gen 1 to gen 3 or from gen 2 to gen 3 so that no overloads occur? If so, how much shift?

11.3 Given the three-bus network shown in Figure 11.12 (see Example 4B),

where $x_{12} = 0.2$ pu
$x_{13} = 0.4$ pu
$x_{23} = 0.25$ pu

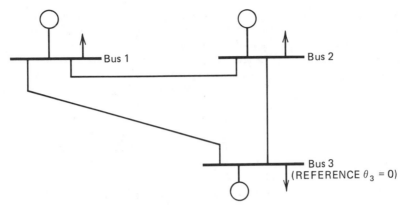

FIG. 11.12 Three-bus system for Problem 11.3.

The [X] matrix is

$$
\begin{bmatrix}
0.2118 & 0.1177 & 0 \\
0.1177 & 0.1765 & 0 \\
0 & 0 & 0
\end{bmatrix}
$$

Use a 100 MVA base. The base loads and generations are

Bus	Load (MW)	Gen (MW)	Gen min (MW)	Gen max (MW)
1	100	150	50	250
2	300	180	60	250
3	100	170	60	300

a. Find base power flows on the transmission lines.

b. Calculate the generation shift factors for line 1-2. Calculate the shift in generation on bus 1 and 2 so as to force the flow on line 1-2 to zero MW. Assume for economic reasons that any shift from base conditions are more expensive for shifts at the generator on bus 1 than for shifts on bus 2 and that the generator on bus 3 can be shifted without any penalty.

FURTHER READING

The subject of power system security and especially state estimation has received a great deal of attention in the engineering literature since the middle 1960s. The list of references presented here is therefore large but also quite limited nonetheless.

Reference 1 is a key paper on the topic of system security and energy control system philosophy. References 2 and 3 provide the basic theory for contingency assessment of power systems using impedance [Z] matrix methods. References 4 and 5 cover contingency analysis using DC load flow methods. Reference 6 is a broad overview of security assessment and contains an excellent bibliography covering the literature on security assessment up to 1975.

The use of AC load flows in contingency analysis is posssible with any AC load flow algorithm. However, the *fast-decoupled Newton* load flow algorithm is generally recognized as the best for this purpose since its Jacobian matrix is constant and single-line outages can be modeled using the matrix inversion lemma. Reference 7 covers the fast-decoupled load flow algorithm and its application.

Correcting the generation dispatch by sensitivity methods is covered by reference 8. The use of linear programming to solve power systems problems is covered in references 9–11. Reference 12 is an excellent overview of LP applications to power system rescheduling and contains an extensive bibliography on this subject.

The full "optimal power flow" solution to power system rescheduling differs from the methods covered in this chapter in that the optimization process is carried out on the AC model rather than on the DC model of the network. A great deal of research has gone into this topic and, as of the writing this text, no one algorithm seems to have clearly emerged as the "best." References 13–18 cover this topic.

1. DyLiacco, T. E., "The Adaptive Reliability Control System," *IEEE Transactions*, Vol. PAS-86, May 1967, pp. 517–531.

2. El-Abiad, A. H., Stagg, G. W., "Automatic Evaluation of Power System Performance—Effects of Line and Transformer Outages," *AIEE Transactions*, Vol. PAS-81, February 1963, pp. 712–716.

3. Brown, H. E., "Contingencies Evaluated by a Z Matrix Method," *IEEE Transactions*, Vol. PAS-88, April 1969, pp. 409–412.

4. Limmer, H., "Techniques and Applications of Security Calculations Applied to Dispatching Computers," Papers STY 4, 3rd Power System Computation Conference, Rome, Italy, 1969.

5. Baughman, M., Schweppe, F. C., "Contingency Evaluation: Real Power Flows from a Linear Model," IEEE Sumer Power Meeting, 1970, Paper CP 689-PWR.

6. Debs, A. S., and Benson, A. R., "Security Assessment of Power Systems," *Systems Engineering For Power: Status and Prospects*, U.S. Government Document CONF-750867, 1967, pp. 1–29.

7. Stott, B., Alsac, O., "Fast Decoupled Load Flow", *IEEE Transactions*, Vol. PAS-93, May/June 1974, pp. 859–869.

8. Ku, W. S., Van Olinda, P., "Security and Voltage Applications of the Public Service Dispatch Computer," PICA Conference Proceedings, 1969, pp. 201–207.

9. Thanikachalam, A., Tudor, J. R., "Optimal Rescheduling of Power for System Reliability," *IEEE Transactions*, Vol. PAS-90, July/August 1971, pp. 1776–1781.

10. Chan, S. M., Yip, E., "A Solution of the Transmission Limited Dispatch Problem by Sparse Linear Programming," *IEEE Transactions on Power Apparatus and Systems*, Vol. PAS-98, May/June 1979 pp. 1044–1053.

11. Wollenberg, B. F., Stadlin, W. O., "A Real Time Optimizer for Security Dispatch," *IEEE Transactions*, Vol. PAS-93, Sept/October 1974, pp. 1640–1644.

12. Stott, B., Marinho, J. L., Alsac, O., "Review of Linear Programming Applied to Power System Rescheduling," Proceedings 1979 PICA Conference, IEEE Document No. 79CH1381-3-PWR, pp. 142–154.

13. Dommel, H. W., Tinney, W. F., "Optimal Power Flow Solutions," *IEEE Transactions*, Vol. PAS-87, October 1968, pp. 1866–1876.

14. Sasson, A. M., "Nonlinear Programming Solutions for Load Flow, Minimum-Loss, and Economic Dispatch Problems," *IEEE Transactions*, Vol. PAS-88, April 1969, pp. 379–409.

15. Peschon, J., Bree, D. W., Hajdu, L. P., "Optimal Solutions Involving System Security" Proceedings, Seventh PICA Conference, Boston, Mass., 1971, pp. 210–218.

16. Carpentier, J. W., "Differential Injections Model, a General Method for Secure and Optimal Load Flows," 1973 PICA Conference Proceedings, pp. 255–262.

17. Alsac, O., Stott, B., "Optimal Load Flow With Steady-State Security," *IEEE Transactions*, Vol. PAS-93, May/June, 1974, pp. 745–751.

18. Barcelo, W. R., Lemmon, W. W., Koen, H. R., "Optimization of the Real Time Dispatch with Constraints for Secure Operation of Bulk Power Systems," *IEEE Transactions*, Vol. PAS-96, May/June, 1977. pp. 741–757.

chapter 12

An Introduction to State Estimation in Power Systems

12.1 INTRODUCTION

State estimation is the process of assigning a value to an unknown system state variable based on measurements from that system according to some criteria. Usually, the process involves imperfect measurements that are redundant and the process of estimating the system states is based on a statistical criteria that estimates the true value of the state variables to minimize or maximize the selected criteria. A commonly used and familiar criteria is that of minimizing the sum of the squares of the differences between the estimated and "true" (i.e., measured) values of a function.

The ideas of least squares estimation have been known and used since the early part of the nineteenth century. The major developments in this area have taken place in the twentieth century in applications in the aerospace field. In these developments the basic problems have involved the location of an aerospace vehicle (i.e., missile, airplane, or space vehicle) and the estimation of its trajectory given redundant and imperfect measurements of its position and velocity vector. In many applications these measurements are based on optical observations and/or radar signals that may be contaminated with noise and may contain system measurement errors. State estimators may be both static and dynamic. Both types of estimators have been developed for power systems. This chapter will introduce the basic development of a static state estimator.

In a power system the state variables are the voltage magnitudes and relative phase angles at the system nodes. Measurements are required in order to estimate the system performance in real time for both system security control and constraints on economic dispatch. The inputs to an estimator are selected, imperfect power system measurements of voltage magnitudes and power, VAR, or ampere-flow quantities. The estimator is designed to produce the "best estimate" of the system voltage and phase angles recognizing that there are errors in the measured quantities and that there may be redundant measurements. The output data are then used in system control centers in the implementation of the security constrained dispatch and control of the system as discussed in the previous chapter.

12.2 POWER SYSTEM STATE ESTIMATION

As introduced in Chapter 11, the problem of monitoring the power flows and voltages on a transmission system is very important in maintaining system security. By simply checking each measured value against its limit, the power system operators can tell where problems exist in the transmission system—and, it is hoped, they can take corrective actions to relieve overloaded lines or out-of-limit voltages.

Many problems are encountered in monitoring a transmission system. These problems come primarily from the nature of the measurement transducers and from communications problems in transmiting the measured values back to the operations control center.

Transducers from power system measurements, like any measurement device, will be subject to errors. If the errors are small, they may go undetected and can cause misinterpretation by those reading the measured values. In addition, transducers may have gross measurement errors that render their output useless. An example of such a gross error might involve having the transducer connected up backward, thus, giving the negative of the value being measured. Finally, the telemetry equipment often experiences periods when communications channels are completely out, thus depriving the system operator of any information about some part of the power system network.

It is for these reasons that power system state estimation techniques have been developed. A state estimator, as we will see shortly, can "smooth out" small random errors in meter readings, detect and identify gross measurement errors, and "fill in" meter readings that have failed due to communications failures.

To begin, we will use a simple DC load flow example to illustrate the principles of state estimation. Suppose the three-bus DC load flow of Example 4B were operating with the load and generation shown in Figure 12.1. The only information we have about this system is provided by three MW power flow meters located as shown in Figure 12.2.

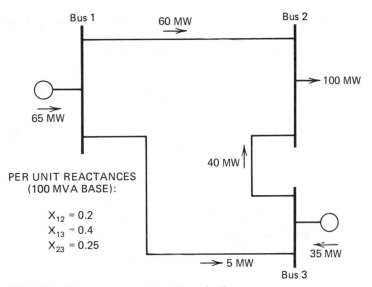

FIG. 12.1 Three-bus system from Example 4B.

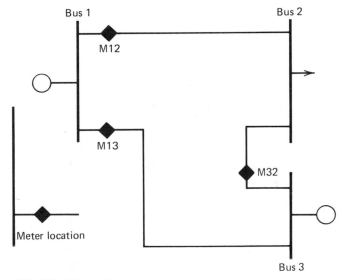

FIG. 12.2 Meter placement.

Only two of these meter readings are required to calculate the bus-phase angles and all load and generation values fully. Suppose we use M_{13} and M_{32} and further suppose that M_{13} and M_{32} give us perfect readings of the flows on their respective transmission lines.

$$M_{13} = 5 \text{ MW} = 0.05 \text{ pu}$$

$$M_{32} = 40 \text{ MW} = 0.40 \text{ pu}$$

Then the flows on lines 1-3 and 3-2 can be set equal to these meter readings.

$$f_{13} = \frac{1}{x_{13}} (\theta_1 - \theta_3) = M_{13} = 0.05 \text{ pu}$$

$$f_{32} = \frac{1}{x_{23}} (\theta_3 - \theta_2) = M_{32} = 0.40 \text{ pu}$$

Since we know that $\theta_3 = 0$ rad, we can solve the f_{13} equation for θ_1 and the f_{32} equation for θ_2 resulting in

$$\theta_1 = 0.02 \text{ rad}$$

$$\theta_2 = -0.10 \text{ rad}$$

We will now investigate the case where all three meter readings have slight errors. Suppose the readings obtained are

$$M_{12} = 62 \text{ MW} = 0.62 \text{ pu}$$

$$M_{13} = 6 \text{ MW} = 0.06 \text{ pu}$$

$$M_{32} = 37 \text{ MW} = 0.37 \text{ pu}$$

If we use only the M_{13} and M_{32} readings as before, we will calculate the phase angles as follows.

$$\theta_1 = 0.024 \text{ rad}$$

$$\theta_2 = -0.0925 \text{ rad}$$

$$\theta_3 = 0 \text{ rad (still assumed to equal zero)}$$

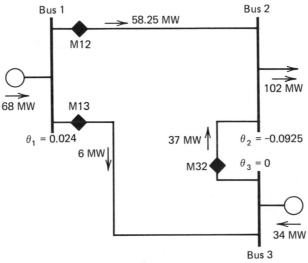

FIG. 12.3   Flows resulting from use of meters M_{13} and M_{32}.

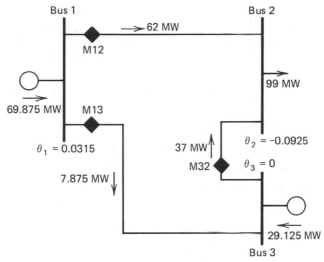

FIG. 12.4 Flows resulting from use of meters M_{12} and M_{32}.

This results in the system flows as shown in Figure 12.3. Note that the predicted flows match at M_{13} and M_{32}, but the flow on line 1-2 does not match the reading of 62 MW from M_{12}. If we were to ignore the reading on M_{13} and use M_{12} and M_{32}, we could obtain the flows shown in Figure 12.4.

All we have accomplished is to match M_{12} but at the expense of no longer watching M_{13}. What we need is a procedure that uses the information available from all three meters to produce the best estimate of the actual angles, line flows, and bus load and generations.

Before proceeding, let's discuss what we have been doing. Since the only thing we know about the power system comes to us from the measurements, we must use the measurements to estimate system conditions. Recall that in each instance the measurements were used to calculate the bus phase angles at bus 1 and 2. Once these phase angles were known, all unmeasured power flows, loads, and generations could be determined. We call θ_1 and θ_2 the *state variables* for the three-bus system since knowing them allows all other quantities to be calculated. In general the state variables for a power system consist of the bus voltage magnitude at all buses and the phase angles at all but one bus. The swing or reference bus phase angle is usually assumed to be zero radians. Note that we could use real and imaginary components of bus voltage if desired. If we can use measurements to estimate the "states" (i.e., voltage magnitudes and phase angles) of the power system, then we can go on to calculate any power flows, generation, loads, and so forth we desire. This presumes that the network configuration (i.e., breaker and disconnect switch statuses) is known and that the impedances in the network are also known. Automatic load tap changing autotransformers or phase angle regulators are often included in a network, and their tap positions may be telemetered to the control center as a measured quantity. Strictly speaking, the transformer taps and phase angle regulator positions should also be considered as states since they must be known in order to calculate the flows through the transformers and regulators.

To return to the three-bus DC power flow model, we have three meters providing us with a set of redundant readings with which to estimate the two states θ_1 and θ_2. We say that the readings are redundant since, as we saw earlier, only two readings are necessary to calculate θ_1 and θ_2, the other reading is always "extra." However, the "extra" reading does carry useful information and ought not to be discarded summarily.

This simple example serves to introduce the subject of *static state estimation*, which is the art of estimating the exact system state given a set of imperfect measurements made on the power system. We will digress at this point to develop the theoretical background for static state estimation. We will return to our three-bus system in Section 12.4.

12.3 MAXIMUM LIKELIHOOD WEIGHTED LEAST-SQUARES ESTIMATION

12.3.1 Introduction

Statistical estimation refers to a procedure where one uses samples to calculate the value of one or more unknown parameters in a system. Since the samples (or measurements) are inexact, the estimate obtained for the unknown parameter is also inexact. This leads to the problem of how to formulate a "best" estimate of the unknown parameters given the available measurements.

The development of the notions of state estimation may proceed along several lines depending on the statistical criterion selected. Of the many criteria that have been examined and used in various applications, three are perhaps the most commonly encountered.

1. *The maximum likelihood criterion*, where the objective is to maximize the probability that the estimate of the state variable $\hat{\mathbf{x}}$, is the true value of the state variable vector, $\mathbf{x}$. (i.e., maximize $P(\hat{\mathbf{x}}) = \mathbf{x}$).
2. *The weighted least-squares criterion*, where the objective is to minimize the sum of the squares of the weighted deviations of the estimated measurements $\hat{\mathbf{z}}$ from the actual measurements $\mathbf{z}$.
3. *The minimum variance criterion*, where the object is to minimize the expected value of the sum of the squares of the deviations of the estimated components of the state variable vector from the corresponding components of the true state variable vector.

When normally distributed, unbiased meter error distributions are assumed, each of these approaches results in identical estimators. This chapter will utilize the maximum likelihood approach because the method introduces the measurement error weighting matrix [R] in a straightforward manner.

The maximum likelihood procedure asks the following questions: "What is the probability (or likelihood) that I will get the measurements I have obtained?" This probability depends on the random error in the measuring device (transducer) as well as the unknown parameters to be estimated. Therefore, a reasonable procedure would be one that simply chose the estimate as the value that maximizes this probability. As we will see shortly, the maximum likelihood estimator assumes that we know the probability density function (PDF) of the random errors in the measurement. Other estimation schemes could also be used. The "least-squares" estimator does not require that we know the probability density function for the sample or measurement errors. However, if we assume that the probability density function of sample or measurement error is a normal (Gaussian) distribution, we will end up with the same estimation formula. We will proceed to develop our estimation formula using the maximum likelihood criterion assuming normal distributions for measurement errors. The result will be a "least-squares" or more precisely a "weighted least-squares" estimation formula even though we will develop the formulation using the maximum likelihood criteria. We will illustrate this method with a simple electrical circuit and show how the maximum likelihood estimate can be made.

First, we introduce the concept of *random measurement error*. Note that we have dropped the term "sample" since the concept of a measurement is much more appropriate to our discussion. The measurements are assumed to be in error: that is, the value obtained from the measurement device is close to the true value of the parameter being measured but differs by an unknown error. Mathematically this can be modeled as follows.

Let z^{meas} be the value of a measurement as received from a measurement device. Let z^{true} be the true value of the quantity being measured. Finally, let η be the random measurement error. We can then represent our measured value as

$$z^{meas} = z^{true} + \eta \qquad (12.1)$$

The random number, η, serves to model the uncertainty in the measurements. If the measurement error is unbiased, the probability density function of η is usually chosen as a normal distribution with zero mean. Note that other measurement probability density functions will also work in the maximum likelihood method as well. The probability density function of η is

$$\text{PDF}(\eta) = \frac{1}{\sigma\sqrt{2\pi}} \exp(-\eta^2/2\sigma^2) \qquad (12.2)$$

where σ is called the standard deviation and σ^2 is called the variance of the random number. $\text{PDF}(\eta)$ describes the behavior of η. A plot of $\text{PDF}(\eta)$ is shown in Figure 12.5. Note that σ, the standard deviation, provides a way to model the seriousness of the random measurement error. If σ is large, the measurement is relatively inaccurate (i.e., a poor quality measurement device), whereas a small value of σ

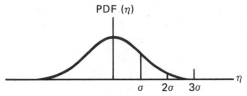

FIG. 12.5 The normal distribution.

denotes a small error spread (i.e., a higher quality measurement device). The normal distribution is commonly used for modeling measurement errors since it is the distribution that will result when many factors contribute to the overall error.

12.3.2 Maximum Likelihood Concepts

The principle of maximum likelihood estimation is illustrated by using a simple DC circuit example as shown in Figure 12.6. In this example, we wish to estimate the value of the voltage source, x^{true}, using an ammeter with an error having a known standard deviation. The ammeter gives a reading of z_1^{meas}, which is equal to the sum of z_1^{true} (the true current flowing in our circuit) and η_1 (the error present in the ammeter). Then we can write

$$z_1^{\text{meas}} = z_1^{\text{true}} + \eta_1 \tag{12.3}$$

Since the mean value of η_1 is zero, we then know that the mean value of z_1^{meas} is equal to z_1^{true}. This allows us to write a probability density function for z_1^{meas} as

$$\text{PDF}(z_1^{\text{meas}}) = \frac{1}{\sigma_1 \sqrt{2\pi}} \exp\left(\frac{-(z_1^{\text{meas}} - z_1^{\text{true}})^2}{2\sigma_1^2}\right) \tag{12.4}$$

where σ_1 is the standard deviation for the random error η_1. If we assume that the value of the resistance, r_1, in our circuit is known, then we can write

$$\text{PDF}(z_1^{\text{meas}}) = \frac{1}{\sigma_1 \sqrt{2\pi}} \exp\left(\frac{-\left(z_1^{\text{meas}} - \frac{1}{r_1}x\right)^2}{2\sigma_1^2}\right) \tag{12.5}$$

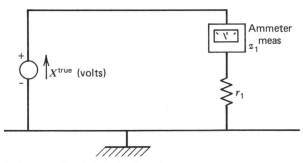

FIG. 12.6 Simple DC circuit with current measurement.

Coming back to our definition of a maximum likelihood estimator, we now wish to find an estimate of x (called x^{est}) that maximizes the probability that the observed measurement z_1^{meas} would occur. Since we have the probability density function of z_1^{meas}, we can write

$$\text{prob}(z_1^{meas}) = \int_{z_1^{meas}}^{z_1^{meas}+dz_1^{meas}} \text{PDF}(z_1^{meas}) \, dz_1^{meas} \qquad \text{as } dz_1^{meas} \to 0$$

$$= \text{PDF}(z_1^{meas}) \, dz_1^{meas} \tag{12.6}$$

The maximum likelihood procedure then requires that we maximize the value of $\text{prob}(z_1^{meas})$, which is a function of x. That is,

$$\max_{x} \text{prob}(z_1^{meas}) = \max_{x} \text{PDF}(z_1^{meas}) \, dz_1^{meas} \tag{12.7}$$

One convenient transformation that can be used at this point is to maximize the natural logarithm of $\text{PDF}(z_1^{meas})$ since maximizing the log of $\text{PDF}(z_1^{meas})$ will also maximize $\text{PDF}(z_1^{meas})$. Then we wish to find

$$\max_{x} \text{Ln}[\text{PDF}(z_1^{meas})]$$

or

$$\max_{x} \left[-\text{Ln}(\sigma_1 \sqrt{2\pi}) - \frac{\left(z_1^{meas} - \dfrac{1}{r_1} x\right)^2}{2\sigma_1^2} \right]$$

Since the first term is constant, it can be ignored. We can maximize the function in brackets by minimizing the second term since it has a negative coefficient, that is,

$$\max_{x} \left[-\text{Ln}(\sigma_1 \sqrt{2\pi}) - \frac{\left(z_1^{meas} - \dfrac{1}{r_1} x\right)^2}{2\sigma_1^2} \right]$$

is the same as

$$\min_{x} \left[\frac{\left(z_1^{meas} - \dfrac{1}{r_1} x\right)^2}{2\sigma_1^2} \right] \tag{12.8}$$

The value of x that minimizes the right-hand term is found by simply taking the first derivative and setting the result to zero.

$$\frac{d}{dx} \left[\frac{\left(z_1^{meas} - \dfrac{1}{r_1} x\right)^2}{2\sigma_1^2} \right] = \frac{-\left(z_1^{meas} - \dfrac{1}{r_1} x\right)}{r_1 \sigma_1^2} = 0 \tag{12.9}$$

or

$$x^{est} = r_1 z_1^{meas}$$

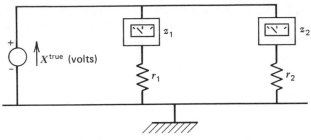

FIG. 12.7 DC circuit with two current measurements.

To most readers this result was obvious from the beginning. All we have accomplished is to declare the maximum likelihood estimate of our voltage as simply the measured current times the known resistance. However, by adding a second measurement circuit, we have an entirely different situation in which the best estimate is not so obvious. Let us now add a second ammeter and resistance as shown in Figure 12.7.

Assume that both r_1 and r_2 are known. As before, model each meter reading as the sum of the true value and a random error.

$$z_1^{\text{meas}} = z_1^{\text{true}} + \eta_1$$
$$z_2^{\text{meas}} = z_2^{\text{true}} + \eta_2$$

(12.10)

Where the errors will be represented as independent zero mean, normally distributed random variables with probability density functions:

$$\text{PDF}(\eta_1) = \frac{1}{\sigma_1\sqrt{2\pi}} \exp\left(\frac{-(\eta_1)^2}{2\sigma_1^2}\right)$$

$$\text{PDF}(\eta_2) = \frac{1}{\sigma_2\sqrt{2\pi}} \exp\left(\frac{-(\eta_2)^2}{2\sigma_2^2}\right)$$

(12.11)

and as before we can write the probability density functions of z_1^{meas} and z_2^{meas} as

$$\text{PDF}(z_1^{\text{meas}}) = \frac{1}{\sigma_1\sqrt{2\pi}} \exp\left(\frac{-\left(z_1^{\text{meas}} - \frac{1}{r_1}x\right)^2}{2\sigma_1^2}\right)$$

$$\text{PDF}(z_2^{\text{meas}}) = \frac{1}{\sigma_2\sqrt{2\pi}} \exp\left(\frac{-\left(z_2^{\text{meas}} - \frac{1}{r_2}x\right)^2}{2\sigma_2^2}\right)$$

(12.12)

The likelihood function must be the probability of obtaining the measurements z_1^{meas} and z_2^{meas}. Since we are assuming that the random errors η_1 and η_2 are independent random variables, the probability of obtaining z_1^{meas} and z_2^{meas} is simply the product of the probability of obtaining z_1^{meas} and the probability of obtaining z_2^{meas}.

$$\text{prob}(z_1^{\text{meas}} \text{ and } z_2^{\text{meas}}) = \text{prob}(z_1^{\text{meas}}) \cdot (\text{prob}(z_2^{\text{meas}}))$$

$$= \text{PDF}(z_1^{\text{meas}}) \, \text{PDF}(z_2^{\text{meas}}) \, dz_1^{\text{meas}} \, dz_2^{\text{meas}}$$

$$= \left[\frac{1}{\sigma_1\sqrt{2\pi}} \exp\left(\frac{-\left(z_1^{\text{meas}} - \dfrac{1}{r_1}x\right)^2}{2\sigma_1^2} \right) \right]$$

$$\times \left[\frac{1}{\sigma_2\sqrt{2\pi}} \exp\left(\frac{-\left(z_2^{\text{meas}} - \dfrac{1}{r_2}x\right)^2}{2\sigma_2^2} \right) \right] dz_1^{\text{meas}} \, dz_2^{\text{meas}} \quad (12.13)$$

To maximize the function we will again take its natural logarithm.

$$\max_x \text{prob}(z_1^{\text{meas}} \text{ and } z_2^{\text{meas}})$$

$$= \max_x \left[-Ln(\sigma_1\sqrt{2\pi}) - \frac{\left(z_1^{\text{meas}} - \dfrac{1}{r_1}x\right)^2}{2\sigma_1^2} - Ln(\sigma_2\sqrt{2\pi}) - \frac{\left(z_2^{\text{meas}} - \dfrac{1}{r_2}x\right)^2}{2\sigma_2^2} \right]$$

$$= \min_x \left[\frac{\left(z_1^{\text{meas}} - \dfrac{1}{r_1}x\right)^2}{2\sigma_1^2} + \frac{\left(z_2^{\text{meas}} - \dfrac{1}{r_2}x\right)^2}{2\sigma_2^2} \right] \quad (12.14)$$

The minimum sought is found by

$$\frac{d}{dx}\left[\frac{\left(z_1^{\text{meas}} - \dfrac{1}{r_1}x\right)^2}{2\sigma_1^2} + \frac{\left(z_2^{\text{meas}} - \dfrac{1}{r_2}x\right)^2}{2\sigma_2^2} \right]$$

$$= \frac{-\left(z_1^{\text{meas}} - \dfrac{1}{r_1}x\right)}{r_1\sigma_1^2} - \frac{\left(z_2^{\text{meas}} - \dfrac{1}{r_2}x\right)}{r_2\sigma_2^2} = 0$$

giving

$$x^{\text{est}} = \frac{\dfrac{z_1^{\text{meas}}}{r_1\sigma_1^2} + \dfrac{z_2^{\text{meas}}}{r_2\sigma_2^2}}{\dfrac{1}{r_1^2\sigma_1^2} + \dfrac{1}{r_2^2\sigma_2^2}} \quad (12.15)$$

If one of the ammeters is of superior quality, its variance will be much smaller than that of the other meter. For example, if $\sigma_2^2 \ll \sigma_1^2$, then the equation for x^{est}

becomes

$$x^{\text{est}} \simeq z_2^{\text{meas}} \cdot r_2$$

Thus, we see that the maximum likelihood method of estimating our unknown parameter gives us a way to weight the measurements properly according to their quality.

It should be obvious by now that we need not express our estimation problem as a maximum of the product of probability density functions. Instead, we can observe a direct way of writing what is needed by looking at Eqs. 12.8 and 12.14. In these equations we see that the maximum likelihood estimate of our unknown parameter is always expressed as that value of the parameter that gives the minimum of the sum of the squares of the difference between each measured value and the true value being measured (expressed as a function of our unknown parameter) with each squared difference divided or "weighted" by the variance of the meter error. Thus, if we are estimating a single parameter, x, using N_m measurements, we would write the expression

$$\min_{x} J(x) = \sum_{i=1}^{N_m} \frac{[z_i^{\text{meas}} - f_i(\hat{x})]^2}{\sigma_i^2} \tag{12.16}$$

where $f_i =$ function that is used to calculate the value being measured by the ith measurement

$\sigma_i^2 =$ variance for the ith measurement

$J(x) =$ measurement residual

$N_m =$ number of independent measurements

$z_i^{\text{meas}} = i^{\text{th}}$ measured quantity

Note that Eq. 12.16 may be expressed in per unit or in physical units such as MW, MVAR, kV.

If we were to try to estimate N_s unknown parameters using N_m measurements, we would write

$$\min_{\{x_1, x_2, \ldots, x_{N_s}\}} J(x_1, x_2, \ldots, x_{N_s}) = \sum_{i=1}^{N_m} \frac{[z_i - f_i(x_1, x_2, \ldots, x_{N_s})]^2}{\sigma_i^2} \tag{12.17}$$

The estimation calculation shown in Eqs. 12.16 and 12.17 is known as a *weighted least squares* estimator, which, as we have shown earlier, is equivalent to a maximum likelihood estimator if the measurement errors are modeled as random numbers having a normal distribution.

12.3.3 Matrix Formulation

If the functions $f_i(x_1, x_2, \ldots, x_{N_s})$ are linear functions, Eq. 12.17 has a closed-form solution. Let us write the function $f_i(x_1, x_2, \ldots, x_{N_s})$ as

$$f_i(x_1, x_2, \ldots, x_{N_s}) = f_i(\mathbf{x}) = h_{i1}x_1 + h_{i2}x_2 + \cdots + h_{iN_s}X_{N_s} \tag{12.18}$$

Then, if we place all the f_i functions in a vector, we may write

$$\mathbf{f(x)} = \begin{bmatrix} f_1(\mathbf{x}) \\ f_2(\mathbf{x}) \\ \vdots \\ f_{N_s}(\mathbf{x}) \end{bmatrix} = [H]\mathbf{x} \tag{12.19}$$

where $[H]$ = an N_m by N_s matrix containing the coefficients of the
           linear functions $f_i(\mathbf{x})$

   N_m = number of measurements

   N_s = number of unknown parameters being estimated

Placing the measurements in a vector:

$$\mathbf{z}^{\text{meas}} = \begin{bmatrix} z_1^{\text{meas}} \\ z_2^{\text{meas}} \\ \vdots \\ z_{N_m}^{\text{meas}} \end{bmatrix} \tag{12.20}$$

We may then write Eq. 12.17 in a very compact form.

$$\min_{\mathbf{x}} J(\mathbf{x}) = [\mathbf{z}^{\text{meas}} - \mathbf{f(x)}]^T [R^{-1}][\mathbf{z}^{\text{meas}} - \mathbf{f(x)}] \tag{12.21}$$

where

$$[R] = \begin{bmatrix} \sigma_1^2 & & & \\ & \sigma_2^2 & & \\ & & \ddots & \\ & & & \sigma_{N_s}^2 \end{bmatrix}$$

$[R]$ is called the *covariance matrix of measurement errors*. To obtain the general expression for the minimum in Eq. 12.21, expand the expression and substitute $[H]\mathbf{x}$ for $\mathbf{f(x)}$ from Eq. 12.19.

$$\min_{\mathbf{x}} J(\mathbf{x}) = \{ \mathbf{z}^{\text{meas}^T} [R^{-1}] \mathbf{z}^{\text{meas}} - \mathbf{x}^T [H]^T [R^{-1}] \mathbf{z}^{\text{meas}}$$

$$- \mathbf{z}^{\text{meas}} [R^{-1}][H]\mathbf{x}$$
$$+ \mathbf{x}^T [H]^T [R^{-1}][H]\mathbf{x} \} \tag{12.22}$$

Similar to the procedures of Chapter 3, the minimum of $J(\mathbf{x})$ is found when $\partial J(\mathbf{x})/\partial x_i = 0$, for $i = 1, \ldots, N_s$; this is identical to the stating that the gradient of $J(\mathbf{x})$, $\nabla J(\mathbf{x})$, is exactly zero.

The gradient of $J(\mathbf{x})$ is (see the appendix to this chapter)

$$\nabla J(\mathbf{x}) = -2[H]^T [R^{-1}] \mathbf{z}^{\text{meas}} + 2[H]^T [R^{-1}][H]\mathbf{x}$$

Then $\nabla J(\mathbf{x}) = \mathbf{0}$ gives

$$\mathbf{x}^{\text{est}} = [[H]^T[R^{-1}][H]]^{-1}[H]^T[R^{-1}]\mathbf{z}^{\text{meas}} \qquad (12.23)$$

Note that Eq. 12.23 holds when $N_s < N_m$, that is, when the number of parameters being estimated is less than the number of measurements being made.

When $N_s = N_m$, our estimation problem reduces to

$$\mathbf{x}^{\text{est}} = [H]^{-1}\mathbf{z}^{\text{meas}} \qquad (12.24)$$

There is also a closed-form solution to the problem when $N_s > N_m$, although in this case we are not estimating $\mathbf{x}$ to maximize a likelihood function since $N_s > N_m$ usually implies that many different values for $\mathbf{x}^{\text{est}}$ can be found that cause $f_i(\mathbf{x}^{\text{est}})$ to equal z_i^{meas} for all $i = 1, \ldots, N_m$ exactly. Rather, the objective is to find $\mathbf{x}^{\text{est}}$ such that the sum of the squares of x_i^{est} are minimized. That is,

$$\min_{\mathbf{x}} \sum_{i=1}^{N_s} x_i^2 = \mathbf{x}^T\mathbf{x} \qquad (12.25)$$

subject to the condition that $\mathbf{z}^{\text{meas}} = [H]\mathbf{x}$. The closed form solution for this case is

$$\mathbf{x}^{\text{est}} = [H]^T[[H][H]^T]^{-1}\mathbf{z}^{\text{meas}} \qquad (12.26)$$

In power system state estimation, underdetermined problems (i.e., where $N_s > N_m$) are not solved as shown in Eq. 12.26. Rather, "pseudomeasurements" are added to the measurement set to give a completely determined or overdetermined problem. We will discuss pseudomeasurements in Section 12.5.3. Table 12.1 summarizes the results for this section.

TABLE 12.1 Estimation Formulas

Case	Description	Solution	Comment
$N_s < N_m$	Overdetermined	$\mathbf{x}^{\text{est}} = [[H]^T[R^{-1}][H]]^{-1}[H]^T[R^{-1}]\mathbf{z}^{\text{meas}}$	$\mathbf{x}^{\text{est}}$ is the maximum likelihood estimate of $\mathbf{x}$ given the measurements $\mathbf{z}^{\text{meas}}$
$N_s = N_m$	Completely determined	$\mathbf{x}^{\text{est}} = [H]^{-1}\mathbf{z}^{\text{meas}}$	$\mathbf{x}^{\text{est}}$ fits the measured quantities to the measurements $\mathbf{z}^{\text{meas}}$ exactly
$N_s > N_m$	Underdetermined	$\mathbf{x}^{\text{est}} = [H]^T[[H][H]^T]^{-1}\mathbf{z}^{\text{meas}}$	$\mathbf{x}^{\text{est}}$ is the vector of minimum norm that fits the measured quantities to the measurements exactly. (The norm of a vector is equal to the sum of the squares of its components)

12.3.4 An Example of Weighted Least-Squares State Estimation

We now return to our three-bus example. Recall from Figure 12.2 that we have three measurements to determine θ_1 and θ_2, the phase angles at buses 1 and 2. From the development in the preceding section we know that the states θ_1 and θ_2 can be estimated by minimizing a residual $J(\theta_1, \theta_2)$ where $J(\theta_1, \theta_2)$ is the sum of the squares of individual measurement residuals divided by the variance for each measurement.

To start, we will assume that all three meters have the following characteristics.

Meter full scale value: 100 MW

Meter accuracy: ± 3 MW

This is interpreted to mean that the meters will give a reading within ± 3 MW of the truc value being measured for approximately 99% of the time. Mathematically, we say that the errors are distributed according to a normal probability density function with a standard deviation, σ, as shown in Figure 12.8.

Notice that the probability of an error decreases as the error magnitude increases. By integrating the PDF between -3σ and $+3\sigma$ we come up with a value of approximately 0.99. We will assume that the meter's accuracy (in our case ± 3 MW) is being stated as equal to the 3σ points on the probability density function. Then ± 3 MW corresponds to a metering standard deviation of $\sigma = 1$ MW $= 0.01$ pu.

The formula developed in the last section for the weighted least-squares estimate is given in Eq. 12.23, which is repeated here.

$$\mathbf{x}^{est} = [[H]^T[R^{-1}][H]]^{-1}[H]^T[R^{-1}]\mathbf{z}^{meas}$$

where $\mathbf{x}^{est}$ = vector of estimated state variables

$[H]$ = measurement function coefficient matrix

$[R]$ = measurement covariance matrix

$\mathbf{z}^{meas}$ = vector containing the measured values themselves

For the three-bus problem we have

$$\mathbf{x}^{est} = \begin{bmatrix} \theta_1^{est} \\ \theta_2^{est} \end{bmatrix} \tag{12.27}$$

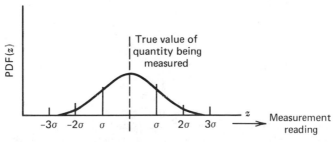

FIG. 12.8   Normal distribution of meter errors.

To derive the $[H]$ matrix we need to write the measurements as a function of the state variables θ_1 and θ_2. These functions are written in per unit as

$$M_{12} = f_{12} = \frac{1}{0.2}(\theta_1 - \theta_2) = 5\theta_1 - 5\theta_2$$

$$M_{13} = f_{13} = \frac{1}{0.4}(\theta_1 - \theta_3) = 2.5\theta_1 \qquad (12.28)$$

$$M_{32} = f_{32} = \frac{1}{0.25}(\theta_3 - \theta_2) = -4\theta_2$$

The reference-bus phase angle, θ_3, is still assumed to be zero. Then

$$[H] = \begin{bmatrix} 5 & -5 \\ 2.5 & 0 \\ 0 & -4 \end{bmatrix}$$

The covariance matrix for the measurements, $[R]$, is

$$[R] = \begin{bmatrix} \sigma^2_{M12} & & \\ & \sigma^2_{M13} & \\ & & \sigma^2_{M32} \end{bmatrix} = \begin{bmatrix} 0.0001 & & \\ & 0.0001 & \\ & & 0.0001 \end{bmatrix}$$

Note that since the coefficients of $[H]$ are in per unit we must also write $[R]$ and $\mathbf{z}^{\text{meas}}$ in per unit.

Our least-squares "best" estimate of θ_1 and θ_2 is then calculated as

$$\begin{bmatrix} \theta_1^{\text{est}} \\ \theta_2^{\text{est}} \end{bmatrix} = \left[\begin{bmatrix} 5 & 2.5 & 0 \\ -5 & 0 & -4 \end{bmatrix} \begin{bmatrix} 0.0001 & & \\ & 0.0001 & \\ & & 0.0001 \end{bmatrix}^{-1} \begin{bmatrix} 5 & -5 \\ 2.5 & 0 \\ 0 & -4 \end{bmatrix} \right]^{-1}$$

$$\times \begin{bmatrix} 5 & 2.5 & 0 \\ -5 & 0 & -4 \end{bmatrix} \begin{bmatrix} 0.0001 & & \\ & 0.0001 & \\ & & 0.0001 \end{bmatrix}^{-1} \begin{bmatrix} 0.62 \\ 0.06 \\ 0.37 \end{bmatrix}$$

$$= \begin{bmatrix} 312500 & -250000 \\ -250000 & 410000 \end{bmatrix}^{-1} \begin{bmatrix} 32500 \\ -45800 \end{bmatrix}$$

$$= \begin{bmatrix} 0.028571 \\ -0.094286 \end{bmatrix}$$

where

$$\mathbf{z}^{\text{meas}} = \begin{bmatrix} 0.62 \\ 0.06 \\ 0.37 \end{bmatrix}$$

From these estimated phase angles we can calculate the power flowing in each transmission line and the net generation or load at each bus. The results are shown in

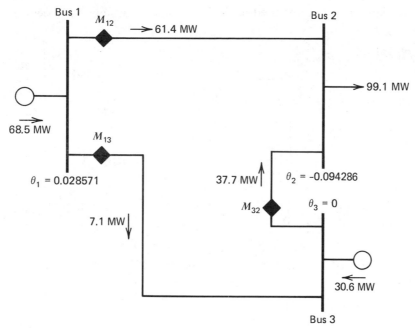

FIG. 12.9 Three-bus example with best estimates of θ_1 and θ_2.

Figure 12.9. If we calculate the value of $J(\theta_1, \theta_2)$, the residual, we get

$$J(\theta_1, \theta_2) = \frac{(z_{12} - f_{12}(\theta_1, \theta_2))^2}{\sigma_{12}^2} + \frac{(z_{13} - f_{13}(\theta_1, \theta_2))^2}{\sigma_{13}^2} + \frac{(z_{32} - f_{32}(\theta_1, \theta_2))^2}{\sigma_{32}^2}$$

$$= \frac{(0.62 - (5\theta_1 - 5\theta_2))^2}{0.0001} + \frac{(0.06 - (2.5\theta_1))^2}{0.0001} + \frac{(0.37 + (4\theta_2))^2}{0.0001}$$

$$= 2.14 \tag{12.29}$$

Suppose the meter on the M_{13} transmission line was superior in quality to those on M_{12} and M_{32}. How will this affect the estimate of the states? Intuitively, we can reason that any measurement reading we get from M_{13} will be much closer to the true power flowing on line 1-3 than can be expected when comparing M_{12} and M_{32} to the flows on lines 1-2 and 3-2, respectively. Therefore, we would expect the results from the state estimator to reflect this if we set up the measurement data to reflect the fact that M_{13} is a superior measurement. To show this, we use the following metering data.

Meters M_{12} and M_{32}:   100 MW full scale
± 3 MW accuracy
($\sigma = 1$ MW $= 0.01$ pu)

Meter M_{13}:   100 MW full scale
± 0.3 MW accuracy
($\sigma = 0.1$ MW $= 0.001$ pu)

The covariance matrix to be used in the least-squares formula now becomes

$$[R] = \begin{bmatrix} \sigma^2_{M12} & & \\ & \sigma^2_{M13} & \\ & & \sigma^2_{M32} \end{bmatrix} = \begin{bmatrix} 1 \times 10^{-4} & & \\ & 1 \times 10^{-6} & \\ & & 1 \times 10^{-4} \end{bmatrix}$$

We now solve Eq. 12.23 again with the new $[R]$ matrix.

$$\begin{bmatrix} \theta_1 \\ \theta_2 \end{bmatrix} = \left[\begin{bmatrix} 5 & 2.5 & 0 \\ -5 & 0 & -4 \end{bmatrix} \begin{bmatrix} 1 \times 10^{-4} & & \\ & 1 \times 10^{-6} & \\ & & 1 \times 10^{-4} \end{bmatrix}^{-1} \begin{bmatrix} 5 & -5 \\ 2.5 & 0 \\ 0 & -4 \end{bmatrix} \right]^{-1}$$

$$\begin{bmatrix} 5 & 2.5 & 0 \\ -5 & 0 & -4 \end{bmatrix} \begin{bmatrix} 1 \times 10^{-4} & & \\ & 1 \times 10^{-6} & \\ & & 1 \times 10^{-4} \end{bmatrix}^{-1} \begin{bmatrix} 0.62 \\ 0.06 \\ 0.37 \end{bmatrix}$$

$$= \begin{bmatrix} 6.5 \times 10^6 & -2.5 \times 10^5 \\ -2.5 \times 10^5 & 4.1 \times 10^5 \end{bmatrix}^{-1} \begin{bmatrix} 1.81 \times 10^5 \\ -0.458 \times 10^5 \end{bmatrix}$$

$$= \begin{bmatrix} 0.024115 \\ -0.097003 \end{bmatrix}$$

From these estimated phase angles we obtain the network conditions shown in Figure 12.10. Compare the estimated flow on line 1-3 as just calculated to the estimated flow calculated on line 1-3 in the previous least-squares estimate. Setting σ_{M13} to 0.01 MW has brought the estimated flow on line 1-3 much closer to the meter reading of 6.0 MW. Also note that the estimates of flow on lines 1-2 and 3-2 are now further from the M_{12} and M_{32} meter readings respectively, which is what we should have expected.

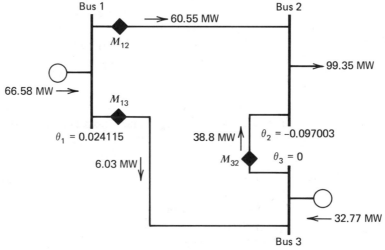

FIG. 12.10 Three-bus example with better meter at M_{13}.

12.4 STATE ESTIMATION OF AN AC NETWORK

12.4.1 Development of Method

We have demonstrated how the maximum likelihood estimation scheme developed in Section 12.3.2 led to a least-squares calculation for measurements from a linear system. In the least-squares calculation we are trying to minimize the sum of measurement residuals.

$$\min_{\mathbf{x}} J(\mathbf{x}) = \sum_{i=1}^{N_m} \frac{(z_i - f_i(\mathbf{x}))^2}{\sigma_i} \tag{12.30}$$

In the case of a linear system, the $f_i(\mathbf{x})$ functions are themselves linear and we solve for the minimum of $J(\mathbf{x})$ directly. In an AC network, the measured quantities are MW, MVAR, MVA, amperes, transformer tap position, and voltage magnitude. The state variables are the voltage magnitude at each bus, the phase angles at all but the reference bus, and the transformer taps. The equation for power flowing over a transmission line is given in Eq. 4.21 and is clearly not a linear function of the voltage magnitude and phase angle at each end of the line. Therefore, the $f_i(\mathbf{x})$ functions will be nonlinear functions, except for a voltage magnitude measurement where $f_i(\mathbf{x})$ is simply unity times the particular x_i that corresponds to the voltage magnitude being measured. For MW and MVAR measurements on a transmission line from bus i to bus j we would have the following terms in $J(\mathbf{x})$ (see Eq. 4.21).

$$\frac{[\mathrm{MW}_{ij}^{\mathrm{meas}} - (|E_i|^2(G_{ij}) - |E_i||E_j|(\cos(\theta_i - \theta_j)G_{ij} + \sin(\theta_i - \theta_j)B_{ij}))]^2}{\sigma_{\mathrm{MW}_{ij}}^2} \tag{12.31}$$

and

$$\frac{[\mathrm{MVAR}_{ij}^{\mathrm{meas}} - (-|E_i|^2(B_{\mathrm{cap}_{ij}} + B_{ij}) - |E_i||E_j|(\sin(\theta_i - \theta_j)G_{ij} - \cos(\theta_i - \theta_j)B_{ij}))]^2}{\sigma_{\mathrm{MVAR}_{ij}}^2}$$

$$\tag{12.32}$$

A voltage magnitude measurement would result in the following term in $J(\mathbf{x})$.

$$\frac{(|E_i|^{\mathrm{meas}} - |E_i|)^2}{\sigma_{|E_i|}^2} \tag{12.33}$$

Similar functions can be derived for MVA or ampere measurements.

If we do not have a linear relationship between the states ($|E|$'s and θ's) and the power flows on a network, we will have to resort to an iterative technique to minimize $J(\mathbf{x})$. A commonly used technique for power system state estimation is to calculate the gradient of $J(\mathbf{x})$ and then force it to zero using Newton's method as was done with the Newton load flow in Chapter 4. We will review how to use Newton's method on multidimensional problems before proceeding to the minimization of $J(\mathbf{x})$.

Given the functions $g_i(\mathbf{x})$, $i = 1, \ldots, n$, we wish to find $\mathbf{x}^{\text{ans}}$ that gives $g_i(\mathbf{x}^{\text{ans}}) = g_i^{\text{des}}$, for $i = 1, \ldots, n$. If we arrange the g_i functions in a vector we can write

$$\mathbf{g}^{\text{des}} - \mathbf{g}(\mathbf{x}) = 0 \text{ for } \mathbf{x} = \mathbf{x}^{\text{ans}} \tag{12.34}$$

by perturbing $\mathbf{x}$ we can write

$$\mathbf{g}^{\text{des}} - \mathbf{g}(\mathbf{x} + \Delta\mathbf{x}) = \mathbf{g}^{\text{des}} - \mathbf{g}(\mathbf{x}) - [\mathbf{g}'(\mathbf{x})]\Delta\mathbf{x} = 0 \tag{12.35}$$

where we have expanded $\mathbf{g}(\mathbf{x} + \Delta\mathbf{x})$ in a Taylor's series about $\mathbf{x}$ and ignored all higher-order terms. The $[\mathbf{g}'(\mathbf{x})]$ term is the Jacobian matrix of first derivatives of $\mathbf{g}(\mathbf{x})$. Then

$$\Delta\mathbf{x} = [\mathbf{g}'(\mathbf{x})]^{-1}(\mathbf{g}^{\text{des}} - \mathbf{g}(\mathbf{x})) \tag{12.36}$$

Note that if $\mathbf{g}^{\text{des}}$ is identically zero we have

$$\Delta\mathbf{x} = [\mathbf{g}'(\mathbf{x})]^{-1}[-\mathbf{g}(\mathbf{x})] \tag{12.37}$$

To solve for $\mathbf{g}^{\text{des}}$ we must solve for $\Delta\mathbf{x}$ using Eq. 12.36. Then calculate $\mathbf{x}^{\text{new}} = \mathbf{x} + \Delta\mathbf{x}$ and reapply Eq. 12.36 until either $\Delta\mathbf{x}$ gets very small or $\mathbf{g}(\mathbf{x})$ comes close to $\mathbf{g}^{\text{des}}$.

Now let us return to the state estimation problem as given in Eq. 12.30.

$$\min_{\mathbf{x}} J(\mathbf{x}) = \sum_{i=1}^{N_m} \frac{[z_i - f_i(\mathbf{x})]^2}{\sigma_i^2}$$

We first form the gradient of $J(\mathbf{x})$ as

$$\nabla_x J(\mathbf{x}) = \begin{bmatrix} \dfrac{\partial J(\mathbf{x})}{\partial x_1} \\[2mm] \dfrac{\partial J(\mathbf{x})}{\partial x_2} \\[2mm] \vdots \end{bmatrix}$$

$$= -2 \begin{bmatrix} \dfrac{\partial f_1}{\partial x_1} & \dfrac{\partial f_2}{\partial x_1} & \dfrac{\partial f_3}{\partial x_1} & \cdots \\[2mm] \dfrac{\partial f_1}{\partial x_2} & \dfrac{\partial f_2}{\partial x_2} & \dfrac{\partial f_3}{\partial x_2} & \cdots \\[2mm] \vdots & \vdots & \vdots & \end{bmatrix} \begin{bmatrix} \dfrac{1}{\sigma_1^2} & & \\ & \dfrac{1}{\sigma_2^2} & \\ & & \ddots \end{bmatrix} \begin{bmatrix} (z_1 - f_1(\mathbf{x})) \\[2mm] (z_2 - f_2(\mathbf{x})) \\[2mm] \vdots \end{bmatrix} \tag{12.38}$$

If we put the $f_i(\mathbf{x})$ functions in a vector form $\mathbf{f}(\mathbf{x})$ and calculate the Jacobian of $\mathbf{f}(\mathbf{x})$, we would obtain

$$\frac{\partial \mathbf{f}(\mathbf{x})}{\partial \mathbf{x}} = \begin{bmatrix} \dfrac{\partial f_1}{\partial x_1} & \dfrac{\partial f_1}{\partial x_2} & \dfrac{\partial f_1}{\partial x_3} & \cdots \\[2mm] \dfrac{\partial f_2}{\partial x_1} & \dfrac{\partial f_2}{\partial x_2} & \dfrac{\partial f_2}{\partial x_3} & \cdots \\[2mm] \vdots & \vdots & \vdots & \end{bmatrix} \tag{12.39}$$

We will call this matrix H. Then,

$$[H] = \begin{bmatrix} \dfrac{\partial f_1}{\partial x_1} & \dfrac{\partial f_1}{\partial x_2} & \dfrac{\partial f_1}{\partial x_3} & \cdots \\[2ex] \dfrac{\partial f_2}{\partial x_1} & \dfrac{\partial f_2}{\partial x_2} & \dfrac{\partial f_2}{\partial x_3} & \cdots \\[2ex] \vdots & \vdots & \vdots & \end{bmatrix} \tag{12.40}$$

And its transpose is

$$[H]^T = \begin{bmatrix} \dfrac{\partial f_1}{\partial x_1} & \dfrac{\partial f_2}{\partial x_1} & \dfrac{\partial f_3}{\partial x_1} & \cdots \\[2ex] \dfrac{\partial f_1}{\partial x_2} & \dfrac{\partial f_2}{\partial x_2} & \dfrac{\partial f_3}{\partial x_2} & \cdots \\[2ex] \vdots & \vdots & \vdots & \end{bmatrix} \tag{12.41}$$

Further, we write

$$\begin{bmatrix} \sigma_1^2 & & \\ & \sigma_2^2 & \\ & & \ddots \end{bmatrix} = [R] \tag{12.42}$$

Equation 12.38 can be written

$$\nabla_x J(\mathbf{x}) = \left\{ -2[H]^T[R]^{-1} \begin{bmatrix} z_1 - f_1(\mathbf{x}) \\ z_2 - f_2(\mathbf{x}) \\ \vdots \end{bmatrix} \right\} \tag{12.43}$$

To make $\nabla_x J(\mathbf{x})$ equal zero, we will apply Newton's method as in Eq. 12.37, then

$$\Delta \mathbf{x} = \left[\frac{\partial \nabla_x J(\mathbf{x})}{\partial \mathbf{x}} \right]^{-1} [-\nabla_x J(\mathbf{x})] \tag{12.44}$$

The Jacobian of $\nabla_x J(\mathbf{x})$ is calculated by treating $[H]$ as a constant matrix.

$$\begin{aligned} \frac{\partial \nabla_x J(\mathbf{x})}{\partial \mathbf{x}} &= \frac{\partial}{\partial \mathbf{x}} \left\{ -2[H]^T[R]^{-1} \begin{bmatrix} z_1 - f_1(\mathbf{x}) \\ z_2 - f_2(\mathbf{x}) \\ \vdots \end{bmatrix} \right\} \\ &= -2[H]^T[R]^{-1}[-H] \\ &= 2[H]^T[R]^{-1}[H] \end{aligned} \tag{12.45}$$

Then

$$\begin{aligned} \Delta \mathbf{x} &= \frac{1}{2} [[H]^{-1}[R]^{-1}[H]]^{-1} \left[2[H]^T[R]^{-1} \begin{bmatrix} z_1 - f_1(\mathbf{x}) \\ \vdots \end{bmatrix} \right] \\ &= [[H]^T[R]^{-1}[H]]^{-1}[H]^T[R]^{-1} \begin{bmatrix} (z_1 - f_1(\mathbf{x})) \\ (z_2 - f_2(\mathbf{x})) \\ \vdots \end{bmatrix} \end{aligned} \tag{12.46}$$

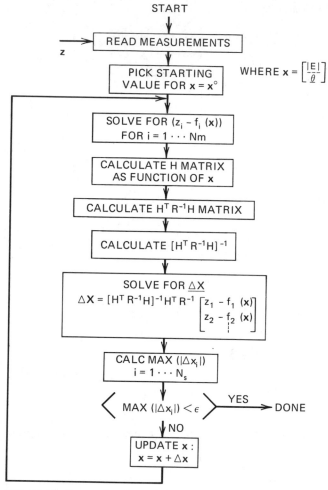

FIG. 12.11 State-estimation solution algorithm.

Equation 12.46 is obviously a close parallel to Eq. 12.23. To solve the AC state estimation problem, apply Eq. 12.46 iteratively as shown in Figure 12.11. Note that this is similar to the iterative process used in the Newton load-flow solution.

12.4.2 Typical Results of State Estimation on an AC Network

Figure 12.12 shows our familiar six-bus system with $P + jQ$ measurements on each end of each transmission line and at each load and generator. Bus voltage is also measured at each system bus.

To demonstrate the use of state estimation on these measurements, the base case conditions shown in Figure 11.1 were used together with a random number

---

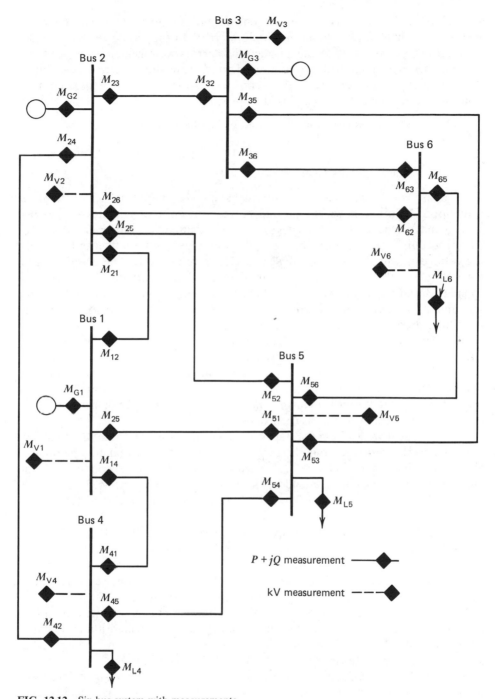

FIG. 12.12 Six-bus system with measurements.

generating algorithm to produce measurements with random errors. The measurements were obtained by adding the random errors to the base-case flows, loads, generations, and bus-voltage magnitudes. The errors were generated so as to be representative of values drawn from a set of numbers having a normal probability density function with zero mean and variance as specified for each measurement type. The measurement variances used were

$P + jQ$ **measurements:** $\quad \sigma = 5$ MW for the P measurement

$\qquad\qquad\qquad\qquad\quad \sigma = 5$ MVAR for the Q measurement

Voltage measurement: $\quad \sigma = 3.83$ kV

The base conditions and the measurements are shown in Table 12.2. The state estimation algorithm shown in Figure 12.11 was run to obtain estimates for the bus-voltage magnitudes and phase angles given the measurements shown in Table 12.2. The procedure took three iterations with $\mathbf{x}^0$ initially being set to 1.0 pu and 0 rad for the voltage magnitude and phase angle at each bus, respectively. At the beginning of each iteration the sum of the measurement residuals, $J(\mathbf{x})$ (see Eq. 12.30), is calculated and displayed. At the end of each iteration, the maximum $\Delta|E|$ and the maximum $\Delta\theta$ are calculated and displayed. The iterative steps for the six-bus system used here produced the results given in Table 12.3.

The value of $J(\mathbf{x})$ at the end of the iterative procedure would be zero if all measurements were without error or if there were no redundancy in the measurements. When there are redundant measurements with errors, the value of $J(\mathbf{x})$ will not normally go to zero. Its value represents a measure of the overall fit of the estimated values to the measurement values. The value of $J(\mathbf{x})$ can in fact be used to detect the presence of bad measurements.

The estimated values from the state estimator are shown in Table 12.4 together with the base-case values and the measured values. Notice that in general the estimated values do a good job of calculating the true (base-case) conditions from which the measurements were made. For example, measurement M23 shows a P flow of 8.6 MW whereas the true flow is 2.9 MW and the estimator predicts a flow of 3.0 MW.

The example shown here started from a base case or "true" state that was shown in Table 12.2. In actual practice we only have the measurements and the resulting estimate of the state, we never know the "true" state exactly and can only compare measurements to estimates. In the presentations to follow, however, we will leave the base case or "true" conditions in our illustrations to aid the reader.

The results in Table 12.4 show one of the advantages of using a state-estimation algorithm in that even with measurement errors, the estimation algorithm calculates quantities that are the "best" possible estimates of the true bus voltages and generator, load, and transmission line MW and MVAR values.

There are, however, other advantages to using a state estimation algorithm. First is the ability of the state estimator to detect and identify bad measurements, and second is the ability to estimate quantities that are not measured and telemetered. These are introduced in the next section.

TABLE 12.2 Base-Case Conditions

Measurement	Base case value			Measured value		
	kV	MW	MVAR	kV	MW	MVAR
M_{V1}	241.5			238.4		
M_{G1}		107.9	16.0		113.1	20.2
M_{12}		28.7	−15.4		31.5	−13.2
M_{14}		43.6	20.1		38.9	21.2
M_{15}		35.6	11.3		35.7	9.4
M_{V2}	241.5			237.8		
M_{G2}		50.0	74.4		48.4	71.9
M_{21}		−27.8	12.8		−34.9	9.7
M_{24}		33.1	46.1		32.8	38.3
M_{25}		15.5	15.4		17.4	22.0
M_{26}		26.2	12.4		22.3	15.0
M_{23}		2.9	−12.3		8.6	−11.9
M_{V3}	246.1			250.7		
M_{G3}		60.0	89.6		55.1	90.6
M_{32}		−2.9	5.7		−2.1	10.2
M_{35}		19.1	23.2		17.7	23.9
M_{36}		43.8	60.7		43.3	58.3
M_{V4}	227.6			225.7		
M_{L4}		70.0	70.0		71.8	71.9
M_{41}		−42.5	−19.9		−40.1	−14.3
M_{42}		−31.6	−45.1		−29.8	−44.3
M_{45}		4.1	−4.9		0.7	−17.4
M_{V5}	226.7			225.2		
M_{L5}		70.0	70.0		72.0	67.7
M_{54}		−4.0	−2.8		−2.1	−1.5
M_{51}		−34.5	−13.5		−36.6	−17.5
M_{52}		−15.0	−18.0		−11.7	−22.2
M_{53}		−18.0	−26.1		−25.1	−29.9
M_{56}		1.6	−9.7		−2.1	−0.8
M_{V6}	231.0			228.9		
M_{L6}		70.0	70.0		72.3	60.9
M_{65}		−1.6	3.9		1.0	2.9
M_{62}		−25.7	−16.0		−19.6	−22.3
M_{63}		−42.8	−57.9		−46.8	−51.1

TABLE 12.3 Iterative Results of State Estimator Solution

Iteration	$J(\mathbf{x})$ at beginning of iteration (pu)	Largest $\Delta|E|$ at end of iteration (puV)	Largest $\Delta\theta$ at end of iteration (rad)
1	3696.86	0.1123	0.06422
2	43.67	0.004866	0.0017
3	40.33	0.0000146	0.0000227

TABLE 12.4 State Estimation Solution

Measurement	Base case value			Measured value			Estimated value		
	kV	MW	MVAR	kV	MW	MVAR	kV	MW	MVAR
M_{V1}	241.5			238.4			240.6		
M_{G1}		107.9	16.0		113.1	20.2		111.9	18.7
M_{12}		28.7	−15.4		31.5	−13.2		30.4	−14.4
M_{14}		43.6	20.1		38.9	21.2		44.8	21.2
M_{15}		35.6	11.3		35.7	9.4		36.8	11.8
M_{V2}	241.5			237.8			239.9		
M_{G2}		50.0	74.4		48.4	71.9		47.5	70.3
M_{21}		−27.8	12.8		−34.9	9.7		−29.4	11.9
M_{24}		33.1	46.1		32.8	38.3		32.4	45.3
M_{25}		15.5	15.4		17.4	22.0		15.6	14.8
M_{26}		26.2	12.4		22.3	15.0		25.9	10.8
M_{23}		2.9	−12.3		8.6	−11.9		3.0	−12.6
M_{V3}	246.1			250.7			244.7		
M_{G3}		60.0	89.6		55.1	90.6		59.5	87.4
M_{32}		−2.9	5.7		−2.1	10.2		−3.0	6.2
M_{35}		19.1	23.2		17.7	23.9		19.2	22.9
M_{36}		43.8	60.7		43.3	58.3		43.3	58.3
M_{V4}	227.6			225.7			226.1		
M_{L4}		70.0	70.0		71.8	71.9		70.2	70.2
M_{41}		−42.5	−19.9		−40.1	−14.3		−43.6	−20.7
M_{42}		−31.6	−45.1		−29.8	−44.3		−30.9	−44.4
M_{45}		4.1	−4.9		0.7	−17.4		4.3	−5.1
M_{V5}	226.7			225.2			225.3		
M_{L5}		70.0	70.0		72.0	67.7		71.8	69.4
M_{54}		−4.0	−2.8		−2.1	−1.5		−4.2	−2.5
M_{51}		−34.5	−13.5		−36.6	−17.5		−35.6	−13.6
M_{52}		−15.0	−18.0		−11.7	−22.2		−15.1	−17.4
M_{53}		−18.0	−26.1		−25.1	−29.9		−18.1	−25.8
M_{56}		1.6	−9.7		−2.1	−0.8		1.3	−10.1
M_{V6}	231.0			228.9			230.1		
M_{L6}		70.0	70.0		72.3	60.9		68.9	65.8
M_{65}		−1.6	3.9		1.0	2.9		−1.2	4.4
M_{62}		−25.7	−16.0		−19.6	−22.3		−25.4	−14.5
M_{63}		−42.8	−57.9		−46.8	−51.1		−42.3	−55.7

12.5 AN INTRODUCTION TO ADVANCED TOPICS IN STATE ESTIMATION

12.5.1 Detection and Identification of Bad Measurements

The ability to detect and identify bad measurements is extremely valuable to a power system's operations department. Transducers may have been wired incorrectly or the transducer itself may be malfunctioning so that it simply no longer gives accurate readings. The statistical theory required to understand and analyze bad measurement detection and identification is straightforward but lengthy. We are going to open the door to this subject in this chapter. The serious student who wishes to pursue this subject should start with the chapter references. For the rest, we present results of these theories and indicate application areas.

To detect the presence of bad measurements, we will rely on the intuitive notion that for a given configuration the residual, $J(\mathbf{x})$, calculated after the state-estimator algorithm converges will be smallest if there are no bad measurements. When $J(\mathbf{x})$ is small, a vector $\mathbf{x}$ (i.e., voltages and phase angles) has been found that causes all calculated flows, loads, generations, and so forth to closely match all the measurements. Generally, the presence of a bad measurement value will cause the converged value of $J(\mathbf{x})$ to be larger than expected. We then need to ask

> What magnitude of $J(\mathbf{x})$ indicates the presence of bad measurements?

The measurement errors are random numbers so that the value of $J(\mathbf{x})$ is also a random number. If we assume that all the errors are described by their respective normal probability density functions, then we can show that $J(\mathbf{x})$ has a probability density function known as a *chi-squared distribution* which is written as $\chi^2(K)$. The parameter K is called the degrees of freedom of the chi-squared distribution. This parameter is defined as follows.

$$K = N_m - N_s$$

where N_m = number of measurements (note that a $P + jQ$ measurement counts as two measurements)

N_s = number of states = $(2n - 1)$

n = number of buses in the network

It can be shown that

The mean value of $J(\mathbf{x})$ equals K

The standard deviation, $\sigma_{J(\mathbf{x})}$, equals $\sqrt{2K}$

When one or more measurements are bad, their errors are frequently much larger than the assumed $\pm 3\sigma$ error bound for the measurement. However, even under normal circumstances (i.e., all errors within $\pm 3\sigma$), $J(\mathbf{x})$ can get to be large— although the chance of this happening is small. If we simply set up a threshold for $J(\mathbf{x})$, which we will call t_J, we could declare that bad measurements are present when $J(\mathbf{x}) > t_J$. This threshold test might be wrong in one of two ways. If we set t_J to a small value, we would get many "false alarms." That is, the test would indicate the presence of bad measurements when in fact there were none. If we set t_J to be a large value, the test would often indicate that "all is well" when in fact bad measurements were present. This can be put on a formal basis by writing the following equation.

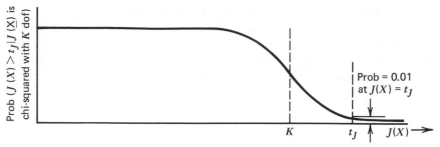

FIG. 12.13 Threshold test probability function.

$$\text{prob}(J(\mathbf{x}) > t_J | J(\mathbf{x}) \text{ is chi-squared}) = \alpha \qquad (12.47)$$
$$\text{with } K \text{ degrees of}$$
$$\text{freedom}$$

This equation says that the probability that $J(\mathbf{x})$ is greater than t_J is equal to α, given that the probability density for $J(\mathbf{x})$ is chi-squared with K degrees of freedom.

This type of testing procedure is formally known as *hypothesis testing*, and the parameter α is called the *significance level* of the test. By choosing a value for the significance level α we automatically know what threshold t_J to use in our test. When using a t_J derived in this manner, the probability of a "false alarm" is equal to α. By setting α to a small number, for example $\alpha = 0.01$, we would say that false alarms would occur in only 1% of the tests made. A plot of the probability function in Eq. 12.47 is shown in Figure 12.13.

In Table 12.3 we saw that the minimum value for $J(\mathbf{x})$ was 40.33. Looking at Figure 12.12 and counting all $P + jQ$ measurements as two measurements, we see that N_m is equal to 62. Therefore, the degrees of freedom for the chi-square distribution of $J(\mathbf{x})$ in our six-bus sample system is

$$K = N_m - N_s = N_m - (2n - 1) = 51$$

where $\quad N_m = 62$
$$n = 6$$

If we set our significance level for this test to 0.01 (i.e., $\alpha = 0.01$ in Eq. 12.47) we get a t_J of 76.6.* Therefore, with a $J(\mathbf{x}) = 40.33$, it seems reasonable to assume that there are no "bad" measurements present.

Now let us assume that one of the measurements is truly bad. To simulate this situation, the state-estimation algorithm was rerun with the M12 measurement reversed. Instead of $P = 31.5$ and $Q = -13.2$, it was set to $P = -31.5$ and $Q = 13.2$. The value of $J(\mathbf{x})$ and the maximum $\Delta|E|$ and $\Delta\theta$ for each iteration for this case are given in Table 12.5. The presence of bad data does not prevent the estimator from converging, but it will increase the value of the residual, $J(\mathbf{x})$.

The calculated flows and voltages for this situation are shown in Table 12.6. Note that the number of degrees of freedom is still 51 but $J(\mathbf{x})$ is now 207.94 at the end of

---

* Standard tables of $\chi^2(K)$ usually only go up to $K = 30$. For $K > 30$ a very close approximation to $\chi^2(K)$ using the normal distribution can be used. The student should consult any standard reference on probability and statistics to see how this is done.

TABLE 12.5 Iterative Results with Bad Measurement

Iteration	$J(\mathbf{x})$ at beginning of iteration (pu)	Largest $\Delta\lvert E\rvert$ at end of iteration (pu V)	Largest $\Delta\theta$ at end of iteration (rad)
1	3701.06	0.09851	0.06416
2	211.13	0.004674	0.001481
3	207.94	0.00002598	0.00004848

TABLE 12.6 State Estimation Solution with Measurement M12 Reversed

Measurement	Base case value			Measured value			Estimated value		
	kV	MW	MVAR	kV	MW	MVAR	kV	MW	MVAR
M_{V1}	241.5			238.4			240.6		
M_{G1}		107.9	16.0		113.1	20.2		99.3	21.9
M_{12}		28.7	-15.4		-31.5	$+13.2$		25.0	-12.2
M_{14}		43.6	20.1		38.9	21.2		40.6	21.9
M_{15}		35.6	11.3		35.7	9.4		33.7	12.3
M_{V2}	241.5			237.8			239.9		
M_{G2}		50.0	74.4		48.4	71.9		54.4	67.0
M_{21}		-27.8	12.8		-34.9	9.7		-24.4	9.2
M_{24}		33.1	46.1		32.8	38.3		35.0	44.1
M_{25}		15.5	15.4		17.4	22.0		16.3	14.7
M_{26}		26.2	12.4		22.3	15.0		25.1	11.3
M_{23}		2.9	-12.3		8.6	-11.9		2.3	-12.2
M_{V3}	246.1			250.7			244.6		
M_{G3}		60.0	89.6		55.1	90.6		61.4	86.3
M_{32}		-2.9	5.7		-2.1	10.2		-2.3	5.8
M_{35}		19.1	23.2		17.7	23.9		-20.5	22.2
M_{36}		43.8	60.7		43.3	58.3		43.2	58.2
M_{V4}	227.6			225.7			226.1		
M_{L4}		70.0	70.0		71.8	71.9		69.0	70.0
M_{41}		-42.5	-19.9		-40.1	-14.3		-39.6	-21.9
M_{42}		-31.6	-45.1		-29.8	-44.3		-33.5	-43.1
M_{45}		4.1	-4.9		0.7	-17.4		4.1	-5.0
M_{V5}	226.7			225.2			225.3		
M_{L5}		70.0	70.0		72.0	67.7		71.8	69.3
M_{54}		-4.0	-2.8		-2.1	-1.5		-4.1	-2.6
M_{51}		-34.5	-13.5		-36.6	-17.5		-32.7	-14.7
M_{52}		-15.0	-18.0		-11.7	-22.2		-15.8	-17.2
M_{53}		-18.0	-26.1		-25.1	-29.9		-19.3	-25.1
M_{56}		1.6	-9.7		-2.1	-0.8		0.1	-9.6
M_{V6}	231.0			228.9			230.0		
M_{L6}		70.0	70.0		72.3	60.9		66.9	66.7
M_{65}		-1.6	3.9		1.0	2.9		-0.1	3.9
M_{62}		-25.7	-16.0		-19.6	-22.3		-24.6	-15.0
M_{63}		-42.8	-57.9		-46.8	-51.1		-42.3	-55.6

our calculation. Since t_J is 76.6, we would immediately expect bad measurements at our 0.01 significance level. If we had not known ahead of running the estimation algorithm that a bad measurement was present, we would certainly have had good reason to suspect its presence when so large a $J(\mathbf{x})$ resulted.

So far, we can say that by looking at $J(\mathbf{x})$ we can detect the presence of bad measurements. But if bad measurements are present, how can one tell which measurements are bad? Without going into the statistical theory, we give the following explanation of how this is accomplished.

Suppose we are interested in the measurement of megawatt flow on a particular line. Call this measured value z_i. In Figure 12.14(a) we have a plot of the normal probability density function of z_i. Since we assume that the error in z_i is normally distributed with zero mean value, the probability density function is centered on the true value of z_i. Since the errors on all the measurements are assumed normal, we will assume that the estimate, $\mathbf{x}^{\mathrm{est}}$ are approximately normally distributed and that any quantity that is a function of $\mathbf{x}^{\mathrm{est}}$ is also an approximately normally distributed quantity. In Figure 12.14(b) we show the probability density function for the calculated megawatt flow, f_i, which is a function of the estimated state, $\mathbf{x}^{\mathrm{est}}$. We have drawn the density function of f_i as having a smaller deviation from its mean than the measurement z_i to indicate that due to redundancy in measurements, the estimate is more accurate.

The difference between the estimate, f_i, and the measurement, z_i, is called the *measurement residual* and is designated y_i. The probability density function for y_i is also normal and is shown in Figure 12.14(c) as having a zero mean and a standard deviation of σ_{y_i}. If we divide the difference between the estimate f_i and the measurement z_i by σ_{y_i}, we obtain what is called a *normalized measurement residual*. The normalized measurement residual is designated y_i^{norm} and is shown in Figure 12.14(d) along with its probability density function, which is normal and has a standard deviation of unity. If the absolute value of y_i^{norm} is greater than 3, we have good reason to suspect that z_i is a bad measurement value. The usual procedure in identifying bad measurements is to calculate all f_i values for the N_m measurements once $\mathbf{x}^{\mathrm{est}}$ is available from the state estimator. Using the z_i's that were used in the estimator and the f_i's, a measurement residual y_i can be calculated for each measurement. Also, using information from the state estimator we can calculate σ_{y_i} (see references for details of this calculation). Using y_i and σ_{y_i}, we can calculate a normalized residual for each measurement. Measurements having the largest absolute normalized residual are labeled as prime suspects. These prime suspects are removed from the state-estimator calculation one at a time starting with the measurement having the largest normalized residual. After a measurement has been removed, the state-estimation calculation (see Figure 12.11) is rerun. This results in a different $\mathbf{x}^{\mathrm{est}}$ and therefore a different $J(\mathbf{x})$. The chi-squared probability density function for $J(\mathbf{x})$ will have to be recalculated assuming that we use the same significance level for our test. If the new $J(\mathbf{x})$ is now less than the new value for t_J, we can say that the measurement that was removed has been identified as bad. If, however, the new $J(\mathbf{x})$ is greater than the new t_J, we must proceed to recalculate $f_i(\mathbf{x}^{\mathrm{est}})$, σ_{y_i}, and then y_i^{norm} for each of the remaining measurements. The measurement

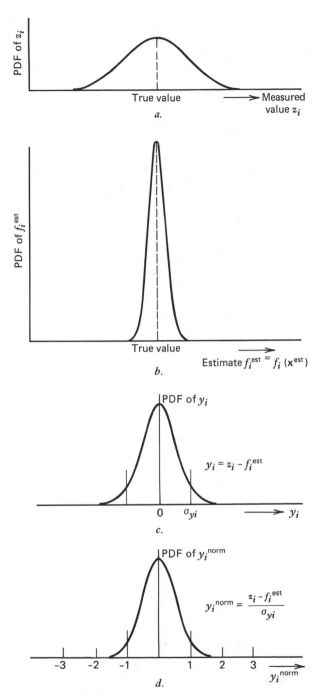

FIG. 12.14 Probability density function of the normalized measurement residual.

with the largest absolute y_i^{norm} is then again removed and the entire procedure repeated successively until $J(\mathbf{x})$ is less than t_J. The references at the end of this chapter discuss a problem that the identification process may encounter wherein several measurements may need to be removed to eliminate one "bad" measurement. That is, the identification procedure often cannot pinpoint a single bad measurement but instead identifies a group of measurements one of which is bad. In such cases the group must be eliminated to eliminate the bad measurement.

The ability to detect (using the chi-squared statistic) and identify (using normalized residuals) are extremely useful features of a state estimator. Without the state-estimator calculation using the system measurement data, those measurements whose values are not obviously wrong have little chance of being detected and identified. With the state estimator, the operations personnel have a greater assurance that quantities being displayed are not grossly in error.

12.5.2 Estimation of Quantities Not Being Measured

The other useful feature of a state-estimator calculation is the ability to calculate (or estimate) quantities not telemetered. This is most useful in cases of failure of communication channels connecting operations centers to remote data gathering equipment or when the remote data gathering equipment fails. Often data from some network substation are simply unavailable because no transducers or data-gathering equipment were ever installed.

An example of this might be the failure of all telemetry from buses 3, 4, 5, and 6 in our six-bus system. Even with the loss of these measurements, we can run the state-estimation algorithm on the remaining measurements at buses 1 and 2, calculate the bus-voltage magnitudes and phase angles at all six buses, and then calculate all network generations, loads, and flows. The results of such a calculation are given in Table 12.7. Notice that the estimate of quantities at the untelemetered buses are not as close to the base case as when using the full set of measurements (i.e., compare Table 12.7 to Table 12.4).

12.5.3 Network Observability and Pseudomeasurements

What happens if we continue to loose telemetry so that fewer and fewer measurements are available? Eventually, the state-estimation procedure breaks down completely. Mathematically, the matrix

$$[[H]^T[R^{-1}][H]]$$

in Eq. 12.46 becomes singular and cannot be inverted. There is also a very interesting engineering interpretation of this phenomena that allows us to alter the situation so that the state-estimation procedure is not completely disabled.

If we take the three-bus example used in the beginning of Section 12.2, we note that when all three measurements are used, we have a redundant set and we can use a least-squares fit to the measurement values. If one of the measurements is lost, we have just enough measurements to calculate the states. If, however, two

Measurement	Base-case value			Measured value			Estimated value		
	kV	MW	MVAR	kV	MW	MVAR	kV	MW	MVAR
M_{V1}	241.5			238.4			238.8		
M_{G1}		107.9	16.0		113.1	20.2		112.4	20.5
M_{12}		28.7	−15.4		31.5	−13.2		30.6	−13.4
M_{14}		43.6	20.1		38.9	21.2		44.7	19.4
M_{15}		35.6	11.3		35.7	9.4		37.1	14.6
M_{V2}	241.5			237.8			237.6		
M_{G2}		50.0	74.4		48.4	71.9		48.2	71.7
M_{21}		−27.8	12.8		−34.9	9.7		−29.6	11.1
M_{24}		33.1	46.1		32.8	38.3		30.5	40.2
M_{25}		15.5	15.4		17.4	22.0		16.1	16.8
M_{26}		26.2	12.4		22.3	15.0		22.4	15.2
M_{23}		2.9	−12.3		8.6	−11.9		8.8	−11.7
M_{V3}	246.1						241.4		
M_{G3}		60.0	89.6					27.2	94.9
M_{32}		−2.9	5.7					−8.7	5.5
M_{35}		19.1	23.2					15.1	25.3
M_{36}		43.8	60.7					20.9	64.0
M_{V4}	227.6						225.0		
M_{L4}		70.0	70.0					67.6	61.2
M_{41}		−42.5	−19.9					−43.6	−18.9
M_{42}		−31.6	−45.1					−29.3	−39.7
M_{45}		4.1	−4.9					5.3	−2.6
M_{V5}	226.7						221.4		
M_{L5}		70.0	70.0					71.9	76.7
M_{54}		−4.0	−2.8					−5.2	−4.8
M_{51}		−34.5	−13.5					−35.9	−15.9
M_{52}		−15.0	−18.0					−15.5	−19.0
M_{53}		−18.0	26.1					−14.0	−28.0
M_{56}		1.6	−9.7					−1.4	−9.0
M_{V6}	231.0						226.2		
M_{L6}		70.0	70.0					40.5	77.2
M_{65}		−1.6	3.9					1.4	3.4
M_{62}		−25.7	−16.0					−21.9	−18.8
M_{63}		−42.8	−57.9					−20.0	−61.8

measurements are lost, we are in trouble. For example, suppose M_{13} and M_{32} were lost leaving only M_{12}. If we now apply Eq. 12.23 in a straightforward manner, we get

$$M_{12} = f_{12} = \frac{1}{0.2}(\theta_1 - \theta_2) = 5\theta_1 - 5\theta_2$$

Then

$$[H] = [5 \quad -5]$$
$$[R] = [\sigma_{M12}^2] = [0.0001]$$

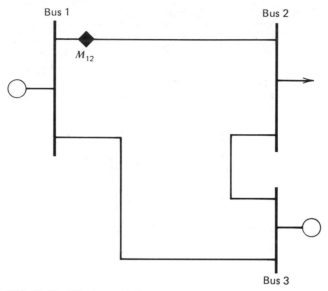

Bus 1

M_{12}

Bus 2

Bus 3

FIG. 12.15 "Unobservable" measurement set.

and

$$\begin{bmatrix} \theta_1^{est} \\ \theta_2^{est} \end{bmatrix} = \left[\begin{bmatrix} 5 \\ -5 \end{bmatrix} [0.0001]^{-1}[5 \quad -5] \right]^{-1} [5 \quad -5][0.0001]^{-1}(0.55)$$

$$= \begin{bmatrix} 2500 & -2500 \\ -2500 & 2500 \end{bmatrix}^{-1} [5 \quad -5][0.0001]^{-1}(0.55) \qquad (12.48)$$

The matrix to be inverted in Eq. 12.48 is clearly singular and, therefore, we have no way of solving for θ_1^{est} and θ_2^{est}. Why is this? The reasons become quite obvious when we look at the one-line diagram of this network as shown in Figure 12.15. With only M12 available, all we can say about the network is that the phase angle across line 1-2 must be 0.11 rad, but with no other information available, we cannot tell what relationship θ_1 or θ_2 has to θ_3, which is assumed to be 0 rad. If we write down the equations for the net injected power at bus 1 and bus 2, we have

$$P_1 = 7.5\,\theta_1 - 5\,\theta_2$$
$$P_2 = -5\,\theta_1 + 9\,\theta_2 \qquad (12.49)$$

If measurement M12 is reading 55 MW (0.55 pu), we have

$$\theta_1 - \theta_2 = 0.11 \qquad (12.50)$$

by substituting Eq. 12.49 into Eq. 12.50 and eliminating θ_1, we obtain

$$P_2 = 1.6\,P_1 - 1.87 \qquad (12.51)$$

Furthermore,

$$P_3 = -P_1 - P_2 = -0.6 P_1 + 1.87 \qquad (12.52)$$

Equations 12.51 and 12.52 give a relationship between P_1, P_2, and P_3, but we still do not know their correct values. The technical term for this phenomena is to say that the network is *unobservable*, that is, with only M12 available, we cannot observe (calculate) the state of the system.

It is very desirable to be able to circumvent this problem. Often a large power-system network will have missing data that render the network unobservable. Rather than just stop the calculations, a procedure is used that allows the estimator calculation to continue. The procedure involves the use of what are called *pseudo-measurements*. If we look at Eqs. 12.51 and 12.52, it is obvious that θ_1 and θ_2 could be estimated if the value of any one of the bus injections (i.e., P_1, P_2, or P_3) could be determined by some means other than direct measurement. This value, the pseudomeasurement, is used in the state estimator just as if it were an actual measured value.

To determine the value of an injection without measuring it we must have some knowledge about the power system beyond the measurements currently being made. For example, it is customary to have access to the generated MW and MVAR values at generating stations through telemetry channels (i.e., the generated MW and MVAR would normally be measurements available to the state estimator). If these channels are out and we must have this measurement for observability, we can probably communicate with the operators in the plant control room by telephone and ask for the MW and MVAR values and enter them into the state-estimator calculation manually. Similarly, if we needed a load-bus MW and MVAR for a pseudomeasurement, we could use historical records that show the relationship between an individual load and the total system load. We can estimate the total system load fairly accurately by knowing the total power being generated and estimating the network losses. Finally, if we have just experienced a telemetry failure, we could use the most recently estimated values from the estimator (assuming that it is run periodically) as pseudomeasurements. Therefore, if needed, we can provide the state estimator with a reasonable value to use as a pseudomeasurement at any bus in the system.

The three-bus sample system in Figure 12.16 requires one pseudomeasurement. Measurement M_{12} allows us to estimate the voltage magnitude and phase angle at

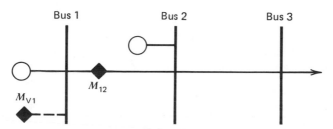

FIG. 12.16 Unobservable system showing importance of location of pseudomeasurements.

bus 2 (bus 1's voltage magnitude is measured and its phase angle is assumed to be zero). But without knowing the generation output at the generator unit on bus 2 or the load on bus 3, we cannot tell what voltage magnitude and phase angle to place on bus 3; hence, the network is unobservable. We can make this three-bus system observable by adding a pseudomeasurement of the net bus injected MW and MVAR at bus 2 or bus 3, but not at bus 1. That is, a pseudomeasurement at bus 1 will do no

TABLE 12.8 State-Estimation Solution with Measurements at Bus 1 and Pseudomeasurements at Buses 2, 3, and 6

Measurement	Base-case value			Measured value			Estimated value		
	kV	MW	MVAR	kV	MW	MVAR	kV	MW	MVAR
M_{V1}	241.5			238.4			238.4		
M_{G1}		107.9	16.0		113.1	20.2		111.4	19.5
M_{12}		28.7	−15.4		31.5	−13.2		33.3	−12.5
M_{14}		43.6	20.1		38.9	21.2		40.7	21.9
M_{15}		35.6	11.3		35.7	9.4		37.4	10.1
M_{V2}	241.5						236.2		
M_{G2}		50.0	74.4	Pseudo:	50.0	74.4		37.5	67.7
M_{21}		−27.8	12.8					−32.1	10.5
M_{24}		33.1	46.1					19.5	44.9
M_{25}		15.5	15.4					14.1	11.5
M_{26}		26.2	12.4					30.0	12.7
M_{23}		2.9	−12.3					6.0	−11.9
M_{V3}	246.1						240.5		
M_{G3}		60.0	89.6	Pseudo:	60.0	89.6		52.6	86.6
M_{32}		−2.9	5.7					−6.0	5.7
M_{35}		19.1	23.2					14.3	19.5
M_{36}		43.8	60.7					44.2	61.4
M_{V4}	227.6						223.8		
M_{L4}		70.0	70.0					51.9	73.3
M_{41}		−42.5	−19.9					−39.6	−21.8
M_{42}		−31.6	−45.1					−18.3	−44.6
M_{45}		4.1	−4.9					6.0	−6.9
M_{V5}	226.7						224.0		
M_{L5}		70.0	70.0					63.9	55.5
M_{54}		−4.0	−2.8					−5.9	−0.4
M_{51}		−34.5	−13.5					−36.3	−11.8
M_{52}		−15.0	−18.0					−13.7	−14.4
M_{53}		−18.0	−26.1					−13.6	−22.9
M_{56}		1.6	−9.7					5.5	−5.9
M_{V6}	231.0						224.9		
M_{L6}		70.0	70.0	Pseudo:	70.0	70.0		77.9	73.4
M_{65}		−1.6	3.9					−5.5	0.3
M_{62}		−25.7	−16.0					−29.3	−15.6
M_{63}		−42.8	−57.9					−43.2	−58.1

good at all because it tells nothing about the relationship of the phase angles between bus 2 and bus 3.

When adding a pseudomeasurement to a network, we simply write the equation for the pseudomeasurement injected power as a function of bus voltage magnitudes and phase angles as if it were actually measured. However, we do not wish to have the estimator treat the pseudomeasurement the same as a legitimate measurement, since it is often quite inaccurate and is little better than a guess. To circumvent this difficulty, we assign a large standard deviation to this measurement. The large standard deviation allows the estimator algorithm to treat the pseudomeasurement as if it were a measurement from a very poor quality metering device.

To demonstrate the use of pseudomeasurements on our six-bus test system, all measurements were removed from buses 2, 3, 4, 5, and 6 so that bus 1 had all remaining measurements. This rendered the network unobservable and required adding pseudomeasurements at buses 2, 3, and 6. In this case, the pseudo-measurements were just taken from our base-case load flow. The results are shown in Table 12.8. Notice that the resulting estimates are quite close to the measured values for bus 1 but that the remaining buses have large measurement residuals. The net injections at buses 2, 3, and 6 do not closely match the pseudomeasurements since the pseudomeasurements were weighted much less than the legitimate measurements.

12.6 APPLICATION OF POWER SYSTEMS STATE ESTIMATION

In this last section we will try to present the "big picture" showing how state estimation, contingency analysis, and generator corrective action fit together in a modern operations control center. Figure 12.17 is a schematic diagram showing the information flow between the various functions to be performed in an operations control center computer system. The system gets its information about the power system from remote terminal units that encode measurement transducer outputs and opened/closed status information into digital signals that are transmitted to the operations center over communications circuits. In addition, the control center can transmit control information such as raise/lower commands to generators and open/close commands to circuit breakers and switches. We have broken down the information coming into the control center as breaker/switch status indications and analog measurements. The analog measurements of generator output must be used directly by the AGC program (see Chapter 9) whereas all other data will be processed by the state estimator before being used by other programs.

In order to run the state estimator, we must know how the transmission lines are connected to the load and generation buses. We call this information the *network topology*. Since the breakers and switches in any substation can cause the network topology to change, a program must be provided that reads the telemetered breaker/switch status indications and restructures the electrical model of the system.

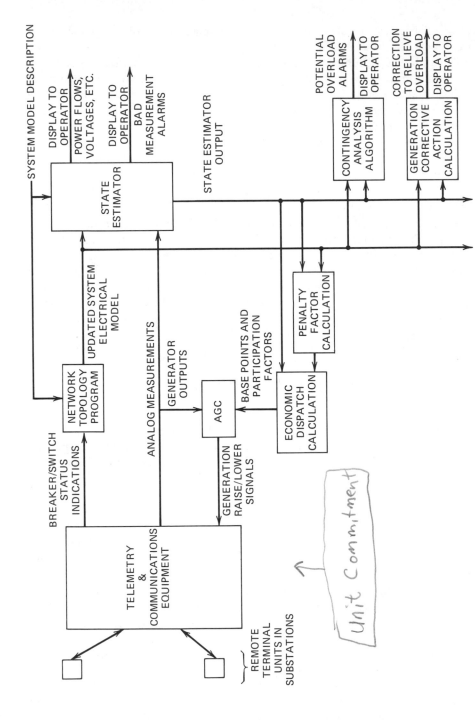

FIG. 12.17 Energy control center system security schematic.

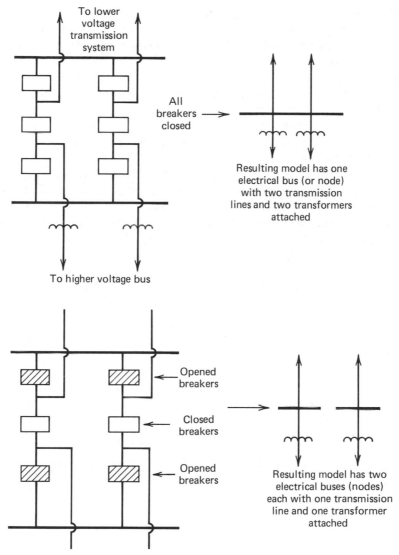

FIG. 12.18 Example of network topology updating.

An example of this is shown in Figure 12.18 where the opening of four breakers requires two electrical buses to represent the substation instead of one electrical bus. We have labeled the program that reconfigures the electrical model the network topology* program. The network topology program must have a complete description of each substation and how the transmission lines are attached to the substation equipment. Bus sections that are connected to other bus sections through

* Alternative names that are often used for this program are "system status processor" and "network configurator."

closed breakers or switches are designated as belonging to the same electrical bus. Thus, the number of electrical buses and the manner in which they are interconnected can be changed in the model to reflect breaker and switch status changes on the power system itself.

As seen in Figure 12.17, the electrical model of the power system's transmission system is sent to the state-estimator program together with the analog measurements. The output of the state estimator consists of all bus voltage magnitudes and phase angles, transmission line MW and MVAR flows calculated from the bus voltage magnitude and phase angles, and bus loads and generations calculated from the line flows. These quantities, together with the electrical model developed by the network topology program provide the basis for the economic dispatch program, congtingency analysis program, and generation corrective action program. Note that since the complete electrical model of the transmission system is available, we can directly calculate bus penalty factors as shown in Section 4.2.4.2.

APPENDIX
Derivation of Least-Squares Equations

One is often confronted with problems wherein data have been obtained by making measurements or taking samples on a process. Furthermore, the quantities being measured are themselves functions of other variables that we wish to estimate. These other variables will be called the state variables and designated $\mathbf{x}$, where the number of state variables is N_s. The measurement values will be called $\mathbf{z}$. We will assume here that the process we are interested in can be modeled using a linear model. Then we say that each measurement z_i is a linear function of the states x_i, that is,

$$z_i = h_i(\mathbf{x}) = h_{i1}x_1 + h_{i2}x_2 + \cdots + h_{iN_s}x_{N_s} \tag{A.1}$$

We can also write this equation as a vector equation if we place the h_{ij} coefficients into a vector $\mathbf{h}$, that is,

$$\mathbf{h_i} = \begin{bmatrix} h_{i1} \\ h_{i2} \\ \vdots \\ h_{iN_s} \end{bmatrix} \tag{A.2}$$

Then Eq. A.1 becomes

$$z_i = \mathbf{h}_i^T \mathbf{x} \tag{A.3}$$

where

$$\mathbf{x} = \begin{bmatrix} x_1 \\ x_2 \\ \vdots \\ x_{N_s} \end{bmatrix}$$

Finally, we can write all the measurement equations in a compact form

$$\mathbf{z} = [H]\mathbf{x} \tag{A.4}$$

where
$$\mathbf{z} = \begin{bmatrix} z_1 \\ z_2 \\ \vdots \\ z_{N_m} \end{bmatrix}$$

$$[H] = \begin{bmatrix} h_{11} & h_{12} & \cdots & h_{1N_s} \\ h_{21} & h_{22} & \cdots & \\ \vdots & & & \\ h_{N_m 1} & & \cdots & h_{N_m N_s} \end{bmatrix}$$

where row i of $[H]$ is equal to vector $\mathbf{h}_i^T$ (see Eq. A.2).

With N_m measurements we can have three possible cases to solve. That is, N_s, the number of states, is either less than N_m, equal to N_m, or greater than N_m. We will deal with each case separately.

The Overdetermined Case ($N_m > N_s$)

In this case we have more measurements or samples than state variables, therefore, we can write more equations, $h_i(\mathbf{x})$, than we have unknowns x_j. One way to estimate the x_i's is to minimize the sum of the squares of difference between the measurement values z_i and the estimate of z_i that is in turn a function of the estimates of x_i. That is, we wish to minimize

$$J(\mathbf{x}) = \sum_{i=1}^{N_m} [z_i - h_i(x_1, x_2, \ldots, x_{N_s})]^2 \tag{A.5}$$

Equation A.5 can be written as

$$J(\mathbf{x}) = \sum_{i=1}^{N_m} (z_i - \mathbf{h}_i^T \mathbf{x})^2 \tag{A.6}$$

and this can be written in a still more compact form as

$$J(\mathbf{x}) = (\mathbf{z} - [H]\mathbf{x})^T (\mathbf{z} - [H]\mathbf{x}) \tag{A.7}$$

If we wish to find the value of $\mathbf{x}$ that minimizes $J(\mathbf{x})$, we can take the first derivative of $J(\mathbf{x})$ with respect to each x_j ($j = 1, \ldots, N_s$) and set these derivatives to zero. That is,

$$\frac{\partial J(\mathbf{x})}{\partial x_j} = 0 \qquad \text{for } j = 1, \ldots, N_s \tag{A.8}$$

If we place these derivatives into a vector, we have what is called the gradient of $J(\mathbf{x})$, which is written $\nabla_x \mathbf{J}(\mathbf{x})$. Then,

$$\nabla_x \mathbf{J}(\mathbf{x}) = \begin{bmatrix} \dfrac{\partial J(\mathbf{x})}{\partial x_1} \\ \dfrac{\partial J(\mathbf{x})}{\partial x_2} \\ \vdots \end{bmatrix} \tag{A.9}$$

Then the goal of forcing each derivative to zero can be written as

$$\nabla_x \mathbf{J}(\mathbf{x}) = \mathbf{0} \tag{A.10}$$

where $\mathbf{0}$ is a vector of N_s elements each of which is zero. To solve this problem, we will first expand Eq. A.7.

$$J(\mathbf{x}) = (\mathbf{z} - [H]\mathbf{x})^T (\mathbf{z} - [H]\mathbf{x})$$
$$= \mathbf{z}^T\mathbf{z} - \mathbf{x}^T[H]^T\mathbf{z} - \mathbf{z}^T[H]\mathbf{x} + \mathbf{x}^T[H]^T[H]\mathbf{x} \tag{A.11}$$

The second and third term in Eq. A.11 are identical so that we can write

$$J(\mathbf{x}) = \mathbf{z}^T\mathbf{z} - 2\mathbf{z}^T[H]\mathbf{x} + \mathbf{x}^T[H]^T[H]\mathbf{x} \tag{A.12}$$

Before proceeding, we will derive a few simple relationships.

The gradient is always a vector of first derivatives of a scalar function that is itself a function of a vector. Thus, if we define $F(\mathbf{y})$ to be a scalar function, then its gradient $\nabla_y \mathbf{F}$ is:

$$\nabla_y \mathbf{F} = \begin{bmatrix} \dfrac{\partial F}{\partial y_1} \\ \dfrac{\partial F}{\partial y_2} \\ \vdots \\ \dfrac{\partial F}{\partial y_n} \end{bmatrix} \tag{A.13}$$

Then, if we define F as follows:

$$F = \mathbf{y}^T\mathbf{b} = \begin{bmatrix} y_1 & y_2 & \cdots \end{bmatrix} \begin{bmatrix} b_1 \\ b_2 \\ \vdots \end{bmatrix} \tag{A.14}$$

where $\mathbf{b}$ is a vector of constants b_i, $i = 1, \ldots, n$, then, F can be expanded as

$$F = y_1 b_1 + y_2 b_2 + y_3 b_3 + \cdots \tag{A.15}$$

and the gradient of F is

$$\nabla_y \mathbf{F} = \begin{bmatrix} \dfrac{\partial F}{\partial y_1} \\[1.2em] \dfrac{\partial F}{\partial y_2} \\[0.5em] \vdots \\[0.5em] \dfrac{\partial F}{\partial y_n} \end{bmatrix} = \begin{bmatrix} b_1 \\ b_2 \\ \vdots \\ b_n \end{bmatrix} = \mathbf{b} \tag{A.16}$$

It ought to be obvious that writing F with $\mathbf{y}$ and $\mathbf{b}$ reversed makes no difference. That is,

$$F = \mathbf{b}^T \mathbf{y} = \mathbf{y}^T \mathbf{b} \tag{A.17}$$

and therefore: $\nabla_y(\mathbf{b}^T \mathbf{y}) = \mathbf{b}$

Suppose we now write the vector $\mathbf{b}$ as the product of a matrix $[A]$ and a vector $\mathbf{u}$.

$$\mathbf{b} = [A]\mathbf{u} \tag{A.18}$$

Then if we take F as shown in Eq. A.14,

$$F = \mathbf{y}^T \mathbf{b} = \mathbf{y}^T [A]\mathbf{u} \tag{A.19}$$

we can say that

$$\nabla_y \mathbf{F} = [A]\mathbf{u} \tag{A.20}$$

Similarly, we can define

$$\mathbf{b}^T = \mathbf{u}^T [A] \tag{A.21}$$

If we can take F as shown in Eq. A.17,

$$F = \mathbf{b}^T \mathbf{y} = \mathbf{u}^T [A]\mathbf{y}$$

Then

$$\nabla_y \mathbf{F} = [A]^T \mathbf{u} \tag{A.22}$$

Finally, we will look at a scalar function F that is quadratic, namely,

$$F = \mathbf{y}^T [A]\mathbf{y}$$

$$= \begin{bmatrix} y_1 & y_2 & \cdots & y_n \end{bmatrix} \begin{bmatrix} a_{11} & a_{12} & \cdots \\ a_{21} & a_{22} & \cdots \\ \vdots & \vdots & \end{bmatrix} \begin{bmatrix} y_1 \\ y_2 \\ \vdots \\ y_n \end{bmatrix}$$

$$= \sum_{i=1}^{n} \sum_{j=1}^{n} y_i a_{ij} y_j \tag{A.23}$$

Then

$$\nabla_y \mathbf{F} = \begin{bmatrix} \dfrac{\partial F}{\partial y_1} \\[2mm] \dfrac{\partial F}{\partial y_2} \\[2mm] \vdots \\[2mm] \dfrac{\partial F}{\partial y_n} \end{bmatrix} = \begin{bmatrix} 2a_{11}y_1 + 2a_{12}y_2 + \cdots \\ 2a_{21}y_1 + 2a_{22}y_2 + \cdots \\ \vdots \end{bmatrix}$$

$$= 2[A]\mathbf{y} \tag{A.24}$$

Then in summary

$$
\begin{array}{ll}
1.\ F = \mathbf{y}^T\mathbf{b} & \nabla_y\mathbf{F} = \mathbf{b} \\
2.\ F = \mathbf{b}^T\mathbf{y} & \nabla_y\mathbf{F} = \mathbf{b} \\
3.\ F = \mathbf{y}^T[A]\mathbf{u} & \nabla_y\mathbf{F} = [A]\mathbf{u} \\
4.\ F = \mathbf{u}^T[A]\mathbf{y} & \nabla_y\mathbf{F} = [A]^T\mathbf{u} \\
5.\ F = \mathbf{y}^T[A]\mathbf{y} & \nabla_y\mathbf{F} = 2[A]\mathbf{y}
\end{array}
\tag{A.25}
$$

We will now use Eq. A.25 to derive the gradient of $J(x)$, that is $\nabla_x \mathbf{J}$, as shown in Eq. A.12. The first term, $\mathbf{z}^T\mathbf{z}$ is not a function of $\mathbf{x}$, so we can discard it. The second term is of the same form as (4) in Eq. A.25, so that,

$$\nabla_x(-2\mathbf{z}^T[H]\mathbf{x}) = -2[H]^T\mathbf{z} \tag{A.26}$$

The third term is the same as (5) in Eq. A.25 with $[H]^T[H]$ replacing $[A]$, then,

$$\nabla_x(\mathbf{x}^T[H]^T[H]\mathbf{x}) = 2[H]^T[H]\mathbf{x} \tag{A.27}$$

Then from Eqs. A.26 and A.27 we have

$$\nabla_x\mathbf{J} = -2[H]^T\mathbf{z} + 2[H]^T[H]\mathbf{x} \tag{A.28}$$

But, as stated in Eq. A.10, we wish to force $\nabla_x\mathbf{J}$ to zero. Then

$$-2[H]^T\mathbf{z} + 2[H]^T[H]\mathbf{x} = 0$$

or

$$\mathbf{x} = [[H]^T[H]]^{-1}[H]^T\mathbf{z} \tag{A.29}$$

If we had wanted to put a different weight, w_i, on each measurement, we could have written Eq. A.6 as

$$J(\mathbf{x}) = \sum_{i=1}^{N_m} w_i(z_i - h_i^T\mathbf{x})^2 \tag{A.30}$$

which can be written as

$$J(\mathbf{x}) = (\mathbf{z} - [H]\mathbf{x})^T[W](\mathbf{z} - [H]\mathbf{x})$$

where $[W]$ is a diagonal matrix. Then

$$J(\mathbf{x}) = \mathbf{z}^T[W]\mathbf{z} - \mathbf{x}^T[H]^T[W]\mathbf{z} - \mathbf{z}^T[W][H]\mathbf{x} + \mathbf{x}^T[H]^T[W][H]\mathbf{x}$$

If we once again use Eq. A.25, we would obtain

$$\nabla_x \mathbf{J} = -2[H]^T[W]\mathbf{z} + 2[H]^T[W][H]\mathbf{x}$$

and

$$\nabla_x \mathbf{J} = 0$$

gives

$$\mathbf{x} = ([H]^T[W][H])^{-1}[H]^T[W]\mathbf{z} \qquad (A.31)$$

The Fully-Determined Case ($N_m = N_s$)

In this case the number of measurements is equal to the number of state variables and we can solve for $\mathbf{x}$ directly by inverting $[H]$.

$$\mathbf{x} = [H]^{-1}\mathbf{z} \qquad (A.32)$$

Underdetermined Case ($N_m < N_s$)

In this case we have fewer measurements than state variables. In such a case it is possible to solve for many solutions $\mathbf{x}^{est}$ that cause $J(\mathbf{x})$ to equal zero. The usual solution technique is to find $\mathbf{x}^{est}$ that minimizes the sum of the squares of the solution values. That is, we find a solution such that

$$\sum_{j=1}^{N_s} x_j^2 \qquad (A.33)$$

is minimized while meeting the condition that the measurements will be solved for exactly. To do this, we treat the problem as a constrained minimization problem and use Lagrange multipliers as shown in the appendix of Chapter 3.

We formulate the problem as

Minimize:
$$\sum_{j=1}^{N_s} x_j^2$$
$$(A.34)$$

Subject to: $\qquad z_i = \sum_{j=1}^{N_s} h_{ij}x_j \qquad \text{for } i = 1, \ldots, N_m$

This optimization problem can be written in vector-matrix form as

$$\begin{aligned} \min & \quad \mathbf{x}^T\mathbf{x} \\ \text{subject to} & \quad \mathbf{z} = [H]\mathbf{x} \end{aligned} \qquad (A.35)$$

The Lagrangian for this problem is

$$\mathcal{L} = \mathbf{x}^T\mathbf{x} + \lambda^T(\mathbf{z} - [H]\mathbf{x}) \qquad (A.36)$$

Following the rules set down in the appendix of Chapter 3, we must find the gradient of $\mathcal{L}$ with respect to $\mathbf{x}$ and with respect to λ. Using the identities found in Eq. A.25 we get

$$\nabla_x \mathscr{L} = 2\mathbf{x} - [H]^T \lambda = 0$$

which gives

$$\mathbf{x} = \frac{1}{2}[H]^T \lambda$$

and

$$\nabla_\lambda \mathscr{L} = \mathbf{z} - [H]\mathbf{x} = 0$$

which gives

$$\mathbf{z} = [H]\mathbf{x}$$

Then

$$\mathbf{z} = \frac{1}{2}[H][H]^T \lambda$$

or

$$\lambda = 2[[H][H]^T]^{-1}\mathbf{z}$$

and finally,

$$\mathbf{x} = [H]^T[[H][H]^T]^{-1}\mathbf{z} \qquad (A.37)$$

The reader should be aware that the matrix inversion shown in Eqs. A.29, A.32, and A.37 may not be possible. That is, the $[[H]^T[H]]$ matrix in Eq. A.29 may be singular, or $[H]$ may be singular in Eq. A.32, or $[[H][H]^T]$ may be singular in Eq. A.37. In the overdetermined case $(N_m > N_s)$ whose solution is Eq. A.29 and the fully determined case $(N_m = N_s)$ whose solution is Eq. A.32, the singularity implies what is known as an "unobservable" system. By unobservable we mean that the measurements do not provide sufficient information to allow a determination of the states of the system. In the case of the underdetermined case $(N_m < N_s)$ whose solution is Eq. A.37, the singularity simply implies that there is no unique solution to the problem.

PROBLEMS

12.1 Using the three-bus sample system shown in Figure 12.1, assume that the three meters have the following characteristics.

Meter	Full Scale	Accuracy	$\sigma(pu)$
M_{12}	100 MW	±6 MW	0.02
M_{13}	100 MW	±3 MW	0.01
M_{32}	100 MW	±0.6 MW	0.002

a. Calculate the best estimate for the phase angles θ_1 and θ_2 given the following measurements.

Meter	Measured Value (MW)
M_{12}	68.0
M_{13}	4.0
M_{32}	40.5

b. Calculate the residual $J(\mathbf{x})$. For a significance level, α, of 0.01 does $J(\mathbf{x})$ indicate the presence of bad data? Explain.

12.2 Given a single transmission line with a generator at one end and a load at the other, two measurements are available as shown in Figure 12.19. Assume that we can model this circuit with a DC load flow using the line reactance shown. Also assume that the phase angle at bus 1 is 0 rad. Given the meter characteristics and meter readings telemetered from the meters, calculate the best estimate of the power flowing through the transmission line.

Meter	Full Scale	Meter Standard Deviation (σ) in pu	Meter Reading
M_{12}	200 MW	0.01	62 MW
M_{21}	200 MW	0.05	-52 MW

Note: M_{12} measures power flowing from bus 1 to bus 2; M_{21} measures power flowing from bus 2 to bus 1.)

Use 100 MVA as base.

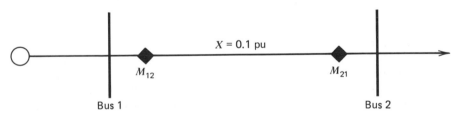

FIG. 12.19 Measurement configuration for Problem 12.2.

12.3 You are given the following network with meters at locations as shown in Figure 12.20.

Branch Impedances (pu)	Bus Conditions
$X_{12} = 0.25$	Load 1 = 50 MW
$X_{13} = 0.50$	Load 2 = 120 MW
$X_{24} = 0.40$	Generation on bus 3 = 90 MW
$X_{34} = 0.10$	Generation on bus 4 = 80 MW

Measurement Values	Measurement Errors
$M_{13} = -70.5$	$\sigma_{13} = 0.01$
$M_{31} = 72.1$	$\sigma_{31} = 0.01$
$M_{12} = 21.2$	$\sigma_{12} = 0.02$

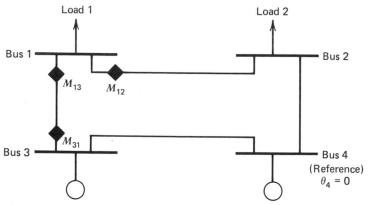

FIG. 12.20 Four-bus system with measurements for Problem 12.3.

a. Is this network observable? Set up the least-squares equations and try to invert $[H^T R^{-1} H]$.

b. Suppose we had a measurement of generation output at bus 3 and included it in our measurement set. Let this measurement be the following.

$$M_{3\,\text{gen}} = 92 \text{ MW} \quad \text{with } \sigma = 0.015$$

Repeat part **a** including this measurement.

FURTHER READING

State estimation originated in the aerospace industry and only came to be of interest to power systems engineers since the late 1960s. Since then, state estimators have been installed on a regular basis in new energy control centers and have proved quite useful. References 1–4 provide a good introduction to this topic. Reference 4 in particular is a carefully written overview with a good bibliography of literature up to 1974. References 5–8 show the variety of algorithms used to solve the state-estimation problem. Reference 8 covers a study undertaken to compare different algorithm types and is quite revealing as to the number of different ways that state estimation algorithms can be written.

The remaining references cover some of the subtopics of state estimation. The use of the state estimator to detect bad measurements and model parameter errors is covered in references 9–12. Network observability determination is covered in references 13 and 14. Finally, methods of automatically updating the network model topology to match switching status are covered in references 15 and 16.

1. Schweppe, F. C., Wildes, J., "Power System Static State Estimation, Part I: Exact Model," *IEEE Transactions*, Vol. PAS-89, January 1970, pp. 120–125.

2. Larson, R. E., Tinney, W. F., Peschon, J., "State Estimation in Power Systems, Part I: Theory and Feasibility," *IEEE Transactions*, Vol. PAS-89, March 1970, 345–352.

3. Larson, R. E., Tinney, W. F., Hajdu, L. P., Piercy, P. S., "State Estimation in Power Systems, Part II: Implementation and Applications," *IEEE Transactions*, Vol. PAS-89, March 1970, pp. 353–359.

4. Schweppe, F. C., Handschin, E., "Static State Estimation in Power Systems," *IEEE Proceedings*, Vol. 62, July 1974.

5. Dopazo, J. F., Klitin, O. A., Stagg, G. W., VanSlyck, L. S., "State Calculation of Power Systems from Line Flow Measurement" *IEEE Transactions*, Vol. PAS-89, September/October 1970, pp. 1698–1708.

6. Dopazo, J. F., Klitin, O. A., VanSlyck, L. S., "State Calculation of Power Systems from Line Flow Measurements, Part II" *IEEE Transcations*, Vol. PAS-91, January/February 1972, pp. 145–151.

7. Simões-Costa, A., Quintana V. H., "A Robust Numerical Technique for Power System State Estimation," *IEEE Transactions*, Vol. PAS-100, February/1981, pp. 691–698.

8. Allemong, J. J., Radu, L., Sasson, A. M., "A Fast and Reliable State Estimation Algorithm for AEP's New Control Center," *IEEE Transactions*, Vol. PAS-101, April 1982, pp. 933–944.

9. Dapozo, J. F., Klitin, O. A., Sasson, A. M., "State Estimation for Power Systems: Detection and Identification of Gross Measurement Errors," Proceedings 8th PICA Conference, Minneapolis, June 1973.

10. Merrill, H. D., Schweepe, F. C., "On-Line System Model Error Correction," IEEE Winter Power Meeting, 1973, Paper C73-106-2.

11. Debs, A. S., "Estimation of Steady-State Power System Model Parameters," *IEEE Transactions*, Vol. PAS-93, No. 5, 1974.

12. Handschin, E., Schweppe, F. C., Kohlar, J., Fiechter, A., "Bad Data Analysis for Power System State Estimation," *IEEE Transactions*, Vol. PAS-94, March/April 1975, pp. 329–337.

13. Clements, K. A., Wollenberg, B. F., "An Algorithm for Observability Determination in Power System State Estimation," IEEE Summer Power Meeting, 1975, Paper A75-447-3.

14. Krumpholz, G. R., Clements, K. A., Davis, P. W., "Power System Observability: A Practical Algorithm Using Network Topology," *IEEE Transactions on Power Apparatus and Systems* Vol. 99, July/August 1980, pp. 1534–1542.

15. Sasson, A. M., Ehrman, S. T., Lynch, P., VanSlyck, L. S., "Automatic Power System Network Topology Determination," *IEEE Transactions* Vol. PAS-92, March/April 1973, pp. 610–618.

16. DyLiacco, T. E., Ramarao, K., Weiner, A., "Network Status Analysis for Real Time Systems," Proceedings 8th PICA Conference, Minneapolis, June 1973.

Index